# NanoRevolution

As the global push toward energy accelerates, *NanoRevolution: Unveiling the Future of Energy through Advanced Materials and Digital Technologies* brings forward a powerful vision of how advanced materials and digital technologies are redefining energy solutions. This comprehensive resource navigates the frontiers of nanotechnology, artificial intelligence (AI), and IoT applications within the renewable energy sector, presenting innovative strategies that address today's most pressing energy challenges.

Across 15 chapters, *NanoRevolution* covers transformative advancements in photovoltaic and energy storage technologies. It explores the application of density functional theory (DFT) in developing high-efficiency, lead-free solar cells, new methodologies for enhancing MXenes-based materials, and emerging AI-powered solutions for energy optimization and predictive maintenance. The book also examines IoT-driven innovations in photovoltaic systems and provides a high-level overview of blockchain's potential in secure energy transactions.

This book is an invaluable guide for researchers, engineers, and industry professionals focused on next-generation energy solutions, offering both theoretical depth and practical applications to drive progress toward a sustainable energy future.

**Abderrahman Abbassi** holds a PhD in Physics, specializing in Engineering and Materials Science (Applied Physics). His extensive scientific research and academic training have provided him with substantial professional experience in Materials Physics, Electrical Engineering, and Renewable Energy. Since December 2020, he has been serving as a Professor and Researcher at Sultan Moulay Slimane University, within the Polydisciplinary Faculty of Beni Mclal.

His research encompasses a wide range of areas, including TCO semiconductors for photovoltaic applications, clean energy, material properties for energy, double perovskite oxides, spintronic and optoelectronics, and 2D systems and layers with applications in energy. He is also engaged in the study of molecular dynamics and digital technologies.

Dr. Abbassi is skilled in simulating materials using quantum physics codes such as WIEN2k and Quantum Espresso, calculating properties of materials for applications in solar energy and optoelectronics, scintillators, batteries and elaborating and fabricating thin films through various methods, including spin coating, laser ablation, and cathodic sputtering. His expertise further extends to solar energy, electrical circuits, IoT, and proficiency in electronic simulation.

## Advances in Digital Technologies for Smart Applications

*Series Editor: Saad Motahhir*

The *Advances in Digital Technologies for Smart Applications* series publishes leading-edge research on innovative digital technologies and their application in smart systems. Key topics include AI, IoT, blockchain, and their integration into various sectors, including finance, healthcare, and public governance.

**Data Analytics for Finance Using Python**
*Nitin Jaglal Untwal, Utku Kose*

**Big Data and Blockchain Technology for Secure IoT Applications**
*Shitharth Selvarajan, Gouse Baig Mohammad, Sadda Bharath Reddy, Praveen Kumar Balachandran*

**Technology-Based Teaching and Learning in Pakistani English Language Classrooms**
*Muhammad Mooneeb Ali*

**Medical Knowledge Paradigms for Enabling the Digital Health Ecosystem**
*Usha Desai, Vivek P Chavda, Ankit Vijayvargiya, Ravichander Janapati*

**Soft Computing in Renewable Energy Technologies**
*Najib El Ouanjli, Mahmoud A. Mossa, Mariya Ouaissa, Sanjeevikumar Padmanaban, Said Mahfoud*

**Leveraging the Potential of Artificial Intelligence in the Real World: Smart Cities and Healthcare**
*Tien Anh Tran, Edeh Michael Onyema, Arij Naser Abougreen*

**eGovernment Whole-of-Government Approach for Good Governance: The Back-Office Integrated Management IT Systems**
*Said Azelmad*

**Advances in Digital Marketing in the Era of Artificial Intelligence: Case Studies and Data Analysis for Business Problem Solving**
*Moez Ltifi*

**Quantum Computing and Artificial Intelligence in Logistics and Supply Chain Management**
*Pushan Kumar Dutta, Pronaya Bhattacharya, Jai Prakash Verma, Ashok Chopra, Neel Kanth Kundu, Khursheed Aurangzeb*

**NanoRevolution: Unveiling the Future of Energy through Advanced Materials and Digital Technologies**
*Abderrahman Abbassi*

For more information about this series, please visit: https://www.routledge.com/Advances-in-Digital-Technologies-for-Smart-Applications/book-series/ADT

# NanoRevolution

## Unveiling the Future of Energy through Advanced Materials and Digital Technologies

Edited by Abderrahman Abbassi

CRC Press
Taylor & Francis Group
Boca Raton London New York

CRC Press is an imprint of the
Taylor & Francis Group, an **informa** business

Designed cover image: shutterstock



First edition published 2025
by CRC Press
2385 NW Executive Center Drive, Suite 320, Boca Raton FL 33431

and by CRC Press
4 Park Square, Milton Park, Abingdon, Oxon, OX14 4RN

*CRC Press is an imprint of Taylor & Francis Group, LLC*



ISBN: 9781032814100 (hbk)
ISBN: 9781032814117 (pbk)
ISBN: 9781003499732 (ebk)

DOI: 10.1201/9781003499732

Typeset in Minion
by KnowledgeWorks Global Ltd.

Access the Support Material: www.routledge.com/9781032814100

# Contents

# Contributors

**Abderrahman Abbassi**
Laboratory of Research in Physics and Engineering Sciences, Sultan Moulay Slimane University
Beni Mellal, Morocco

**Hassan Abboudi**
Sidi Mohamed Ben Abdullah University
Fez, Morocco

**Samira Achki**
Department of Computer Science, Cadi Ayyad University
Marrakech, Morocco

**Mohamed Agouri**
Laboratory of Research in Physics and Engineering Sciences, Sultan Moulay Slimane University
Beni Mellal, Morocco

**Aumeur El Amrani**
ESTM, Moulay Ismail University of Meknes
Meknes, Morocco

**Mohamed El Amraoui**
LASMAR, Moulay Ismail University of Meknes
Meknes, Morocco

**Layla Aziz**
Faculté Polydisciplinaire, Doukkali University
El Jadida, Morocco

**Ayoub El Baraka**
Sidi Mohamed Ben Abdullah University
Fez, Morocco

**Lahcen Bejjit**
ESTM, Moulay Ismail University of Meknes
Meknes, Morocco

**Hamza Belkhanchi**
Superior School of Technology, Sultan Moulay Slimane University
Beni-Mellal, Morocco

**El Mokhtar Darkaoui**
Laboratory of Research in Physics and Engineering Sciences, Sultan Moulay Slimane University
Beni Mellal, Morocco

**Faycal Djeffal**
Department of Electronics, University of Batna
Batna, Algeria

**Mohamed Driouich**
Laboratory of Research in Physics and Engineering Sciences, Sultan Moulay Slimane University
Beni Mellal, Morocco

**Redouane En-Nadir**
Sidi Mohamed Ben Abdullah University
Fez, Morocco

**Ilyass Ez-Zejjari**
ENSAM, University Hassan-II University
Casablanca, Morocco

**Hmad Fatihi**
Laboratory of Research in Physics and Engineering Sciences, Sultan Moulay Slimane University
Beni Mellal, Morocco

**Hichem Ferhati**
Department of Electronics, University of Batna
Batna, Algeria

**Ketankumar Rameshbhai Gayakvad**
Department of Physics, K. J. Somaiya College of Science and Commerce
Mumbai, India

**Haddou El Ghazi**
ENSAM, University Hassan-II University
Casablanca, Morocco

**Abdelali El Gorari**
Department of Computer Science, Cadi Ayyad University
Marrakech, Morocco

**Sana Handor**
Energy and Materials Engineering Laboratory, Sultan Moulay Slimane University
Beni-Mellal, Morocco

**Laura Hrostea**
RAMTECH, Department of Exact and Natural Sciences, University of Lasi
Lasi, Romania

**Moha El Idrissi**
Sultan Moulay Slimane University
Beni Mellal, Morocco

**Mohamed A. Basyooni-M. Kabatas**
DPME, Delft University of Technology
Delft, The Netherlands

**Sanaa Kabil**
Department of Computer Sciences, Private University of Marrakech
Marrakech, Morocco

**Karisma Karisma**
Tunku Abdul Rahman University of Management and Technology
Pulau Pinang, Malaysia

**Aman Kumar**
Faculty of Science, Swami Vivekanand Subharti University
Uttar Pradesh, India

**Anuj Kumar**
Department of Physics, Mahamaya Government Degree College
Uttar Pradesh, India

**Charaf Laghlimi**
MASTSD
Beni-Mellal, Morocco

**Liviu Leontie**
Faculty of Physics, University of Lasi
Lasi, Romania

**Raja Mahmou**
LABSTI, Private University of Marrakech
Marrakech, Morocco

**Bouzid Manaut**
Laboratory of Research in Physics and Engineering Sciences, Sultan Moulay Slimane University
Beni Mellal, Morocco

**Abdelhak Maoucha**
Department of Electronics, University of Batna
Batna, Algeria

**Abdelmalek Mimouni**
ESTM, Moulay Ismail University of Meknes
Meknes, Morocco

**Hakima Ouhenou**
Laboratory of Research in Physics and Engineering Sciences, Sultan Moulay Slimane University
Beni Mellal, Morocco

**Ketaki Ketan Patankar**
Department of Physics, Rajaram College
Kolhapur, India

**Abdelati Razouk**
Energy and Materials Engineering Laboratory, Sultan Moulay Slimane University
Beni-Mellal, Morocco

**Mohamed Sahlaoui**
Energy and Materials Engineering Laboratory, Sultan Moulay Slimane University
Beni-Mellal, Morocco

**Mohammed Sajieddine**
Energy and Materials Engineering Laboratory, Sultan Moulay Slimane University
Beni-Mellal, Morocco

**Souad Taj**
Laboratory of Research in Physics and Engineering Sciences, Sultan Moulay Slimane University
Beni Mellal, Morocco

**Mohamed Tihtih**
ICPE, University of Miskolc
Miskolc, Hungary

**Abderrahman Waqdim**
Laboratory of Research in Physics and Engineering Sciences, Sultan Moulay Slimane University
Beni Mellal, Morocco

**Ait Lahcen Yassine**
Department of Computer Science, Cadi Ayyad University
Marrakech, Morocco

**Abderrahman Zaghrane**
Laboratory of Research in Physics and Engineering Sciences, Sultan Moulay Slimane University
Beni Mellal, Morocco

**Zakaryaa Zarhri**
CONAHCYT, The Autonomous University of Morelos State
Morelos, Mexico

**Younes Ziat**
Superior School of Technology, Sultan Moulay Slimane University
Beni-Mellal, Morocco

**Izeddine Zorkani**
Sidi Mohamed Ben Abdullah University
Fez, Morocco

Please note that color versions of figures found in the book may be found on the following website: www.routledge.com/9781032814100.

CHAPTER 1

# Artificial Intelligence Applications in Energy

Abderrahman Zaghrane, Hakima Ouhenou, Abderrahman Abbassi, Souad Taj, and Bouzid Manaut

## 1.1 INTRODUCTION

The energy industry, a crucial sector with far-reaching impacts, stands to gain immensely from advancements in artificial intelligence (AI) and machine learning (ML). This chapter seeks to explore the potential of AI applications in energy, focusing on the exciting opportunities they present for the discovery, characterization, and production of novel materials essential for the development of energy-related technologies. From solar panels to batteries and beyond, the energy sector heavily relies on sophisticated materials whose discovery and optimization have historically been time-consuming and resource-intensive.

By harnessing the power of AI, it is now possible to accelerate the identification and development of these specialized materials, thus transforming the energy landscape [1].

The sections that follow will delve into the transformative potential of AI and ML in energy materials science. First, Section 1.2 will explore how AI and ML techniques accelerate the discovery, design, and optimization of materials tailored for energy applications. This section will highlight AI-driven simulations and predictive models that revolutionize material discovery processes, showcasing breakthroughs in solar cells, battery electrodes, catalysts, and other critical components. Next, Section 1.3 expands on AI's role in enhancing the speed and accuracy of material characterization through techniques like computer vision, spectroscopy, and sensor data analysis. It will also examine how AI-driven simulations predict

DOI: 10.1201/9781003499732-1

material performance under varied conditions, supporting the development and quality control of advanced energy materials. Following this, Section 1.4 discusses AI's impact on automating and optimizing manufacturing processes for energy materials. It covers AI-driven predictive maintenance, real-time quality control, and scalable, sustainable production methods crucial for meeting the demand for clean energy solutions. Finally, Section 1.5 addresses persistent obstacles such as data availability and computational complexity in integrating AI into energy systems. It also explores anticipated breakthroughs and the broader societal impact of AI-driven material science in advancing the global energy transition.

## 1.2 AI FOR ADVANCED ENERGY MATERIAL DISCOVERY AND DESIGN

The energy sector's relentless pursuit of advanced, high-performance materials has long been a critical driver of technological progress. From the development of efficient solar cells to the innovation of next-generation battery chemistries, breakthroughs in materials science have been pivotal in advancing clean energy solutions [2–6]. However, the traditional materials research and development process has often been arduous, costly, and time-consuming, with researchers having to meticulously explore vast chemical and structural design spaces through painstaking experimentation [7–9]. The rapid advancement of AI and ML is proving to be a pivotal catalyst in transforming the materials discovery and design process for the energy sector [7]. By leveraging the pattern-recognition capabilities of ML algorithms and the predictive power of atomistic simulations enhanced by AI, researchers are now able to explore vast chemical and structural design spaces in ways that were previously unimaginable [7, 10].

With ML models being trained on extensive datasets of known material structures and their measured or computed properties, researchers can leverage these algorithms to rapidly screen and evaluate the predicted performance of thousands, or even millions, of hypothetical new compositions [11, 12].This high-throughput computational screening represents a stark contrast to the time-consuming experimental approach, allowing research teams to quickly identify the most promising material candidates for further investigation and development [13, 14].

One compelling example of ML-driven materials screening for the energy sector is a study published in the journal *Energy & Environmental Science* by Sendek et al. [15].

The flowchart in Figure 1.1 illustrates a comprehensive approach to screening and evaluating lithium-containing structures for potential use in ionic

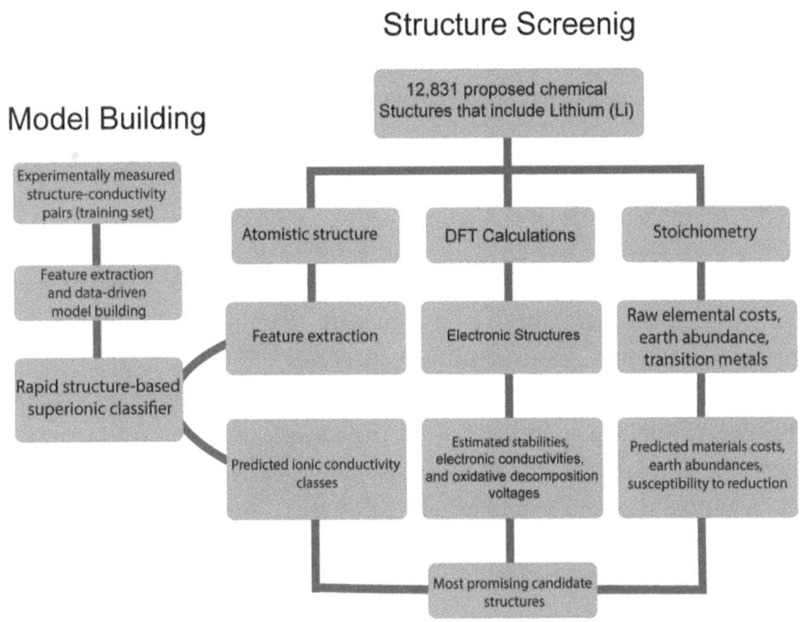


FIGURE 1.1 Approach for screening Li-containing structures and building ionic conductivity models to identify promising superionic conductor candidates for battery applications.

conductivity applications, particularly in battery materials. This two-pronged strategy combines structure screening and model building. Initially, the structure screening process begins with 12,831 Li-containing candidate structures, analyzed through three main pathways: stoichiometry, Density Functional Theory (DFT)-based electronic structure, and atomistic structure. These pathways yield calculations and predictions regarding materials costs, electronic conductivities, and ionic conductivity classes, which converge to identify the most promising candidate structures. By focusing on materials that meet specific technological requirements, the researchers reduced the pool from 12,831 candidates to approximately 300. Concurrently, the model-building component utilized experimentally measured structure-conductivity pairs to develop a rapid structure-based superionic classifier, which feeds back into the screening process to enhance the prediction of ionic conductivity classes. This integration of prerequisite property evaluation and ML-powered predictive modeling allowed the researchers to identify the highest confidence superionic conductivity predictions, ultimately shortlisting the 21 best candidate materials, along with several additional notable options.

The overall goal was to efficiently identify materials with high potential for superionic conductivity, considering factors such as stability, cost, and performance, which is crucial for advancing lithium-ion battery technology. This multi-step process demonstrates the power of AI-driven computational techniques in navigating the vast design space of energy materials, enabling rapid prioritization of the most promising solid electrolyte candidates for further investigation and optimization, beyond traditional trial-and-error experimentation.

Moreover, Sudha Priyanga et al. [16] conducted a comprehensive study exploring the use of ML techniques to predict the nature of band gaps in perovskite oxides with the ABO3 structure. This flowchart outlines the structured ML approach used in their research.

The researchers compiled an extensive dataset by collecting information from various databases and research papers on the structural, compositional, and band gap properties of these materials. They then preprocessed and formatted the data to make it suitable for ML analysis, including splitting it into training and testing sets (see Figure 1.2). Exploring multiple algorithms, the team ultimately selected the random forest model as the most effective for predicting the direct or indirect nature of the band gaps in perovskite oxides. This ML approach proved highly successful, achieving an impressive accuracy of around 91% in correctly classifying the band gap characteristics. Notably, the random forest model outperformed other ML algorithms tested, highlighting its suitability for this specific materials prediction task. Furthermore, the researchers were able to leverage the predictive capabilities of their model to identify specific perovskite

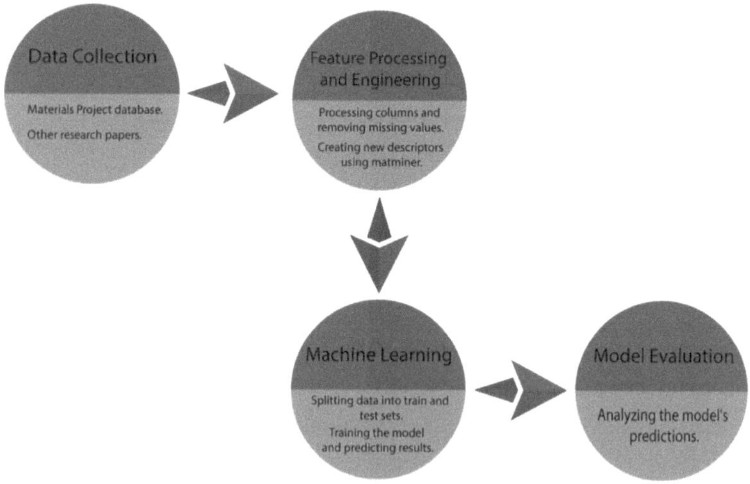


FIGURE 1.2 Machine learning approach flowchart illustrating the sequential steps of Sudha Priyang's study.

compounds, such as Sodium Plutonium Oxide ($NaPuO3$) and Vanadium Lead Oxide ($VPbO3$), as promising candidates for solar cell applications based on their direct band gaps around 1.5 eV. Altogether, this step-by-step process demonstrates the power of integrating ML techniques, particularly the random forest algorithm, into the materials discovery workflow to rapidly and effectively identify novel perovskite compositions with the desired electronic properties for energy applications.

Furthermore, in a compelling study published in the *Journal of Hazardous Materials*, Jaffari et al. [17] have employed the power of ML techniques to predict and optimize the photocatalytic degradation of malachite green (MG) dye using noble metal-doped $BiFeO3$ composites with the following steps mentioned in Figure 1.3.

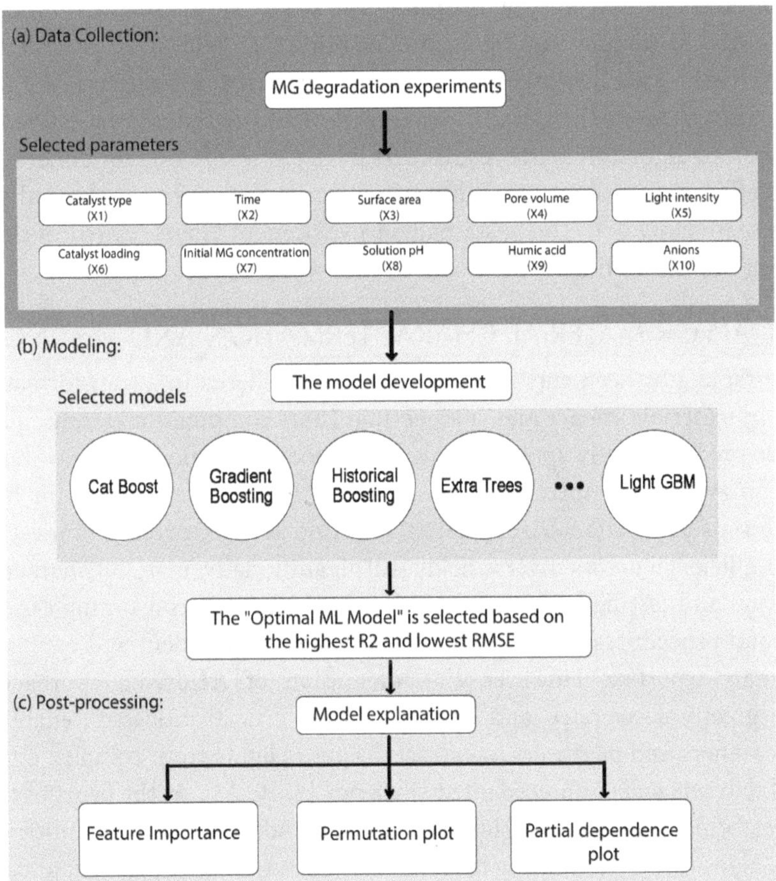


FIGURE 1.3 Workflow of the machine learning approach for analyzing MG degradation experiments.

The researchers conducted a comprehensive experimental investigation, collecting 1200 data points across ten key parameters influencing the degradation efficiency. By comparing various ML models, the team identified the CatBoost algorithm as the optimal approach, demonstrating the highest test coefficient of determination (0.99) and the lowest error metrics [18]. Through the CatBoost model, the authors were able to gain invaluable insights into the factors governing the photocatalytic process. Notably, the model revealed that the photocatalytic reaction conditions were more critical than the material properties in influencing the MG dye degradation efficiency. Further analysis of feature importance highlighted the pivotal role of reaction time as the most crucial input variable, significantly impacting the overall dye removal process. The implications of Jaffari et al.'s work extend far beyond the specific application of MG dye degradation. The study showcases the immense potential of integrating ML techniques into the field of environmental engineering, particularly in the realm of wastewater treatment. By taking advantage of the pattern-recognition and predictive capabilities of advanced algorithms, the researchers have paved the way for enhancing the efficiency and sustainability of pollutant removal processes, aligning seamlessly with the Sustainable Development Goals related to clean water and sanitation.

## 1.3 AI FOR MATERIAL CHARACTERIZATION AND TESTING

The rapid advancements in AI and ML have ushered in a transformative era in materials science and engineering. These computational techniques, once predominantly applied to energy-related applications, are now revolutionizing a wide spectrum of industries. AI and ML are redefining how materials are characterized, tested, and understood, driving innovation and efficiency across diverse fields [10]. Traditionally, material characterization and testing have been heavily reliant on time-consuming experimental procedures, often requiring specialized equipment and extensive human expertise. However, the integration of AI-driven approaches has greatly accelerated and enhanced these crucial processes, enabling researchers and engineers to extract deeper insights from complex material datasets with unprecedented efficiency [8, 19, 20]. At the heart of this transformation lies the pattern-recognition and predictive capabilities of ML algorithms. By training these models on extensive experimental data, scientists can now rapidly identify and correlate the relationships between material composition, structure, and performance, often uncovering hidden patterns and unexpected connections that would be challenging to discern through conventional analytical methods [21–23].

This is the focus of the work by Anke Stoll and Peter Benner [24], who present an application of ML techniques to predict the tensile properties of materials based on data from simpler and faster small punch tests (SPT) [25–28]. Their work demonstrates a strong correlation between the SPT data and the more complex and costly tensile test results, allowing them to develop predictive ML models that can replace the need for traditional tensile testing. Specifically, they applied various ML regression models to correlate SPT data with the ultimate tensile strength of different materials. The results confirmed the reliability of SPT measurements in predicting tensile test outcomes. This case study highlights the potential of ML methods to assist in materials characterization by extracting relevant property information from basic experimental data. It represents a significant advancement, enabling the replacement of complex and expensive tensile tests with simpler SPT combined with ML-based prediction. Faramarz Bagherzadeh, from the Alfred Wegener Institute, Helmholtz Centre for Polar and Marine Research, and Torkan Shafighfard, the corresponding author from the Institute of Fluid Flow Machinery at the Polish Academy of Sciences [29], conducted a study in 2022 that aimed to provide a time and cost-efficient technique for material characterization of carbon nanotube (CNT)-reinforced cementitious composites.

They employed ensemble ML methods, specifically random forest and gradient boosting machine, to predict the flexural and compressive strengths of these advanced composites (see Figures 1.4 and 1.5). By utilizing feature selection techniques, they were able to identify the key parameters, such as water-cement ratio, curing temperature, and CNT content, that most significantly influence the mechanical properties of the material. The team also analyzed the data distributions, feature correlations, and model evaluation metrics like mean squared error and R-squared to optimize the predictive accuracy. Sensitivity analysis further revealed the parameters with the highest contribution to the model predictions. From the characterization of mechanical properties and thermal behavior to the evaluation of optical and electronic characteristics, AI-powered techniques have revolutionized the way materials are scrutinized and understood. By automating the data acquisition and analysis processes, these computational tools have significantly reduced the time and resources required for material testing, ultimately accelerating the development and optimization of novel materials for a wide range of applications.

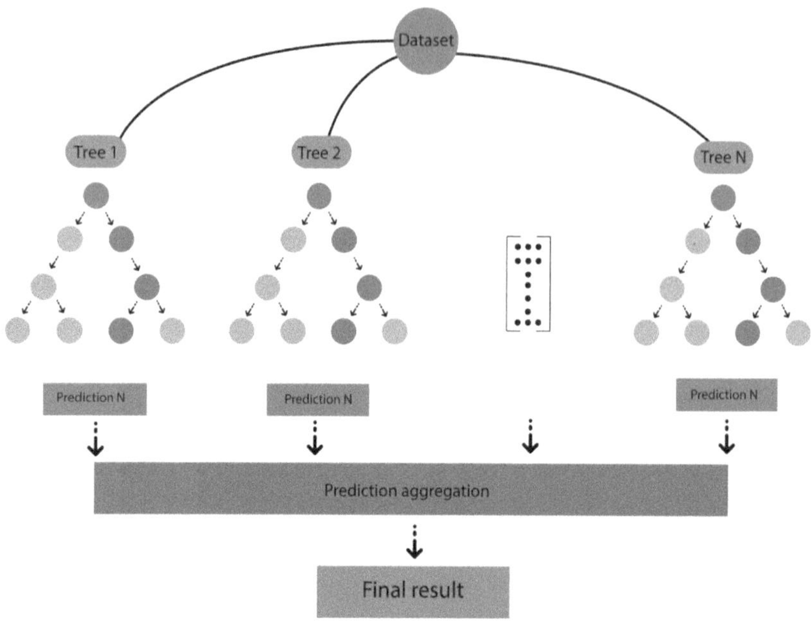


FIGURE 1.4 Random forest technique.

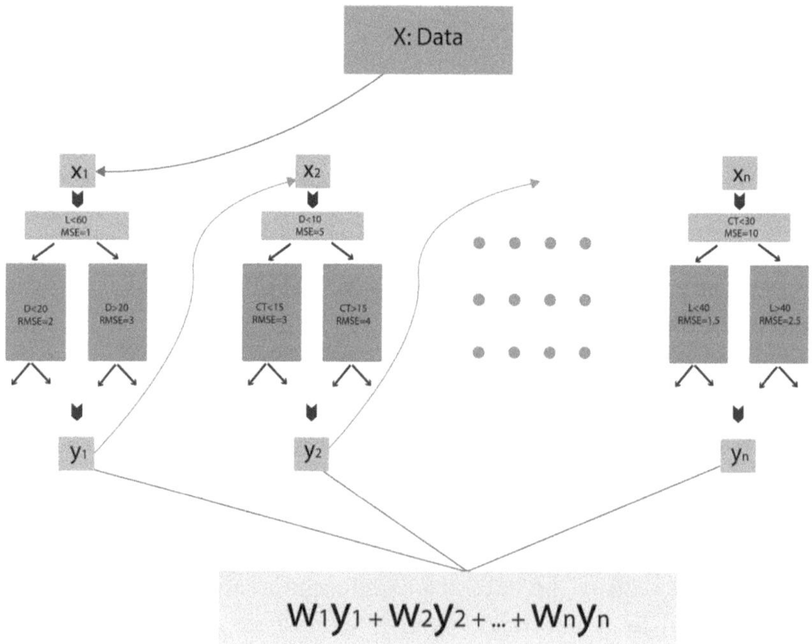


FIGURE 1.5 Gradient boosting machine.

## 1.4 AI FOR ENERGY MATERIAL MANUFACTURING AND PRODUCTION

The development of advanced energy materials is not only dependent on the discovery of novel chemical compositions and structural designs but also on the ability to reliably and efficiently manufacture these materials at scale. Traditional energy material production methods, however, are often rigid, time-consuming, and resource-intensive, presenting a significant bottleneck in the materials innovation lifecycle. To address this challenge, there is growing interest in applying AI and ML techniques to transform the manufacturing and production of energy materials [30, 31]. By integrating ML-powered intelligence into every stage of the production process, from process optimization to real-time quality control, manufacturers can unlock unprecedented levels of efficiency, adaptability, and product performance. This section will explore the emerging applications of AI and ML in energy material manufacturing, outlining how these advanced computational methods are enabling the development of smart, automated, and simulation-driven production systems that can accelerate the commercialization of innovative energy technologies.

Recent studies have demonstrated the significant impact of AI and ML on material science and engineering. For instance, Butler et al. [32] highlighted how ML algorithms can predict the properties of materials with remarkable accuracy, thereby speeding up the discovery and development process. Similarly, a study by Rajan [33] discussed the potential of AI-driven frameworks to optimize material design and manufacturing processes, reducing both time and cost. Figure 1.6 presents a

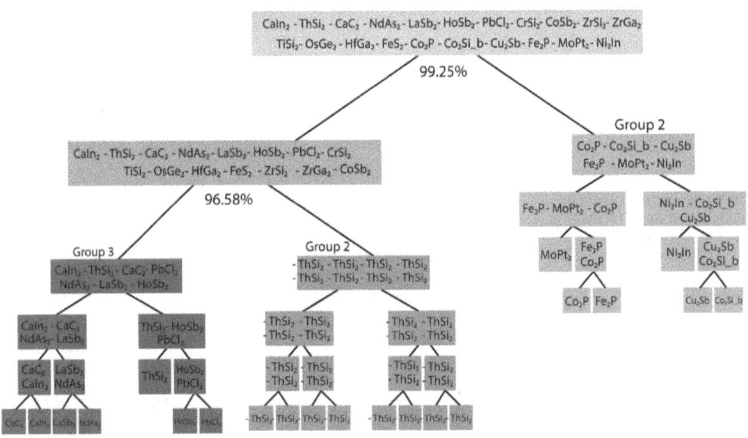


FIGURE 1.6 Discovery of new subclassifications in binary intermittencies.

diagram that depicts the discovery of new subclassifications in binary intermittencies, achieved through the use of unsupervised learning techniques. The diagram organizes the subclassifications into various groups and subgroups, each labeled with chemical element symbols and formulas. The key finding highlighted is the ability to identify a class of planar defects known as twins, which are characteristic of the compounds in Group 1. The subgroups within each group appear to be differentiated by the presence or absence of certain elements and their configurations. This discovery demonstrates the power of the analytical methods employed, as it reveals insights into the structural and compositional characteristics of the materials under investigation that were not readily apparent before. The data presented in this figure is from Reference [34].

Additionally, Kalidindi and De Graef [35] explored the role of computational methods in materials science, emphasizing the ability of ML to handle complex data and uncover patterns that traditional methods might miss. The integration of AI and ML in manufacturing is not limited to process optimization; it extends to real-time monitoring and quality control. A study by Lookman et al. [14] demonstrated how ML-powered active learning methods can accelerate the discovery of new materials with desirable properties. By building surrogate models that provide uncertainty estimates and strategically sampling the most informative new data points, active learning can guide researchers to promising material candidates more efficiently than traditional trial-and-error. This accelerated discovery process could have significant benefits for energy applications, where developing high-performance, cost-effective materials is critical for advancing technologies like solar cells, batteries, and renewable energy. The ability to rapidly identify optimal material compositions and structures through active learning could also translate to improvements in manufacturing workflows, as ML models integrated into production lines could help predict and mitigate defects, optimize processes, and ensure quality – reducing waste, improving yields, and enabling more efficient large-scale energy materials manufacturing.

Furthermore, in their 2019 [36] paper published in the journal *Nature Communications*, researchers from the University of Chicago, including Xie, Zhang, and Ong, developed an innovative unsupervised learning approach called graph dynamical networks (GDyNets) for extracting insights from molecular dynamics simulations of materials. This ML-driven technique has important implications for materials manufacturing, particularly for the production of advanced energy technologies. Understanding atomic-scale dynamics is critical for optimizing the performance of functional materials used in applications like energy storage, conversion, and transport [37–39]. By leveraging deep learning methods, including graph convolutional neural networks and Koopman analysis, the GDyNets framework can automatically identify the key dynamical modes and coordination patterns governing material behavior at the nanoscale [40, 41].

## 1.5 CHALLENGES AND FUTURE OUTLOOK

ML is transforming the field of materials science, enabling the discovery and development of new materials with unprecedented speed and efficiency. However, the integration of ML into materials research is not without its challenges. This discussion addresses key issues related to model interpretation, integration, ethics, data representation, data quality, and model selection [42].

## 1.6 INTERPRETATION AND EXPLANATION

One significant challenge with ML models is their complexity and opacity, often described as the "black-box" nature of these models [42]. This lack of transparency can limit trust in their outputs and hinder understanding of their decision-making processes. To address this issue, it is essential to employ tools such as feature importance and sensitivity analysis [43, 44]. Feature importance techniques help identify which input variables most influence the model's predictions, providing insights into the model's behavior. Sensitivity analysis assesses how changes in input variables affect the output, offering a clearer picture of the model's internal workings. Visualizing these aspects can further enhance interpretability, making the model's decision-making process more transparent and understandable to researchers and stakeholders [45].

## 1.7 MODEL INTEGRATION AND DEPLOYMENT

Integrating ML models into existing systems and deploying them effectively is another critical challenge [46]. For ML models to be useful in practical applications, they must be compatible, interoperable, and robust. This involves ensuring that models can communicate with other components of the system, handle diverse data inputs, and maintain performance under varying conditions. Deployment also requires a robust infrastructure that supports real-time processing and decision-making.

Continuous monitoring and maintenance are necessary to ensure that models remain effective and adapt to any changes in the data or operating environment [47].

## 1.8 MODEL ETHICS AND SOCIAL IMPACT

The application of ML models in materials science also raises ethical concerns and potential social impacts. These models can inadvertently introduce biases or be used in ways that have negative societal consequences [48]. It is crucial to ensure that ML models are developed and applied responsibly, adhering to ethical principles such as fairness, transparency, and accountability. This includes conducting thorough impact assessments, implementing bias mitigation strategies, and involving diverse stakeholders in the decision-making process. By doing so, researchers can ensure that ML models contribute positively to society and advance materials science in an ethical manner [49–51].

## 1.9 DATA REPRESENTATION AND FEATURE ENGINEERING

Materials data often possesses complex structures, making data representation and feature engineering particularly challenging. Selecting the appropriate data representation and feature extraction methods is critical for the performance and interpretability of ML models [52, 53]. Effective feature engineering involves transforming raw data into meaningful features that capture the underlying patterns and relationships. Techniques such as dimensionality reduction, domain-specific feature extraction, and the use of domain knowledge can significantly enhance model performance [52]. Proper data representation also aids in better generalization and robustness of the models.

## 1.10 DATA QUALITY AND AVAILABILITY

High-quality data is the cornerstone of effective ML models. However, materials data can be inaccessible, incomplete, inconsistent, or of poor quality. ML practitioners must invest significant effort in curating, cleaning, and augmenting data from various sources to improve its quality and diversity [54]. This involves dealing with missing values, correcting errors, standardizing formats, and integrating data from different experiments or simulations. Data augmentation techniques, such as synthetic data generation and transfer learning, can also help address data scarcity and enhance model performance.

## 1.11 MODEL SELECTION AND VALIDATION

The choice of the appropriate ML model for a specific task is another critical challenge. Different models have varying strengths and weaknesses, and selecting the best model requires careful consideration of the problem domain, data characteristics, and performance criteria. Model validation is equally important to ensure that the selected model generalizes well to new, unseen data. Techniques such as cross validation, bootstrapping, and rigorous performance metrics can help assess model accuracy, robustness, and reliability [55–58]. It is also essential to consider the computational efficiency and scalability of the model, especially for large-scale applications in materials science.

## 1.12 CONCLUSION

AI is fundamentally transforming the landscape of materials science and the energy industry. Through advanced computational techniques and innovative algorithms, AI accelerates the discovery, characterization, and manufacturing of energy materials, driving unprecedented efficiencies and breakthroughs. By optimizing the molecular structure and properties of materials, AI helps in developing high-performance solar cells, batteries, and other critical energy technologies. Moreover, AI's role in material characterization and testing enables faster, more accurate assessments, ensuring that materials meet the rigorous demands of modern energy applications. In manufacturing, AI-driven automation and predictive maintenance contribute to scalable, efficient, and sustainable production processes, addressing the growing need for clean and renewable energy sources. Despite the remarkable progress, challenges remain in data availability, computational demands, and integration of AI into existing systems. However, the ongoing advancements in AI and ML hold promise for overcoming these obstacles, leading to further breakthroughs in energy materials science. As AI continues to evolve, its impact on the energy transition will become even more profound, driving the development of innovative materials and technologies that are essential for a sustainable future. This chapter highlights the transformative potential of AI in the energy sector, emphasizing its critical role in advancing material science and contributing to global energy solutions.

## REFERENCES

1. Szczepaniuk, H., and E. K. Szczepaniuk. "Applications of artificial intelligence algorithms in the energy sector." Energies 16.1 (2022) https://doi.org/10.3390/en16010347

2. Enrichi, Francesco, and Giancarlo Righini, eds. Solar Cells and Light Management: Materials, Strategies and Sustainability. Elsevier, 2019.
3. Abbassi, A., et al. "Magneto-thermal, mechanical, and opto-electronic properties of S R 2 MWO 6 (M = V, R h, R u): Ab initio study." Journal of Superconductivity and Novel Magnetism 36.3 (2023) https://doi.org/10.1007/s10948-023-06537-0
4. Agouri, M., et al. "Computational study of stability, photovoltaic, and thermoelectric properties of new inorganic lead-free halide perovskites." Europhysics Letters 146.1 (2024) https://doi.org/10.1209/0295-5075/ad2cb7
5. Shamoon, A., et al. "Understanding the role of advanced materials for energy infrastructure and transmission." Materials Today: Proceedings 62 (2022) https://doi.org/10.1016/j.matpr.2022.04.756
6. He, Q., et al. "Density functional theory for battery materials." Energy & Environmental Materials 2.4 (2019) https://doi.org/10.1002/eem2.12056
7. Badini, S., S. Regondi, and R. Pugliese. "Unleashing the power of artificial intelligence in materials design." Materials 16.17 (2023) https://doi.org/10.3390/ma16175927
8. Narula, C. K., et al. "Materials chemistry issues related to advanced materials applications in the automotive industry." Chemistry of Materials 8.5 (1996) https://doi.org/10.1021/cm950588m
9. Ceder, G. "Opportunities and challenges for first-principles materials design and applications to Li battery materials." MRS Bulletin 35.9 (2010) https://doi.org/10.1557/mrs2010.681
10. DeCost, B. L., et al. "Scientific AI in materials science: a path to a sustainable and scalable paradigm." Machine Learning: Science and Technology 1.3 (2020) https://doi.org/10.1088/2632-2153/ab9a20
11. Jain, A., et al. "Commentary: The materials project: a materials genome approach to accelerating materials innovation." APL Materials 1.1 (2013) https://doi.org/10.1063/1.4812323
12. Ward, L., et al. "Matminer: An open source toolkit for materials data mining." Computational Materials Science 152 (2018) https://doi.org/10.1016/j.commatsci.2018.05.018
13. Raccuglia, P., et al. "Machine-learning-assisted materials discovery using failed experiments." Nature 533.7601 (2016) https://doi.org/10.1038/nature17439
14. Lookman, T., et al. "Active learning in materials science with emphasis on adaptive sampling using uncertainties for targeted design." npj Computational Materials 5.1 (2019) https://doi.org/10.1038/s41524-019-0153-8
15. Sendek, A. D., et al. (2017). Holistic computational structure screening of more than 12,000 candidates for solid lithium-ion conductor materials. Energy & Environmental Science, 10(1), 306–320.
16. Sudha Priyanga, G., et al. "Prediction of nature of band gap of perovskite oxides (ABO3) using a machine learning approach." Journal of Materiomics 8.5 (2022) https://doi.org/10.1016/j.jmat.2022.04.006
17. Jaffari, Z. H., et al. "Machine learning approaches to predict the photocatalytic performance of bismuth ferrite-based materials in the removal of malachite green." Journal of Hazardous Materials 442 (2023) https://doi.org/10.1016/j.jhazmat.2022.130031

18. Hancock, J. T., and T. M. Khoshgoftaar. "CatBoost for big data: an interdisciplinary review." Journal of Big Data 7.1 (2020) https://doi.org/10.1186/s40537-020-00369-8
19. Darkaoui, E., et al. "First-principles investigation of MXenes M4C3 (M= Sc, Cr, andMn) for clean energy." Chinese Journal of Physics 89 (2024) https://doi.org/10.1016/j.cjph.2024.03.041
20. Ouhenou, H., et al. "First-principles examination of structural, electronic, magnetic, and optical properties of a free lead scintillating double perovskites A2NATAX6 (A= C s, Rb; X= C l, B r, I)." Optical and Quantum Electronics 56.6 (2024) https://doi.org/10.1007/s11082-024-06804-x
21. Mahesh, B. "Machine learning algorithms—a review." International Journal of Science and Research (IJSR). [Internet] 9.1 (2020).
22. Morgan, D., and R. Jacobs. "Opportunities and challenges for machine learning in materials science." Annual Review of Materials Research 50.1 (2020) https://doi.org/10.1002/9781119148739.ch4
23. Wagner, N., and J. M. Rondinelli. "Theory-guided machine learning in materials science." Frontiers in Materials 3 (2016) https://doi.org/10.3389/fmats.2016.00028
24. Stoll, A., and P. Benner. "Machine learning for material characterization with an application for predicting mechanical properties." GAMM-Mitteilungen 44.1 (2021) https://doi.org/10.1002/gamm.202100003
25. Fleury, E., and J. S. Ha. "Small punch tests to estimate the mechanical properties of steels for steam power plant: I. Mechanical strength." International Journal of Pressure Vessels and Piping 75.9 (1998) https://doi.org/10.1016/S0308-0161(98)00074-X
26. Mao, X., and H. Takahashi. "Development of a further-miniaturized specimen of 3 mm diameter for tem disk (ø 3 mm) small punch tests." Journal of Nuclear Materials 150.1 (1987) https://doi.org/10.1016/0022-3115(87)90092-4
27. Misawa, T., et al. "Small punch tests for evaluating ductile-brittle transition behavior of irradiated ferritic steels." Journal of Nuclear Materials 150. 2 (1987): 194–202. https://doi.org/10.1016/0022-3115(87)90075-4
28. Misawa, T., et al. "Fracture toughness evaluation of fusion reactor structural steels at low temperatures by small punch tests." Journal of Nuclear Materials 169 (1989): 225–232. https://doi.org/10.1016/0022-3115(89)90538-2
29. Bagherzadeh, F., and T. Shafighfard. "Ensemble machine learning approach for evaluating the material characterization of carbon nanotube-reinforced cementitious composites." Case Studies in Construction Materials 17 (2022) https://doi.org/10.1016/j.cscm.2022.e01537
30. Liu, Y., et al. "Machine learning for advanced energy materials." Energy and AI 3 (2021) https://doi.org/10.1016/j.egyai.2021.100049
31. Hemavathi, B., et al. "Machine learning in the era of smart automation for renewable energy materials." e-Prime-Advances in Electrical Engineering, Electronics and Energy (2024) https://doi.org/10.1016/j.prime.2024.100458
32. Butler, K. T., et al. "Machine learning for molecular and materials science." Nature 559.7715 (2018) https://doi.org/10.1038/s41586-018-0337-2

33. Rajan, K. "Materials informatics: the materials "gene" and big data." Annual Review of Materials Research 45.1 (2015) https://doi.org/10.1146/annurev-matsci-070214-021132
34. Kong, C. S., et al. "Mapping the 'materials gene' for binary intermetallic compounds—a visualization schema for crystallographic databases." Computational Science & Discovery 5.1 (2012) https://doi.org/10.1088/1749-4699/5/1/015004
35. Kalidindi, S. R., and M. De Graef. "Materials data science: current status and future outlook." Annual Review of Materials Research 45.1 (2015) https://doi.org/10.1146/annurev-matsci-070214-020844
36. Xie, T., et al. "Graph dynamical networks for unsupervised learning of atomic scale dynamics in materials." Nature Communications 10.1 (2019) 10.24435/materialscloud:2019.0017/v1
37. Rangel-Martinez, D., K. D. P. Nigam, and L. A. Ricardez-Sandoval. "Machine learning on sustainable energy: a review and outlook on renewable energy systems, catalysis, smart grid and energy storage." Chemical Engineering Research and Design 174 (2021) https://doi.org/10.1016/j.cherd.2021.08.013
38. Zaghrane, A., et al. "First-principles investigation of structural, electronic, optical, and magnetic properties of a scintillating double perovskite halides (Cs2LiCeX6) with (X= F, Br, and I)." Chinese Journal of Physics (2023) https://doi.org/10.1016/j.cjph.2023.08.001
39. Artrith, N. "Machine learning for the modeling of interfaces in energy storage and conversion materials." Journal of Physics: Energy 1.3 (2019) https://doi.org/10.1088/2515-7655/ab2060
40. Mardt, A., et al. "VAMPnets for deep learning of molecular kinetics." Nature Communications 9.1 (2018) https://doi.org/10.1038/s41467-017-02388-1
41. Wu, H., and F. Noé. "Variational approach for learning Markov processes from time series data. arXiv." arXiv preprint arXiv:1707.04659 (2017) https://doi.org/10.48550/arXiv.1707.04659
42. Meredig, B. "Five high-impact research areas in machine learning for materials science." Chemistry of Materials 31.23 (2019) https://doi.org/10.1021/acs.chemmater.9b04078
43. Yeung, D. S., et al. Sensitivity Analysis for Neural Networks. Springer, 2010 https://doi.org/10.1007/978-3-642-02532-7
44. Zhang, P. "A novel feature selection method based on global sensitivity analysis with application in machine learning-based prediction model." Applied Soft Computing 85 (2019) https://doi.org/10.1016/j.asoc.2019.105859
45. Brownlee, J. "Sensitivity analysis of dataset size vs. model performance." Python Machine Learning (2021).
46. Paleyes, A., R.-G. Urma, and N. D. Lawrence. "Challenges in deploying machine learning: a survey of case studies." ACM Computing Surveys 55.6 (2022) https://doi.org/10.1145/3533378
47. Garg, S., et al. "On continuous integration/continuous delivery for automated deployment of machine learning models using MLOps." 2021 IEEE fourth international conference on artificial intelligence and knowledge engineering (AIKE). IEEE, 2021 https://doi.org/10.1109/AIKE52691.2021.00010.

48. Lo Piano, S. "Ethical principles in machine learning and artificial intelligence: cases from the field and possible ways forward." Humanities and Social Sciences Communications 7.1 (2020) https://doi.org/10.1057/s41599-020-0501-9
49. Saltz, J., et al. "Integrating ethics within machine learning courses." ACM Transactions on Computing Education (TOCE) 19.4 (2019) https://doi.org/10.1145/3341164
50. Collectif, C. Research ethics in machine learning. Diss. CERNA; ALLISTENE, 2018 https://hal.science/hal-01724307/
51. Shadowen, N. "Ethics and bias in machine learning: A technical study of what makes us "good"." The Transhumanism Handbook (2019) https://doi.org/10.1007/978-3-030-16920-6_12
52. Wang, H., et al. "Machine learning basics." Deep learning (2016) http://whdeng.cn/Teaching/PPT_01_Machine%20learning%20Basics.pdf
53. Bengio, Y., A. Courville, and P. Vincent. "Representation learning: a review and new perspectives." IEEE Transactions on Pattern Analysis and Machine Intelligence 35.8 (2013) https://doi.org/10.1109/TPAMI.2013.50
54. Jain, A., et al. "Overview and importance of data quality for machine learning tasks." Proceedings of the 26th ACM SIGKDD international conference on knowledge discovery & data mining. 2020 https://doi.org/10.1145/3394486.3406477.
55. Raschka, S. "Model evaluation, model selection, and algorithm selection in machine learning." arXiv preprint arXiv:1811.12808 (2018) https://doi.org/10.48550/arXiv.1811.12808
56. Bartlett, P. L., S. Boucheron, and G. Lugosi. "Model selection and error estimation." Machine Learning 48 (2002) https://doi.org/10.1023/A:1013999503812
57. Yang, W., et al. "A novel machine learning-based electricity price forecasting model based on optimal model selection strategy." Energy 238 (2022) https://doi.org/10.1016/j.energy.2021.121989
58. Farahmand, A.-M., and C. Szepesvári. "Model selection in reinforcement learning." Machine Learning 85.3 (2011) https://doi.org/10.1007/s10994-011-5254-7

CHAPTER 2

# Density Functional Theory for 2D Anode Materials

Mohamed Agouri, Abderrahman Waqdim, Abderrahman Abbassi, Souad Taj, Bouzid Manaut, and Mohamed Driouich

## 2.1 INTRODUCTION

The increasing demand for energy and the escalation of environmental pollution have made the shift from traditional energy sources to green energy an urgent and complex challenge. Renewable energy sources, including solar, wind, tidal, and geothermal energies, have great potential to enable and facilitate this transition [1–4]. They are typically collected as electrical energy and then distributed/delivered to users. Consequently, energy storage strategies play a crucial role in determining the efficiency and effectiveness of energy utilization [5]. With regard to this, secondary batteries, specifically rechargeable lithium-ion batteries (LIBs), are recognized as the largest and most widely used advanced energy storage technology [6]. However, the energy density, safety, and long lifetime of LIB devices hardly fail to meet the demands arising from recent societal and industrial developments [7].

Since the inception of LIB systems, graphite has been considered an ideal and perfect anode material owing to its unmatched combination of relatively low cost, abundance, power density, long cycle life, and high energy density. However, conventional batteries face limitations in meeting the growing demand for large-scale applications [8, 9]. Recent research indicates that the lithium storage performance of graphite can be


DOI: 10.1201/9781003499732-2

improved, highlighting the promising future of advanced LIBs for electric vehicles and grid-scale energy storage stations. Consequently, the development and design of new negative electrode materials with high specific capacity, safety, and excellent cycle performance are critical tasks for LIB devices [10, 11]. Despite extensive research efforts and rapid advancements in characterization techniques, many complex challenges remain for the commercialization of these battery systems. One significant hurdle is the difficulty in experimentally investigating the reaction mechanisms of new batteries at the atomic level [12]. Therefore, gaining a deeper understanding of the reaction mechanisms in these kinds of battery systems is a fundamental role for designing new and advanced battery materials suitable for industrial applications.

Since the discovery of graphene in 2004, a vast array of two-dimensional (2D) materials has garnered significant attention and been extensively studied and used in various fields, including energy storage systems. Their proper chemical and physical characteristics compared to their bulk counterparts offer distinct advantages, opening the door for enhanced performance and efficiency. This has led to novel technological advancements in energy storage applications, providing potential solutions to global challenges related to energy storage [13, 14]. Many 2D materials, including borophene, silicene, blue/black phosphorene, MXenes, and transition metal dichalcogenides, have been intensively discovered and explored [15–21]. These materials have been examined theoretically and experimentally for their applicability as anode materials in LIB systems. However, their practical use is principally limited by weak electrochemical characteristics, such as high diffusion energy barriers, low storage capacity, and poor stability [22]. Ongoing research focuses on discovering new 2D materials, understanding their fundamental properties, and developing scalable synthesis methods.

Advancements in computational capabilities have made computer simulations crucial for materials discovery and design. Atomistic simulations, using density functional theory (DFT), offer significant value by revealing processes at the atomic and electronic scales. These simulations aid in interpreting experimental observations and predicting outcomes. DFT, developed by Kohn and Sham, has become the predominant tool in computational material science and solid-state physics for numerically solving the Schrodinger equation [23, 24]. DFT plays a crucial role in guiding the application of materials for energy storage, aiding in the design and exploration of electrode materials with optimal electrochemical properties, especially for negative electrode anodes [25, 26]. Specifically, DFT provides detailed insights into atomic and molecular interactions, energy calculations, and potential energy surfaces, which are challenging to achieve experimentally. By rapidly assessing the physical and chemical properties of novel electrode materials, DFT helps identify the most promising candidates, saving time and resources. Recently, many studies concentrated and focused on experiments with DFT investigations used for verification, while some relied solely on DFT for comprehensive theoretical analysis, and others emphasized a balanced integration of both approaches [27]. This synergistic strategy between experimental methods and DFT calculations enhances the overall understanding and development of advanced electrode anode materials [28–31].

In this chapter, our primary focus is to discuss some theoretical strategies, challenges, and opportunities using DFT in the design of novel 2D anode materials. The chapter is organized into seven main sections. Section 2.2 addresses the theoretical stability estimations of negative electrode materials through DFT investigations, including thermodynamic, mechanical, and dynamic stabilities. Section 2.3 deals with the adsorption process of ions with surfaces providing detailed insights into atomic-level interactions. This process serves as a predictive tool for optimizing material designs and understanding the electrochemical reaction mechanisms. Section 2.4 presents electronic analyses of anode materials, such as electronic conductivity, band gap, density of states (DOS), and charge transfer from ion-adsorbed on surfaces. Additionally, it also discusses ionic diffusion barriers in anode materials, crucial for understanding ion mobility of surfaces and its influence on the rate capacity of anode materials. Section 2.5 describes theoretical capacity and open-circuit voltage (OCV) estimations of anode materials. Section 2.6 presents an example of the electrochemical properties of the o-$Al_2N_2$ monolayer using DFT investigations. The last section summarizes key insights and findings discussed throughout the chapter.

## 2.2 THEORETICAL STABILITY

Ensuring the stability of materials prior to their integration and application into industrial applications, such as batteries, is paramount for safeguarding operational integrity, safety, and performance. Designing and exploring novel anode materials with predictable characteristics, including electrochemical stability, performance, and excellent ionic mobility, requires thorough stability checks, which significantly impact battery efficiency. Cohesive energy, formation energy, phonon dispersion, and Gibbs free energy are crucial methods for predicting and assessing the stability of negative electrode materials. This section introduces these methods and their application in designing new anode materials using DFT calculations.

### 2.2.1 Formation Energy

Formation energy refers to the energy change when a solid or material is formed from its constituent elements in their standard states. It indicates the stability of a material by showing how much energy is absorbed or released during its formation. The formation energy can be calculated using the following equation [32]:

$$\Delta H_f = \frac{E_{tot}\left(X_x Y_y\right) - xE_X - yE_Y}{x + y} \tag{2.1}$$

where $E_{tot}\left(X_x Y_y\right)$ is the total energy of the $X_x Y_y$ material, and $E_{i=X/Y}$ represents the individual energy of each atom in its bulk phase. The formation energy is principally measured at standard pressure and temperature, ensuring the stability of the compound relative to its elements. A positive formation energy value implies that the formation process is endothermic, meaning energy is absorbed, and the material is less stable. Conversely, a negative formation energy value suggests that the formation process is exothermic, indicating energy is released, and the material is stable.

### 2.2.2 Cohesive Energy

Cohesive energy of a material refers to the energy released when completely separating its constituent atoms into isolated atomic species. It measures the strength of the bonds holding the atoms together, providing insights into how atoms interact to form a stable structure within the material. The cohesive energy per atom was calculated via the equation [33]:

$$E_c = \frac{E_{tot}\left(X_x Y_y\right) - xE_X - yE_Y}{x + y} \tag{2.2}$$

where $E_{tot}\left(X_x Y_y\right)$ is the total energy of the $X_x Y_y$ material, and $E_{i=X/Y}$ is the total energy of each constituent atom in its isolated state. As is well known, greater stability of a material with strong bonding is directly related to a high cohesive energy value. This information is crucial for understanding a material's properties, including mechanical strength and thermal stability.

### 2.2.3 Gibbs Free Energy

Gibbs free energy is a crucial concept in thermodynamics, offering valuable insights into defining equilibrium and predicting whether a process will occur spontaneously under constant pressure and temperature conditions. Moreover, it aids in determining cell parameters and atomic positions to predict the most stable crystal structure in a given state, thereby minimizing the relevant thermodynamic quantities [34]. For a given system, the Gibbs free energy is defined by the following equation:

$$G = H - TS = E + PV - TS \tag{2.3}$$

where $H$, $T$, $S$, $E$, $P$, and $V$ represent the enthalpy, temperature of the system, and entropy, internal energy, pressure, and volume, respectively. Gibbs free energy plays a fundamental role in assessing the electrochemical stability and phase transitions of anode materials in battery systems. It dictates whether the intercalation of metal ions (Li, Na, K, Ca, etc.) into the anode materials occurs spontaneously, which directly impacts the operating voltage and overall performance of the battery [35]. Therefore, understanding and controlling the Gibbs free energy is important in developing new advanced negative electrode materials. In the framework of DFT calculations, Gibbs free energy calculations are highly relevant to experimental work on LIBs. Several packages can be utilized to predict and calculate various thermal properties, including Gibbs energy, at a constant pressure. For example, Phonopy [36] is a powerful open-source package that offers a robust interface for several simulation codes, such as Quantum ESPRESSO [37], WIEN2k [38], etc. Phonopy is used for calculating phonon dispersion and thermal properties based on the quasi-harmonic approximation. Recently, researchers have utilized Phonopy to comprehensively understand and predict thermodynamic stability, phase transitions, and reaction kinetics of materials within LIB systems.

### 2.2.4 Phonon Dispersion

Understanding and optimizing advanced anode materials for LIBs relies significantly on phonon dispersion analysis, which provides crucial insights into thermal conductivity, phase stability, and electrochemical performance. By examining the quantized normal modes of lattice vibrations known as the phonon spectrum, researchers can characterize harmonic lattice vibrations in a material. Assessing phonon dispersion is essential for determining dynamic stability; the presence of imaginary vibrational modes indicates structural instability [39]. This analysis is particularly important for anode materials as it sheds light on phase transformations and lattice distortion occurring during the battery's charge and discharge processes. The insights enable the development of new anode materials that offer enhanced stability, efficiency, and performance during battery operation [40]. Moreover, it aids in identifying these changes and predicting material stability, thereby facilitating experimental synthesis.

In theoretical investigations, researchers have recently employed the Phonopy package in conjunction with DFT to predict the dynamic stability of materials by calculating their phonon dispersion [36]. However, experimental confirmation remains crucial to validate these theoretical predictions and understand real-world aspects. Techniques such as X-ray diffraction and Raman spectroscopy offer direct measurements of structural properties, phase transitions, and thermal behaviors under specific conditions. Comparing experimental and theoretical findings allows researchers to validate theoretical models and optimize material designs for practical applications. A prominent example is graphite. L. Wirtz et al. [41] conducted various calculations and measurements of the phonon dispersion relation of graphite, highlighting the importance of integrating theoretical calculations with experimental measurements. This integration refines models and enhances understanding of phonon behavior in graphite and related nanostructures.

## 2.3 ADSORPTION PROCESS

In the electrochemical reactions of batteries, adsorption is the process whereby ions and molecules from a liquid or gas adhere to a solid surface. This phenomenon occurs due to intermolecular forces, including Van der Waals forces, between an adsorbent (ion or molecule) and the surface of a substance. Adsorption significantly impacts the kinetics and efficiency of electrochemical reactions by modifying the surface energy and reactivity of the electrode [42].

In theory, assessing the adsorption energy of a single metal (Li, Na, K, Ca, etc.) can help identify and select materials with potential to serve as anodes or cathodes in specific secondary batteries. For anode materials, ions can spontaneously adsorb onto the surface, enhancing energy storage and release efficiency. Therefore, in DFT calculations, the adsorption energy is a crucial parameter for comprehending and investigating interactions between adsorbates and surfaces [28].

The adsorption energy $\left(E_{ads}\right)$ is generally defined as the change in energy when an adsorbent binds to a surface, calculated using the following formula:

$$E_{ads} = E_{system} - E_{surface} - E_{adsorbate} \tag{2.4}$$

where $E_{system}$ and $E_{surface}$ represent the total energy of systems with and without adsorption. $E_{adsorbate}$ refers to the energy of a single adsorbate ion in its bulk phase. A negative adsorption energy indicates attractive interactions between the adsorbent and surface atom electron density, leading to binding forces. Conversely, a positive adsorption energy suggests repulsive interactions [43].

To better understand the adsorption mechanism using DFT calculations, we start by identifying the adsorption sites on the surface [26]. This is a fundamental technique for a detailed analysis of the adsorption process, aiming to optimize electrode materials in electrochemical applications where high specific capacity and efficient energy storage are desired. These adsorption sites are specific locations where ions attach, varying in energy and availability based on the surface structure, symmetry, and composition. The adsorption energy and kinetics of lithium ions at these different sites significantly impact the charge-discharge rates, cycle life, and energy efficiency.

In general, there are typically three types of adsorption sites, as shown in Figure 2.1. First, the top site (T), where the adsorbate is directly above each surface atom, generally experiences weaker adsorption energies. Second, the bridge site (B), where the adsorbate is located between two adjacent surface atoms, allowing significant interaction and offering moderate adsorption energies. The hallow sites (H) have the adsorbate positioned over a site surrounded by three or more surface atoms, usually providing the strongest adsorption energy owing to multiple bonds with the surface. Optimizing these adsorption sites in materials design through DFT calculations relies on identifying sites with minimum total energies, which provides insights into the energetically favorable sites for adsorption. By comprehensively understanding the distribution and energetics of these adsorption sites, we can facilitate the design and exploration

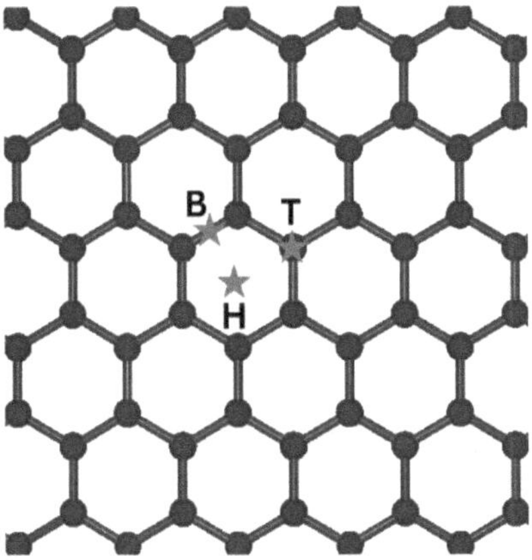


FIGURE 2.1 Three main adsorption sites on 2D materials: hollow (H), bridge (B), and top of atoms (T).

of materials with higher specific capacities, as more adsorbate ions can be efficiently stored on the surface. Graphene is one of the most prominent examples of 2D materials and is currently the most well-known and applicable anode material in batteries [44]. K.T. Chan et al. [45] reported the adsorption of 12 metals on graphene using First-principles studies. Owing to the symmetry of graphene, three available adsorption sites were selected: the T site above a carbon atom, the B site above the center of a C-C bond, and the H site located in the hollow surrounded by six carbon atoms. In the anode part, the half-reaction of the lithiation process can be expressed as follows [46]:

$$C_6 + Li^+ + e^- \rightarrow LiC_6 \tag{2.5}$$

Based on the adsorption energy (Equation 2.4), the most stable site for the adsorption of lithium is the hollow site (H), which presents the lowest adsorption energy (−1.17 eV) compared to the other sites. Table 2.1 shows the lithium adsorption energy at the most suitable site of several well-known 2D active anode materials. These findings highlight significant adsorption energies, suggesting that these materials can effectively adsorb lithium ions. Therefore, all materials listed in Table 2.1 have been theoretically studied as anode materials, showing great promise for developing high-performance batteries and warranting further experimental synthesis.

TABLE 2.1 Lithium Adsorption Energy at the Most Stable Site of Various 2D Materials Reported Recently

| 2D Material | Stable Site | Adsorption Energy (eV) |
|---|---|---|
| Graphene | H | −1.17 [46] |
| Silicene | H | −2.08 [47] |
| Germanene | H | −2.07 [48] |
| Borophosphene | H | −1.06 [49] |
| Blue phosphorus | H | −1.97 [18] |
| Black phosphorus | H | −1.96 [18] |
| Arsenene | H | −2.55 [50] |
| h-Bas | H | −0.42 [51] |
| h-AlC | H | −2.10 [52] |
| o-$Al_2N_2$ | H | −0.39 [53] |
| o-$B_2N_2$ | H | −1.27 [54] |
| o-$B_2P_2$ | H | −0.47 [55] |
| $Si_3C$ | H/B | −0.75 [56] |
| $WB_4$ | H | −2.52 [57] |
| Siliborophene | H | −1.33 [58] |
| $AlB_4$ | H | −2.14 [59] |
| $Ti_2PX_2$ (X = S, Se, Te) | T | −0.43 [60] |
| t-BN | H | −1.51 [61] |
| $VS_2$ | T | −2.13 [62] |
| $TiB_2C_2$ | H | −1.63 [63] |

## 2.4 ELECTRONIC PROPERTIES AND IONIC MOBILITY

### 2.4.1 Electronic Properties

Understanding the electronic properties of 2D anode materials is crucial for optimizing their performance in LIB systems. This includes comprehending their electronic band structure (BS), DOS, and charge transfer processes during lithiation. Through DFT calculations, the BS analysis reveals the availability of electronic states for charge carriers, electronic conductivity, and storage capacity. Additionally, the DOS reveals the distribution of these states across energy levels [64]. The intercalation of lithium ions profoundly impacts the electronic properties of materials, causing shifts in the BS and changes in the band gap energy as lithium transfers charge to the material. These alterations are critical in influencing the electrochemical properties, such as the capacity to store and release lithium ions during the charge-discharge processes. They can also lead to chemical changes on the surface and structural distortions, influencing the electronic properties of the materials. Understanding these electronic information is important for selecting anode materials that can achieve efficiency and stability in practical battery technologies. Graphene is a semi-metal material characterized by its Dirac cone at the K-point of the first Brillouin zone, resulting from its $sp^2$ hybridized carbon structure. This unique BS grants graphene high electronic mobility [65]. However, the intercalation of lithium significantly changes its electronic properties. This intercalation shifts the Fermi level above the Dirac point, alters the DOS, and changes the carrier concentration. These alterations improve its electrical conductivity, making it a good candidate for anode materials in battery devices [66]. Similarly, other 2D materials with semiconductor aspects, such as o-$Al_2N_2$ and h-AlC, exhibit significant physical characteristics, including a notable electronic band gap $\left(E_g < 2\ eV\right)$ and unique optical properties [52, 53]. The adsorption of lithium on these materials changes their electronic nature from semiconducting to metallic, thereby improving their electronic conductivity. These improvements facilitate their use as negative electrodes and present a promising potential to replace graphene in current LIB systems.

### 2.4.2 Ionic Mobility

Understanding the migration mechanism of lithium ions on anode materials is crucial for designing and applying 2D materials effectively in LIB systems. The movement of lithium ions across the surface plays a pivotal role in optimizing novel negative electrodes and improving overall battery functionality. In fact, the rate performance of the charge-discharge process in LIB systems depends on the mobility of Li ions within the active materials [67]. The study of the atomic-level ion diffusion mechanisms in battery systems provides invaluable insights, as ion diffusion directly impacts battery speed, cycling performance, and structural stability. These insights are essential for advancing the development of next-generation materials for energy storage, where enhanced ion mobility can significantly boost device performance and stability under diverse operational conditions [68].

DFT investigations have been extensively used to simulate ion transport processes in batteries by calculating the activation energy along ion diffusion paths. The climbing image nudged elastic band (CI-NEB) method is an essential computational technique for simulating the diffusion process of anode materials [69]. This approach determines the minimum energy pathway between two stable states by creating a series of intermediate images or configurations that form a smooth transition path. This path reveals the energy landscape ions transverse, maintaining equal distances among images using spring forces [70]. By accurately locating the highest energy

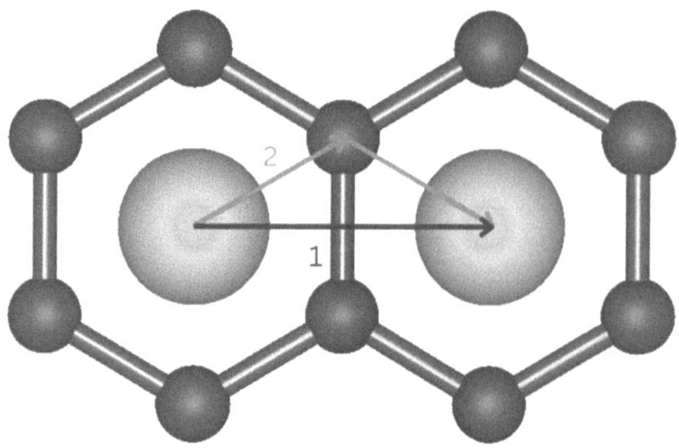


FIGURE 2.2 Two possible pathways for lithium-ion diffusion on graphene.

point (saddle point) along the pathway, which represents the activation energy barrier, the CI-NEB method provides a precise calculation of the activation energy. This technique is crucial for understanding ion mobility in battery systems, guiding the design of novel anode materials with improved performance, faster charging rates, and greater efficiency [71].

Based on the CI-NEB method, E.G. Leggesse et al. [72] reported the mobility of Li ion on graphene by investigating the minimum energy path and identifying the saddle point along two pathways. In these pathways, lithium migrates from the most favorable site to its first neighbor, as presented in Figure 2.2. The first pathway goes on the C-C bond, while the second pathway passes over the top of a C atom. The authors reported that path 1 presents the lowest diffusion barrier, approximately 0.18 eV. Table 2.2 shows the diffusion barrier of several other 2D materials. These

TABLE 2.2 Diffusion Barrier of Various 2D Materials Reported Recently from First Principles Study

| 2D Material | Diffusion Barrier (eV) | 2D Material | Diffusion Barrier (eV) |
|---|---|---|---|
| Graphene | 0.18 [72] | o-$B_2N_2$ | 0.38 [54] |
| Silicene | 0.23 [47] | o-$B_2P_2$ | 0.35 [55] |
| Germanene | 0.21 [48] | $Si_3C$ | 0.18 [56] |
| Borophosphene | 0.50 [49] | $WB_4$ | 0.47 [57] |
| Blue phosphorus | 0.16 [18] | Siliborophene | 0.32 [58] |
| Black phosphorus | 0.12 [18] | $AlB_4$ | 0.47 [59] |
| Arsenene | 0.16 [50] | $Ti_2PX_2$ (X=S, Se, Te) | 0.17–0.21 [60] |
| h-Bas | 0.52 [51] | T-BN | 0.35 [61] |
| h-AlC | 0.78 [52] | $VS_2$ | 0.22 [62] |
| o-$Al_2N_2$ | 0.26 [53] | $TiB_2C_2$ | 0.54 [63] |

materials theoretically present significant diffusion barriers, which make them potential candidates for high-performance battery systems.

## 2.5 OPEN-CIRCUIT VOLTAGE AND THEORETICAL CAPACITY

OCV and specific capacity (C) are important parameters for designing new anode materials, influencing the performance and efficiency of batteries. The OCV provides valuable insight into the thermodynamic stability and potential of the anode materials, indicating its capacity to propel electrons, which affects the energy density and efficiency of the battery. On the other hand, the specific capacity quantifies the maximum charge (lithium) that the anode can store, which is an essential factor impacting the battery's total energy storage capacity. High specific capacity and significant average OCV are crucial for achieving high energy density and long lifetime during the charge/discharge process [73, 74].

To better understand these electrochemical characteristics, the strategy to increase the number of Li atoms adsorbed on anode materials has been utilized, conceptually respecting the adsorption energy of Li atoms [75]. In general, the half-reaction that occurs on the anode part of an LIB can be rearranged as follows:

$$Li_{n_1}XY+(n_2-n_1)Li^{+}+(n_2-n_1)e^{-} \Leftrightarrow Li_{n_2}XY \tag{2.6}$$

where $n_1$ and $n_2$ are two different concentrations of lithium adsorbed ($n_1 < n_2$). Theoretically, the OCV is an essential parameter to understand the applicability of anode materials in battery systems. The Nernst equation to compute OCV is expressed as [76]

$$V=-\frac{\Delta G}{(n_2-n_1)F}=-\frac{G(Li_{n_2}XY)-G(Li_{n_1}XY)-(n_2-n_1)G(Li)}{(n_2-n_1)F} \tag{2.7}$$

where $\Delta G$ and $F$ represent the change in Gibbs free energy and the Faraday constant, respectively. At low temperatures, especially in first principles calculations (0K), the entropy and volume change are small and can be neglected. Thus, the reaction free energy can be approximated by the internal energy $\Delta G \approx \Delta E$. The internal energy of reaction can be computed using DFT calculations, and the equilibrium reaction voltage becomes as [76]

$$V=-\frac{E(Li_{n_2}XY)-E(Li_{n_1}XY)-(n_2-n_1)E(Li)}{(n_2-n_1)} \tag{2.8}$$

where $E\left(Li_{n_{1,2}}XY\right)$ presents the total energy of $XY$ material with the adsorption of $n_1$ and $n_2$ number of Li atoms. $E(Li)$ is the total energy of Li atom in its bulk phase.

In addition to OCV, another factor that describes how much charge can be stored per unit mass of a material is the theortical specific capacity (C). This capacity is directly based on the number of electrons involved in the redox reaction. Principally, the theoretical specific capacity of designed anode materials highly depends on the number of reactive electrons and the molar weight of the material, as expressed by the following equation [76]:

$$C = \frac{n_{max}F}{M(XY)} \tag{2.9}$$

where $n_{max}$, $M(XY)$, and F represent the number of reactive electrons per formula unit, the relative molecular mass of material, and the Faraday constant, respectively. In fact, the number of electrons is linked with the number of Li ions stored in the host lattice. Importantly, this equation indicates that a smaller molecular weight and accommodation of more electrons per formula unit can theoretically lead to a higher capacity.

As is well-known, commercial graphite serves as the primary anode material in LIB systems owing to its wide lithium insertion range and low voltage. It has a significant theoretical capacity of about 372 mAh/g, which is nearly achieved in practice, while its applicability is limited to the growing demand for large scale applications [46]. Consequently, there is a considerable interest in designing alternative anode materials. Researchers are currently focused on designing and predicting novel anode materials with higher specific capacities. Table 2.3 shows recent 2D materials alongside graphite, showing their theoretical capacities. The materials listed in this table are potential candidates for negative electrode materials owing to their high specific capacities. However, challenges including volume expansion and instability during the charge-discharge process can hinder their widespread adoption in battery systems, presenting another challenge for selecting superior anode materials [77]. The materials listed in the table present minimal volume change and excellent stability, making them favorable candidates for experimental synthesis as anode materials.

TABLE 2.3 Theoretical Capacity (mAh/g) of Various 2D Materials Reported Recently from First Principles Study

| 2D Material | Theoretical Capacity | 2D Material | Theoretical Capacity |
|---|---|---|---|
| Graphene | 372 [46] | o-$B_2N_2$ | 3239.74 [54] |
| Silicene | 954 [47] | o-$B_2P_2$ | 1924 [55] |
| Germanene | 369 [48] | $Si_3C$ | 1394.11 [56] |
| Borophosphene | 1282.7 [49] | $WB_4$ | 708 [57] |
| Blue phosphorus | 865 [18] | Siliborophene | 1291.48 [58] |
| Black phosphorus | 216 [18] | $AlB_4$ | 3053.3 [59] |
| Arsenene | 1609.8 [50] | $Ti_2PX_2$ (X=S, Se, Te) | 281 [60] |
| h-Bas | 522.08 [51] | T-BN | 1620 [61] |
| h-AlC | 739.61 [52] | $VS_2$ | 466 [62] |
| o-$Al_2N_2$ | 1144.3 [53] | $TiB_2C_2$ | 1075 [63] |

## 2.6 APPLICATION: o-$AL_2N_2$ MONOLAYER

Orthorhombic-$Al_2N_2$ is a III-V compound, theoretically proposed as a novel 2D inorganic material [78]. It consists of two atoms of aluminum and two atoms of nitrogen, arranged in a planar structure similar to graphene. Figure 2.3a shows the top and side views of the optimized structure

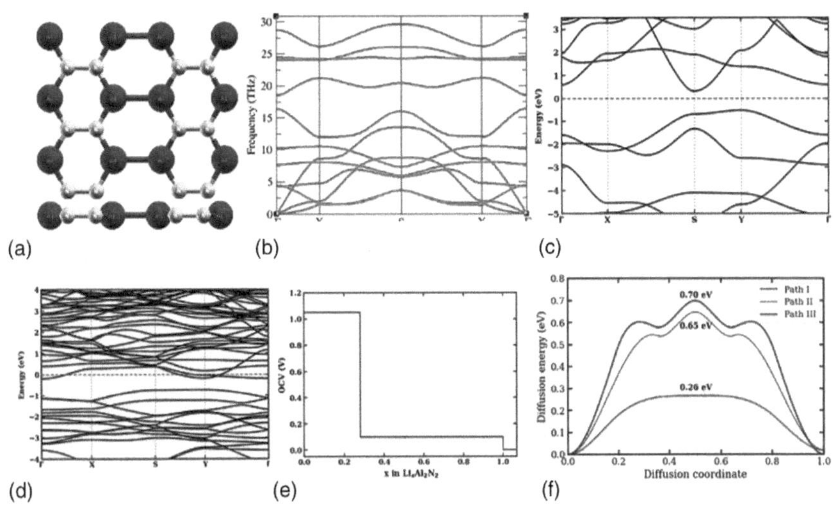


FIGURE 2.3 (a) Top and side views of the optimized structure of o-$Al_2N_2$ monolayer; (b) phonon dispersion spectra of o-$Al_2N_2$ monolayer; (c) electronic band structure of o-$Al_2N_2$ monolayer; (d) electronic band structure after the intercalation of Li on o-$Al_2N_2$ monolayer; (e) voltage profile of o-$Al_2N_2$ at different Li adsorption concentrations; (f) diffusion pathways on o-$Al_2N_2$ and the corresponding energy barriers.

of o-$Al_2N_2$ monolayer. The structural stability of this 2D material has been investigated by computing its cohesive energy and phonon dispersion. The computed cohesive energy was found to be −5.59 eV/atom, ensuring thermodynamic stability and suggesting a strong bonding network in o-$Al_2N_2$ monolayer. Figure 2.3b displays the phonon dispersion curve of the o-$Al_2N_2$ monolayer along the high symmetry points in the 1st Brillouin zone. The absence of negative frequencies confirms the dynamical stability of the o-$Al_2N_2$ phase. Additionally, the electronic properties were calculated by computing its electronic BS, as shown in Figure 2.3c. The o-$Al_2N_2$ monolayer presents a semiconductor aspect with an indirect band gap (S-Y). The performance of the o-$Al_2N_2$ monolayer as an anode material in LIBs has been theoretically investigated. The intercalation of Li ions at the most available site of the o-$Al_2N_2$ monolayer induces a transition from a semiconducting to a metallic aspect, offering better electronic conductivity. Figure 2.3d shows the electronic BS after the intercalation of Li, indicating an overlap between valence and conduction bands. During the storage process, the o-$Al_2N_2$ monolayer demonstrated a favorable average voltage of about 0.575 V (Figure 2.3e) and a high storage capacity of 1144.2913 mAh/g, significantly larger than that of the commercial graphite anode. Additionally, the lithiation process causes a small shrinking in cell volume, approximately 2.87% at maximum storage, indicating excellent cycling performance of the monolayer. Furthermore, the o-$Al_2N_2$ monolayer presents an ultra-low energy barrier of lithium, about 0.26 eV, facilitating the mobility of Li ions on the o-$Al_2N_2$ monolayer (Figure 2.3f). Based on these findings, the 2D o-$Al_2N_2$ monolayer emerges as a novel and promising candidate for anode material in LIBs [53]. To be noted, more experimental studies are needed for this kind of 2D material to fully understand its potential and practical applications.

## 2.7 CONCLUSION

In conclusion, this chapter has summarized various theoretical strategies and tools using DFT in the design of novel 2D anode materials. We have demonstrated the importance of DFT in understanding the fundamental electrochemical mechanisms of anode materials. DFT has proven to be an invaluable technique in developing efficient anode materials by providing insights into structural stability, electronic properties, adsorption process, ionic mobility, specific capacity, and OCV. The o-$Al_2N_2$ monolayer was discussed as a case study, highlighting the importance of theoretical studies in predicting the performance of novel anode material and paving a way for future experimental validation and potential industrial applications.

Continued advancements in computational approaches will accelerate the exploration and optimization of novel generation anode materials.

## REFERENCES


1. M.M. Aman, et al., Renew. Sustain. Energy Rev. **41**, 1190–1204 (2015), DOI: 10.1016/j.rser.2014.08.086.
2. M.R. Islam, et al., Renew. Sustain. Energy Rev. **21**, 456–468 (2013), DOI: 10.1016/j.rser.2013.01.007.
3. S.P. Neill, et al., Renewable Energy **127**, 763–778 (2018), DOI: 10.1016/j.renene.2018.05.007.
4. M. Balat, et al., Energy Sources, B: Econ. Plann. Policy **4**, 295–309 (2009), DOI: 10.1080/15567240701621158.
5. M. Amir, et al., J. Energy Storage **72**, 108694 (2023), DOI: 10.1016/j.est.2023.108694.
6. B. Dunn, et al., Science **334**, 928–935 (2021), DOI: 10.1126/science.1212741.
7. F.M. Nizam Uddin Khan, et al., Energy Rep. **9**, 11–21 (2023), DOI: 10.1016/j.egyr.2023.08.069.
8. H. Zhang, et al., Energy Storage Mater. **36**, 147–170 (2021), DOI: 10.1016/j.ensm.2020.12.027.
9. K.S. Chen, et al., ACS Energy Lett. **2**, 2026–2034 (2017), DOI: 10.1021/acsenergylett.7b00476.
10. Y. Yang, et al., Energy Rev. **3**, 100068 (2024), DOI: 10.1016/j.enrev.2024.100068.
11. X. Yi, et al., J. Energy Storage **95**, 112480 (2024), DOI: 10.1016/j.est.2024.112480.
12. J. Sung, et al., Mater. Chem. Front. **8**, 1861–1887 (2024), DOI: 10.1039/D3QM01171B.
13. Y. Lei, et al., ACS Nanosci. **2**, 450–485 (2022), DOI: 10.1021/acsnanoscienceau.2c00017.
14. P. Kumbhakar, et al., iScience **26**, 106671 (2023), DOI: 10.1016/j.isci.2023.106671.
15. B. Mortazavi, et al., J. Power Sources **329**, 106671 (2016), DOI: 10.1016/j.jpowsour.2016.08.109.
16. J. Zhuang, et al., Adv. Mater. **29**, 1606716 (2017), DOI: 10.1002/adma.201606716.
17. J. Zhu, et al., Nano-Micro Lett. **12**, 120 (2020), DOI: 10.1007/s40820-020-00453-x.
18. Q.-F. Li, et al., J. Phys. Chem. C **119**, 8662–8670 (2015), DOI: 10.1021/jp512411g.
19. S. Sun, et al., J. Chem. Eng. **338**, 27–45 (2018), DOI: 10.1016/j.cej.2017.12.155.
20. E. Darkaoui, et al., Chin. J. Phys. **89**, 884–895 (2024), DOI: 10.1016/j.cjph.2024.03.041.
21. X. Sun, et al., Beilstein J. Nanotechnol. **8**, 2711–2718 (2017), DOI: 10.3762/bjnano.8.270.
22. M.R. Kumar, et al., J. Power Sources **556**, 232256 (2023), DOI: 10.1016/j.jpowsour.2022.232256.

23. E.K.U. Gross, et al., Dens. Funct. Theory (1993), DOI: 10.1007/978-1-4757-9975-0.
24. P. Hohenbergt, et al., Phys. Rev. **136**, B864 (1964), DOI: 10.1103/PhysRev.136.B864.
25. P. Patel, et al., J. Energy Storage **71**, 108074 (2023), DOI: 10.1016/j.est.2023.108074.
26. D.G. Ladha, et al., Mater. Today Chem. **11**, 94e111 (2019), DOI: 10.1016/j.mtchem.2018.10.006.
27. J. Tang, et al., J. Alloys Compd. **730**, 228–233 (2018), DOI: 10.1016/j.jallcom.2017.09.197.
28. S. Kansara, et al., J. Mater. Chem. A **11**, 24482 (2023), DOI: 10.1039/d3ta05042d.
29. J. Lin, et al., Comput. Mol. Sci. **10**, e1473 (2020), DOI: 10.1002/wcms.1473.
30. Q. Bai, et al., Adv. Energy Mater. **8**, 1702998 (2018), DOI: 10.1002/aenm.201702998.
31. D. Chen, et al., J. Am. Chem. Soc. **139**, 7071–7081 (2017), DOI: 10.1021/jacs.7b03141.
32. W. Chen, et al., J. Power Sources **561**, 232720 (2023), DOI: 10.1016/j.jpowsour.2023.232720.
33. Q. He, et al., Energy Environ. Mater. **2**, 1–16 (2019), DOI: 10.1002/eem2.12056.
34. M. Geng, et al., J. Phys. Chem. C **123**, 464–472 (2019), DOI: 10.1021/acs.jpcc.8b09047.
35. C. Liu, et al., Mater. Today **19**, 109–123 (2016), DOI: 10.1016/j.mattod.2015.10.009.
36. A. Togo, et al., Scr. Mater. **108**, 1–5 (2015), DOI: 10.1016/j.scriptamat.2015.07.021.
37. P. Giannozzi, et al., J. Phys.: Condens. Matter. **29**, 465901 (2017), DOI: 10.1088/1361-648X/aa8f79.
38. P. Blaha, et al., Chem. Phys. **152**, 074101 (2020), DOI: 10.1063/1.5143061.
39. I. Pallikara, et al., Electron. Struct. **4**, 033002 (2022), DOI: 10.1088/2516-1075/ac78b3.
40. H. Wang, et al., J. Phys. Chem. C **119**, 10197–10205 (2015), DOI: 10.1021/acs.jpcc.5b00353.
41. L. Wirtz, et al., Solid State Commun. **131**, 141–152 (2004), DOI: 10.1016/j.ssc.2004.04.042.
42. M. Alaqarbeh, RHAZES: Green Appl. Chem. **13**, 43–51 (2021), DOI: 10.48419/IMIST.PRSM/rhazes-v13.28283.
43. S. Stavric, et al., Phys. Rev. Mater. **2**, 114007 (2018), DOI: 10.1103/PhysRevMaterials.2.114007.
44. R. Yazami, et al., J. Power Sources **153**, 312–318 (2006), DOI: 10.1016/j.jpowsour.2005.05.087.
45. K.T. Chan, et al., Phys. Rev. B **77**, 235430 (2008), DOI: 10.1103/PhysRevB.77.235430.
46. C.H. Doh, et al. Bull. Korean Chem. Soc. **32**, 2045 (2011), DOI: 10.5012/bkcs.2011.32.6.2045.

47. G.A. Tritsaris, et al., Nano Lett. **13**, 2258–2263 (2013), DOI: 10.1021/nl400830u.
48. D.K. Sharma, et al., Comput. Condens. Matter. **16**, e00314 (2018), DOI: 10.1016/j.cocom.2018.e00314.
49. S. Wang, et al., Phys. Chem. Chem. Phys. **22**, 21326–21333 (2020), DOI: 10.1039/D0CP03247F.
50. H. Benzidi, et al., Phys. Chem. Chem. Phys. **21**, 19951–19962 (2019), DOI: 10.1039/C9CP03230D.
51. N. Khossossi, et al., Phys. Chem. Chem. Phys. **21**, 18328–18337 (2019), DOI: 10.1039/C9CP03242H.
52. D. Chodvadiya, et al., Appl. Surf. Sci. **593**, 153424 (2022), DOI: 10.1016/j.apsusc.2022.153424.
53. M. Agouri, et al., J. Energy Storage **94**, 112351 (2024), DOI: 10.1016/j.est.2024.112351.
54. N. Khossossi, et al., Nano Energy **96**, 107066 (2022), DOI: 10.1016/j.nanoen.2022.107066.
55. W.C. Sun, et al., J. Mater. Sci. **56**, 13763–13771 (2021), DOI: 10.1007/s10853-021-06174-9.
56. Y. Wang, et al., J. Mater. Chem. A **8**, 4274–4282 (2020), DOI: 10.1039/C9TA11589G.
57. M.K. Masood, et al., J. Phys. Chem. Solids **186**, 111814 (2024), DOI: 10.1016/j.jpcs.2023.111814.
58. Z. Tabandeh, et al., J. Energy Storage **76**, 109640 (2024), DOI: 10.1016/j.est.2023.109640.
59. S. Ma, et al., Appl. Surf. Sci. **648**, 159024 (2024), DOI: 10.1016/j.apsusc.2023.159024.
60. B. Ge, et al., Appl. Surf. Sci. **550**, 149177 (2021), DOI: 10.1016/j.apsusc.2021.149177.
61. X. Xiong, et al., Comput. Mater. Sci. **232**, 112626 (2024), DOI: 10.1016/j.commatsci.2023.112626.
62. Y. Jing, et al., J. Phys. Chem. C **117**, 25409–25413 (2013), DOI: 10.1021/jp410969u.
63. A. Etrini, et al., Phys. Chem. Chem. Phys. **25**, 21699–21707 (2023), DOI: 10.1039/D3CP02724D.
64. S. Lebègue, et al., Phys. Rev. B **97**, 115409 (2009), DOI: 10.1103/PhysRevB.79.115409.
65. K.S. Novoselov, et al., Phys. Status Solidi B **244**, 4106–4111 (2007), DOI: 10.1002/pssb.200776208.
66. L.J. Zhou, et al., J. Phys. Chem. C **116**, 21780–21787 (2012), DOI: 10.1021/jp304861d.
67. J. Asenbauer, et al., Sustainable Energy Fuels **4**, 5387–5416 (2020), DOI: 10.1039/D0SE00175A.
68. G. Wang, et al., Nano-Micro Lett. **16**, 150 (2024), DOI: 10.1007/s40820-024-01363y.
69. G. Henkelman, et al., J. Chem. Phys. **113**, 9901–9904 (2000), DOI: 10.1063/1.1329672.

70. G. Henkelman, et al., J. Chem. Phys. **113**, 9978–9985 (2000), DOI: 10.1063/1.1323224.
71. K. Fan, et al., Mater. Rep.: Energy **3**, 100213 (2023), DOI: 10.1016/j.matre.2023.100213.
72. E.G. Leggesse, et al., Carbon **103**, 209–216 (2016), DOI: 10.1016/j.carbon.2016.03.016.
73. A.M. Yao, et al., J. Phys. Chem. Lett. **15**, 1143–1151 (2024), DOI: 10.1021/acs.jpclett.3c03129.
74. M.d.H.Hossain,et al.,Chem.Eng.J.Adv. **16**, 100569 (2023), DOI: 10.1016/j.ceja.2023.100569.
75. A. Urban, et al., npj Comput. Mater. **2**, 16002 (2016), DOI: 10.1038/npjcompumats.2016.2.
76. S.A. Arote, Fundamentals and perspectives of lithium-ion batteries (2022), DOI: 10.1088/978-0-7503-4881-2ch1.
77. F.K. Tareq, et al., Mater. Today Commun. **39**, 108653 (2024), DOI: 10.1016/j.mtcomm.2024.108653.
78. J. Zhao, et al., Phys. Chem. Chem. Phys. **23**, 3771 (2021), DOI: 10.1039/d0cp05909a.

CHAPTER 3

# Advancements in Solar Cell Technology

## *Integrating DFT and SLME Methods for Lead-Free Perovskite Development*

Hakima Ouhenou, Abderrahman Zaghrane, Abderrahman Abbassi, Souad Taj, and Bouzid Manaut

### 3.1 INTRODUCTION

The technology of materials in solar cells has seen significant advancements, driven by the quest for higher efficiency, stability, and environmentally friendly options [1]. Traditional silicon-based solar cells, while efficient, face limitations in terms of production costs and material availability [2]. This has led to the exploration of alternative materials such as perovskites, which have garnered attention for their remarkable light absorption and charge carrier properties [3].

Perovskite solar cells (PSCs) first appeared with a power conversion efficiency (PCE) of 3.8% in 2009 and quickly advanced to over 10% within a few years [4–8]. After extensive research, their efficiency has now surpassed 29.5%, marking a remarkable achievement in the history of photovoltaics compared to the development of conventional solar cells [9, 10]. PSCs are also attractive due to their simple fabrication processes, low processing costs, and the abundance of raw materials, which reduce their overall production costs to two-thirds that of single-crystalline and polycrystalline silicon solar cells [11]. To successfully commercialize PSC technology in

DOI: 10.1201/9781003499732-3

the near future, several key challenges must be addressed. These include achieving controllable thin film growth and deposition, ensuring high stability and long lifespan, and reducing toxicity. Additionally, to remain competitive, manufacturing companies must focus on cost efficiency [12].

Lead-free halide perovskites have emerged as a promising material in solar technologies due to their exceptional performance characteristics and wide range of applications [13]. They are notable for their use in multilayer ceramic capacitors, their potential in catalytic and magnetic applications, and their effectiveness in optoelectronic devices like solar cells, light-emitting diodes (LEDs), and photodetectors. Their superior attributes, including a favorable bandgap, robust light absorption, high carrier mobility, defect tolerance, and low exciton binding energy, offer significant advantages over silicon-based solar cells, particularly in manufacturing feasibility. These properties collectively make lead-halide perovskites a highly attractive option for sustainable and high-performance solar technologies [14].

Density Functional Theory (DFT) and Spectroscopic Limited Maximum Efficiency (SLME) play pivotal roles in determining the properties of materials, particularly in the context of solar cell development [15, 16]. DFT is a quantum mechanical method used to investigate the electronic structure of many-body systems, allowing researchers to predict and understand material properties at the atomic level [17]. By utilizing DFT, scientists can calculate fundamental properties such as band structure, density of states, and charge distribution, which are crucial for evaluating a material's suitability for photovoltaic applications. SLME, on the other hand, provides a theoretical framework to estimate the maximum efficiency of a solar cell by considering both the electronic properties derived from DFT calculations and the optical absorption spectra [15]. This combined approach enables a comprehensive assessment of a material's potential performance in solar cells, guiding the design and optimization of new, efficient, and stable materials for sustainable energy solutions.

In this chapter, we have discussed the development and advantages of lead-free PSCs. Our analysis is based on several research studies that examine the physical properties of lead-free perovskites using DFT. We have reviewed and compared the results, focusing on their application in solar cells through SLME evaluation.

## 3.2 DFT AND SLME IN SOLAR CELL DEVELOPMENT

DFT is a computational quantum mechanical modeling method used to investigate the electronic structure of many-body systems, including atoms and molecules. It has become an essential tool in material science and has significantly advanced our understanding and development of new materials for various applications, including solar cell technology [17].

DFT plays a pivotal role in advancing solar cell technologies across multiple fronts. It serves as a cornerstone for material screening by predicting and analyzing electronic structures and properties, enabling the identification of materials suited for solar cell applications [18]. Additionally, DFT is instrumental in band gap engineering, crucial for enhancing solar cell efficiency by accurately calculating and optimizing band gaps tailored for optimal solar energy conversion. Moreover, DFT aids in understanding defects within materials, including their formation energies and impacts on electronic and optical properties, which are essential for evaluating and mitigating performance limitations in solar cell devices [17].

The relationship between (DFT) and the SLME of solar cells is pivotal in advancing photovoltaic technology. DFT provides a robust computational framework to predict and optimize the electronic structure and optical properties of materials used in solar cells. By accurately simulating the behavior of electrons within these materials, DFT enables researchers to identify optimal configurations that enhance light absorption, charge carrier mobility, and overall efficiency [19]. This predictive capability aligns closely with SLME, which quantifies the maximum efficiency achievable in converting sunlight into electricity based on these material properties [16]. Integrating DFT with SLME calculations allows for a deeper understanding of how material characteristics influence solar cell performance, thereby guiding the design of more efficient and stable photovoltaic devices. Figure 3.1 represents the workflow from DFT to SLME method. The SLME is theoretically determined based on the optical properties and electronic band structures of materials. The proposed SLME simulation model is designed to assess the PCE of solar cells, leveraging an extended variation of the Shockley-Queisser model, as represented in the following equation [16]:

$$SLME(\eta) = \frac{P_{max}}{P_{in}} \tag{3.1}$$

where $P_{max}$ and $P_{in}$ denote the maximum output power density and the incident power, they are given as:

$$P_{max} = max\left( I_{sc} - I_0 exp\left( \frac{eV}{K_b T} - 1 \right) \right) \tag{3.2}$$

$$P_{in} = \int_0^{\infty} I_{sun}(E)\,dE \tag{3.3}$$

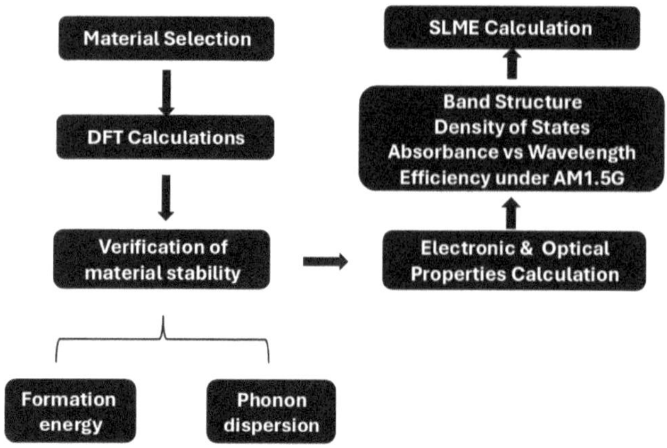


FIGURE 3.1 From DFT simulations to SLME estimation: A comprehensive workflow for solar cell materials.

where $I_{sc}$, $I_0$, V, $K_b$, T, and e denote short-circuit current density, current density at reverse saturation, potential difference across the photo absorber material, Boltzmann's constant, absolute temperature, and elementary charge, respectively.

$I_{sc}$ and $I_0$ can be determined using the diode equation below [20]:

$$J_{sc} = e\int_0^\infty \alpha(E) I_{sun}(E) dE \tag{3.4}$$

$$J_0 = \frac{J_0^\tau}{f_r} = \frac{e\pi}{f_r}\int_0^\infty \alpha(E) I_{bb}(E,T) dE \tag{3.5}$$

where $I_{bb}$ $(E,T)$ represents the blackbody emission spectrum, while $I_{sun}$ represents the Air Mass 1.5 Global solar irradiance spectrum.

$J_0^\tau$, $f_r$ and $\alpha$ denote the radiative recombination current density, the radiative fraction, and the absorption coefficient, respectively.

$$\alpha(E) = 1 - e^{-2\alpha(E)L} \tag{3.6}$$

$$f_r = exp\left(\frac{\Delta}{KT}\right) \tag{3.7}$$

$$\Delta = E_g^{da} - E_g \tag{3.8}$$

where Δ represents the difference between the direct allowed and fundamental bandgaps and L is the thickness of material.

## 3.3 EXPLORING LEAD-FREE PEROVSKITE MATERIALS FOR SOLAR CELL APPLICATION

In all lead-free alternatives, double perovskites have shown the greatest potential in replacing lead-based perovskites in solar cell applications. These materials offer excellent stability, favorable electronic and optical properties, and comparable or even superior SLME [19]. Recent studies on double perovskite halides reveal their promising characteristics for sustainable and efficient solar energy conversion, making them a viable and environmentally friendly option for future solar cell development.

In this section, we have discussed the results of research studies that have utilized DFT to investigate the stability, electronic, optical, and SLME properties of double perovskite halides, as presented in Table 3.1, along with the crystal structure illustrated in Figure 3.2. These studies provide a comprehensive understanding of the material characteristics and potential performance enhancements, offering valuable insights for the development of efficient and lead-free solar cells.

After calculating the properties of any material, it is essential to verify its stability. The tolerance factor (t) are calculated to evaluate the structural stability, using the following equation [21]:

$$t = \frac{R_A + R_X}{\sqrt{2}\left(\frac{R_B + R_{B'}}{2} + R_X\right)} \tag{3.9}$$

Figure 3.3 confirms the structural stability of the double-halide perovskite materials, which were identified within specific ranges for key parameter: $0.71 < \tau < 1.04$ for Goldschmidt's tolerance factor [22]. The calculated tolerance factors confirm the stability of these materials in the cubic phase. Their lattice parameters are presented in Figure 3.4.

TABLE 3.1 Chemical Composition of the Studied Materials

| A | B | B′ | X |
|---|---|---|---|
| Cs | Cu | Al | F |
| Rb | Ga | Ga | Cl |
| | Tl | Sb | Br |
| | Bi | Co | |
| | | Ag | |

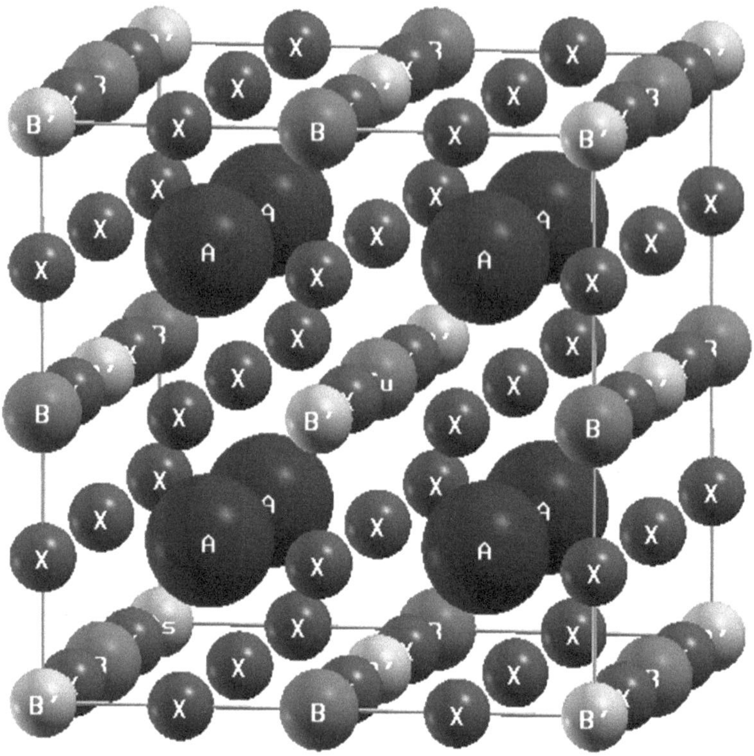


FIGURE 3.2 Crystal structure of $A_2BB'X_6$.

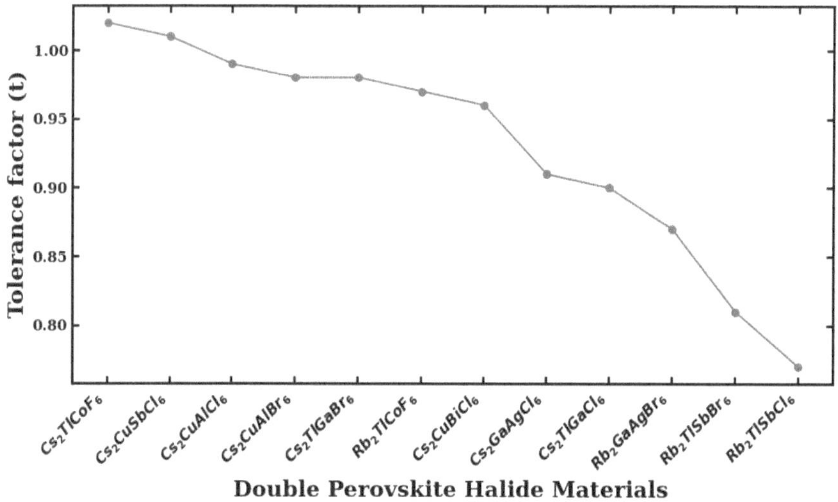


FIGURE 3.3 Goldschmidt factor of studied materials.

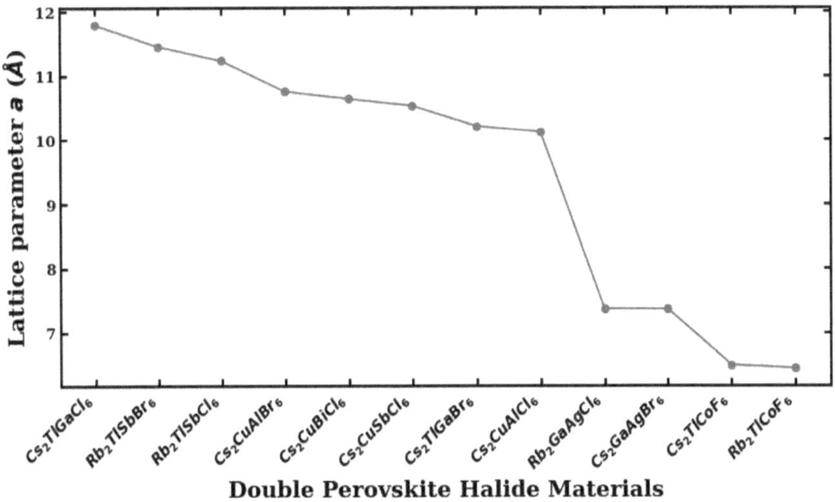


FIGURE 3.4 Lattice parameters for studied double perovskite halides.

To assess thermodynamic stability of materials, the formation energies are calculated using this equation [23]:

$$E_F = \frac{E_{tot}\left(A_2BB'X_6\right) - \left(jE^{A}_{Bulk} + kE^{B}_{Bulk} + lE^{B'}_{Bulk} + mE^{X}_{Bulk}\right)}{j+k+l+m} \tag{3.10}$$

where $E_{tot}(A_2BB'X_6)$ denote the total energy of $A_2BB'X_6$, $E^{A}_{Bulk}$, $E^{B}_{Bulk}$, $E^{B'}_{Bulk}$, $E^{X}_{Bulk}$ are energies of A, B, B′, and X atom, respectively, in their bulk structure.

The negative formation energy values calculated for all these materials confirm their stability, as shown in Table 3.2.

For understanding and evaluating the mechanical behavior and stability of solid materials heavily relies on the calculation of independent

TABLE 3.2 Formation Energies of Studied Materials

| Compound | Formation Energy (eV/Atom) |
|---|---|
| $Cs_2CuAlBr_6$ | −1.437[24] |
| $Rb_2TlSbCl_6$ | −1.51[25] |
| $Cs_2CuAlCl_6$ | −1.549[24] |
| $Cs_2TlGaBr_6$ | −1.58[26] |
| $Rb_2TlSbBr_6$ | −1.63[25] |
| $Cs_2TlGaCl_6$ | −1.82[26] |
| $Rb_2TlCoF_6$ | −4.38[27] |
| $Cs_2TlCoF_6$ | −5.03[27] |

TABLE 3.3 Elastic Constants of Studied Double Halide Perovskites

| Compound | C11 | C12 | C44 |
|---|---|---|---|
| $Rb_2TlCoF_6$[27] | 46.681 | 31.248 | 9.351 |
| $Cs_2TlCoF_6$[27] | 44.797 | 29.681 | 8.328 |
| $Cs_2CuSbCl_6$[30] | 57.21 | 19.56 | 5.05 |
| $Cs_2CuBiCl_6$[30] | 56.27 | 19.23 | 4.05 |
| $Cs_2TlGaCl_6$[26] | 33.42 | 7.39 | 7.84 |
| $Cs_2TlGaBr_6$[26] | 28.30 | 7.47 | 7.98 |
| $Rb_2TlSbCl_6$[25] | 27.50 | 2.57 | 3.51 |
| $Rb_2TlSbBr_6$[25] | 26.05 | 1.86 | 2.41 |
| $Rb_2GaAgCl_6$[31] | 41.02055 | 10.17550 | 19.72355 |
| $Cs_2GaAgCl_6$[31] | 27.45410 | 13.00920 | 24.44670 |

stiffness constants. These constants, commonly known as elastic parameters, provide significant insights into the characteristics of materials and play an essential role in many engineering applications. For materials with cubic symmetry, there are three key elastic constants: $C_{11}$, $C_{12}$, and $C_{44}$ [28]. The calculated values of these elastic constants are presented in Table 3.3. To determine the mechanical stability of a crystalline substance, these calculated elastic constants must satisfy the well-known Born stability conditions. The conditions are as follows [29]:

$$C_{11} - C_{12} > 0, C_{11} > 0, C_{44} > 0 \quad \text{and} \quad C_{11} + 2C_{12} > 0 \tag{3.11}$$

Abeer Mera and Muhammad Awais Rehman [27] utilized DFT to investigate the structural, electronic, optical, and mechanical properties of $Ae_2TlCoF_6$ (Ae = Rb, Cs) double perovskite materials. They confirmed the materials' dynamic stability and feasibility for experimental synthesis through calculations of phonon dispersion curves and formation energies. The electronic band structures revealed semiconductor characteristics, with direct band gap values of 1.35 eV for $Rb_2TlCoF_6$ and 1.64 eV for $Cs_2TlCoF_6$. Optical analysis indicated high absorption coefficients, low reflectivity, and minimal energy loss functions. Furthermore, using the SLME metric, they assessed the efficiency of these perovskite materials. Results showed that $Rb_2TlCoF_6$ achieved a maximum SLME of 25.07% at 0.67 micrometers, while $Cs_2TlCoF_6$ reached 24.59% at 2.17 micrometers.

Similarly, Moussa Kibbou et al. [24] conducted a comprehensive investigation into the electronic, optical, and solar cell characteristics of hybrid double perovskite (HDP) materials, focusing on $Cs_2CuAlCl_6$ and $Cs_2CuAlBr_6$. Their research highlighted these materials' robust structural and thermodynamic stability, underscored by their negative formation energies and high tolerance factors (0.98–0.99). Electronic analysis revealed that $Cs_2CuAgCl_6$ and $Cs_2CuAgBr_6$ exhibit semiconductor behavior with direct bandgaps of 1.35 eV and 0.93 eV, respectively. Particularly noteworthy, $Cs_2CuAgCl_6$ demonstrated an impressive maximum-PCE (32.72%), outperforming $Cs_2CuAgBr_6$ (29.31%). These findings advocate strongly for the practical application of these adjustable direct bandgap double-halide perovskites in solar cells and other optoelectronic devices, positioning them as superior alternatives capable of enhancing power efficiency in energy harvesting technologies.

Muhammad Awais Rehman et al. [26] investigated the double perovskite halides $Cs_2TlGaCl_6$ and $Cs_2TlGaBr_6$, emphasizing their suitability for photovoltaic applications due to their optoelectronic properties. Employing first-principles DFT with the HSE06 hybrid functional, they systematically explored structural stability, electronic structure, optical absorption, and mechanical properties. The study confirmed stable crystal structures based on calculated tolerance and octahedral factors. $Cs_2TlGaCl_6$ exhibited an indirect band gap of 2.50 eV, while $Cs_2TlGaBr_6$ showed 1.54 eV, both conducive to visible-light absorption. Elastic constant calculations supported mechanical stability. SLME estimates suggested promising photovoltaic efficiencies of 12.41% and 29.73% for $Cs_2TlGaCl_6$ and $Cs_2TlGaBr_6$, respectively, highlighting their potential for cost-effective, high-efficiency solar cells and warranting further experimental validation.

Similarly, Abeer Mera et al. [25] utilized DFT to investigate the optoelectronic properties of $Rb_2TlSbM_6$ (M = Cl, Br) halides. The stability of their cubic phases was assessed using Goldsmith's tolerance factor and enthalpy of formation calculations. Electronic characteristics were analyzed, revealing bandgaps of 1.713 eV for both $Rb_2TlSbCl_6$ and $Rb_2TlSbBr_6$ using *Generalized Gradient Approximation using the Perdew-Burke-Ernzerhof functional* GGA-PBE potentials, ensuring accuracy and comparison with prior data. These perovskite halides exhibit high ultraviolet absorption coefficients and low reflectivity (~20%), highlighting their potential for optoelectronics and photovoltaics. SLME analysis at 293 K indicated promising SLME values of 31.04% for $Rb_2TlSbCl_6$ and 27.53% for $Rb_2TlSbBr_6$, underscoring their suitability for solar cell applications.

Moreover, Zia Ur Rehman et al. [31] studied the structural, mechanical, vibrational, electronic, optical, and SLME properties of $A_2GaAgCl_6$ (A = Cs, Rb), a double perovskite material, using DFT. First-principles calculations were performed with CASTEP and Quantum ESPRESSO. $A_2GaAgCl_6$

possesses a cubic structure with space-group symmetry Fm-3m. The lattice parameters of $Cs_2GaAgCl_6$ and $Rb_2GaAgCl_6$ were optimized using energy-volume curves, resulting in values of 7.357 Å and 7.365 Å, respectively. Population analysis confirmed charge transfer among transition metals and halogen atoms. The stability of $A_2GaAgCl_6$ (A = Cs, Rb) was demonstrated by analyzing phonon dispersion curves, which showed no negative frequencies. Band structure calculations indicated the semiconducting nature of the compounds, with energy gaps of 0.96 eV for $Cs_2GaAgCl_6$ and 0.88 eV for $Rb_2GaAgCl_6$. Optical characteristics confirmed that the materials are suitable for devices operating primarily in the visible region of the electromagnetic spectrum. SLME results showed that $Cs_2GaAgCl_6$ has an efficiency of 30% and $Rb_2GaAgCl_6$ has an efficiency of 27%, suggesting their potential use in photovoltaics.

However, Muskan Nabi and Dinesh C. Gupta [30] studied the structural stability, electronic, elastic, and optical properties of inorganic halide double perovskites $Cs_2CuMCl_6$ (M = Sb, Bi). Both materials were found to be stable in a cubic structure, following the space symmetry of the Fm-3m space group. Stability within this group was confirmed using energy optimization and tolerance factor calculations. The positive values of the elastic constants further authenticated the mechanical stability of these materials. The band structure and density of states revealed a semiconducting character with small indirect band gaps, measured at 1.55 eV for $Cs_2CuSbCl_6$ and 2 eV for $Cs_2CuBiCl_6$ using GGA + mBJ + SOC methods. These materials exhibited excellent optical absorption in the visible range, making them suitable for optoelectronic applications. The SLME was calculated for both materials, with $Cs_2CuSbCl_6$ achieving 30% and $Cs_2CuBiCl_6$ achieving 19%, indicating their potential for solar cell applications.

Figure 3.5 illustrates the band gap values of various lead-free perovskites studied, ranging from 0.88 eV to 2.50 eV. These values indicate the semiconductor behavior of these materials. A bandgap in this range suggests that these perovskites can efficiently absorb sunlight and potentially be used in solar cell applications. Understanding their bandgap characteristics is crucial for assessing their suitability and performance in photovoltaic devices.

Figure 3.6 represents the SLME values for various double perovskite halides. The materials included in the study are $Cs_2CuAlCl_6$, $Rb_2TlSbCl_6$, $Cs_2GaAgCl_6$, $Cs_2CuSbCl_6$, $Cs_2TlGaBr_6$, $Cs_2CuAlBr_6$, $Rb_2TlSbBr_6$, $Rb_2GaAgCl_6$, $Rb_2TlCoF_6$, $Cs_2TlCoF_6$, $Cs_2CuBiCl_6$, and $Cs_2TlGaCl_6$. Among

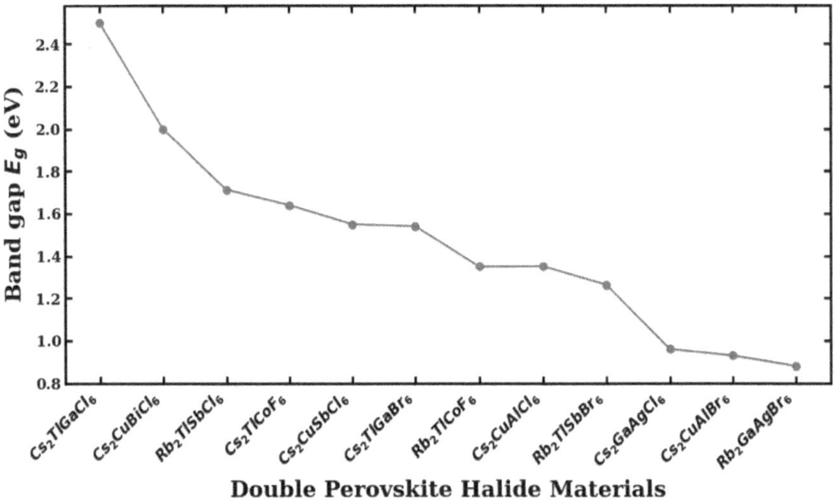


FIGURE 3.5 Band gap values of studied materials.

these, $Cs_2CuAlCl_6$ exhibits the highest SLME value at 32.72%, indicating its potential as a highly efficient material for solar cell applications. Following closely are $Rb_2TlSbCl_6$ and $Cs_2GaAgCl_6$ with SLME values of 31.04% and 30%, respectively. On the lower end of the spectrum, $Cs_2CuBiCl_6$ and $Cs_2TlGaCl_6$ show SLME values of 19% and 12.41%, respectively. This variation in SLME values highlights the diverse potential of double perovskite halides in achieving high-efficiency solar cells, with certain compositions demonstrating significantly higher efficiency potentials.

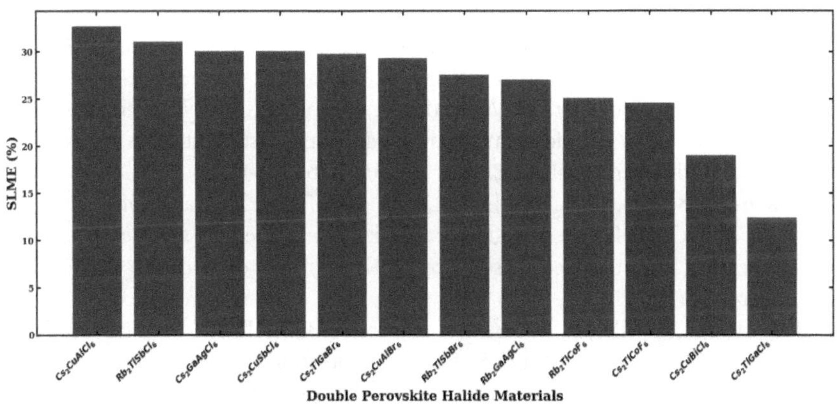


FIGURE 3.6 Spectroscopic Limited Maximum Efficiency (SLME) of the studied materials.

## 3.4 CONCLUSION

This chapter extensively explores the potential of lead-free halide perovskites in solar cell technologies, highlighting their exceptional characteristics and promising applications. The advantageous traits of these materials, including a favorable bandgap, high SLME efficiency, and robust defect tolerance, underscore their significant potential for sustainable and high-performance solar energy solutions. Leveraging DFT analyses provides a comprehensive understanding of the structural, electronic, optical, and mechanical properties of lead-free halide perovskites, further reinforcing their suitability for the solar technology sector. Notably, the SLME results of the studied lead-free perovskite materials reveal a remarkable efficiency of 32.72% for $Cs_2CuAlCl_6$, indicating its strong candidacy for solar cell applications and highlighting the overall excellence of these materials in this specific domain.

## REFERENCES

1. Chu, S., Cui, Y., & Liu, N. (2017). The path towards sustainable energy. *Nature materials*, *16*(1), 16–22.
2. Green, M. A. (2017). Corrigendum to 'Solar cell efficiency tables (version 49)' [Prog. Photovolt: Res. Appl. 2017; 25: 3–13]. *Progress in photovoltaics: Research and applications*, *25*(4), 333–334.
3. Lin, Y., Chen, B., Zhao, F., Zheng, X., Deng, Y., Shao, Y., ..., & Huang, J. (2017). Matching charge extraction contact for wide-bandgap perovskite solar cells. *Advanced materials*, *29*(26), 1700607.
4. Kojima, A., Teshima, K., Shirai, Y., & Miyasaka, T. (2009). Organometal halide perovskites as visible-light sensitizers for photovoltaic cells. *Journal of the American chemical society*, *131*(17), 6050–6051.
5. Im, J. H., Lee, C. R., Lee, J. W., Park, S. W., & Park, N. G. (2011). 6.5% efficient perovskite quantum-dot-sensitized solar cell. *Nanoscale*, *3*(10), 4088–4093.
6. Lee, M. M., Teuscher, J., Miyasaka, T., Murakami, T. N., & Snaith, H. J. (2012). Efficient hybrid solar cells based on meso-superstructured organometal halide perovskites. *Science*, *338*(6107), 643–647.
7. Kim, H. S., Lee, C. R., Im, J. H., Lee, K. B., Moehl, T., Marchioro, A., ..., & Park, N. G. (2012). Lead iodide perovskite sensitized all-solid-state submicron thin film mesoscopic solar cell with efficiency exceeding 9%. *Scientific reports*, *2*(1), 591.
8. Burschka, J., Pellet, N., Moon, S. J., Humphry-Baker, R., Gao, P., Nazeeruddin, M. K., & Grätzel, M. (2013). Sequential deposition as a route to high-performance perovskite-sensitized solar cells. *Nature*, *499*(7458), 316–319.
9. Green, M., Dunlop, E., Hohl-Ebinger, J., Yoshita, M., Kopidakis, N., & Hao, X. (2021). Solar cell efficiency tables (version 57). *Progress in photovoltaics: Research and applications*, *29*(1), 3–15.

10. Kim, J. Y., Lee, J. W., Jung, H. S., Shin, H., & Park, N. G. (2020). High-efficiency perovskite solar cells. *Chemical reviews*, *120*(15), 7867–7918.
11. Park, N. G. (2013). Organometal perovskite light absorbers toward a 20% efficiency low-cost solid-state mesoscopic solar cell. *The journal of physical chemistry letters*, *4*(15), 2423–2429.
12. Yang, W. S., Park, B. W., Jung, E. H., Jeon, N. J., Kim, Y. C., Lee, D. U., …, & Seok, S. I. (2017). Iodide management in formamidinium-lead-halide–based perovskite layers for efficient solar cells. *Science*, *356*(6345), 1376–1379.
13. Saparov, B., & Mitzi, D. B. (2016). Organic–inorganic perovskites: structural versatility for functional materials design. *Chemical reviews*, *116*(7), 4558–4596.
14. Ma, X., Yang, L., Lei, K., Zheng, S., Chen, C., & Song, H. (2020). Doping in inorganic perovskite for photovoltaic application. *Nano energy*, *78*, 105354.
15. Neugebauer, J., & Hickel, T. (2013). Density functional theory in materials science. *Wiley interdisciplinary reviews: Computational molecular science*, *3*(5), 438–448.
16. Yu, L., & Zunger, A. (2012). Identification of potential photovoltaic absorbers based on first-principles<? Format?> Spectroscopic screening of materials. *Physical review letters*, *108*(6), 068701.
17. Orio, M., Pantazis, D. A., & Neese, F. (2009). Density functional theory. *Photosynthesis research*, *102*, 443–453.
18. Giorgi, G., Fujisawa, J. I., Segawa, H., & Yamashita, K. (2013). Small photocarrier effective masses featuring ambipolar transport in methylammonium lead iodide perovskite: a density functional analysis. *The journal of physical chemistry letters*, *4*(24), 4213–4216.
19. Qian, J., Liu, Y., Song, J., Liu, L., Xu, B., Chen, G., & Tian, W. (2018). Spectroscopic Limited Practical Efficiency (SLPE) model for organometal halide perovskites solar cells evaluation. *Organic electronics*, *59*, 389–398.
20. Cuevas, A. (2014). The recombination parameter J0. *Energy procedia*, *55*, 53–62.
21. Ouhenou, H., Zaghrane, A., Darkaoui, E., Agouri, M., Waqdim, A., Abbassi, A., …, & Driouich, M. (2024). First-principles examination of structural, electronic, magnetic, and optical properties of a free lead scintillating double perovskites $A_2NaTaX_6$(A = Cs, Rb; X = Cl, Br, I). *Optical and quantum electronics*, *56*(6), 941.
22. Li, Z., Yang, M., Park, J. S., Wei, S. H., Berry, J. J., & Zhu, K. (2016). Stabilizing perovskite structures by tuning tolerance factor: formation of formamidinium and cesium lead iodide solid-state alloys. *Chemistry of materials*, *28*(1), 284–292.
23. Zaghrane, A., Ouhenou, H., Darkaoui, E., Agouri, M., Abbassi, A., Mekaoui, Y., …, & Manaut, B. (2023). First-principles investigation of structural, electronic, optical, and magnetic properties of a scintillating double perovskite halides ($Cs_2LiCeX_6$) with (X = F, Br, and I). *Chinese journal of physics*, *90*, 911–921.

24. Kibbou, M., Haman, Z., Lahbi, Z., Ouabida, E., Essaoudi, I., Ahuja, R., & Ainane, A. (2024). Advancing photovoltaics and optoelectronics: exploring the superior performance of lead-free halide perovskites. *Optical materials, 147,* 114737.
25. Mera, A., Rehman, M. A., ur Rehman, Z., Sarfraz, Z., Sohaib, M., Fatima, J., & Usman, M. (2024). Exploring the physical properties of $Rb_2TlSbM_6$ (M = Cl, Br) inorganic halide perovskites for solar cell applications: a DFT study. *Inorganic chemistry communications, 165,* 112528.
26. Rehman, M. A., Manzoor, O., Rehman, Z. U., Ifseisi, A. A., Fatima, J., Usman, M., …, & Hamad, A. (2024). Study of physical properties of $Cs_2TlGaX_6$ (X = Cl, Br) halide perovskites via HSE-06 hybrid technique for high efficiency solar cells. *Optical and quantum electronics, 56*(7), 1079.
27. Mera, A., & Rehman, M. A. (2023). Exploring the physical properties of $Ae_2TlCoF_6$ (Ae = Rb, Cs) double perovskites for solar cell applications by first-principles calculations. *Inorganic chemistry communications, 158,* 111656.
28. Abbassi, A., Agouri, M., Iacomi, F., Manaut, B., & Elhadadi, B. (2023). Magneto-thermal, mechanical, and opto-electronic properties of $Sr_2MWO_6(M = V, Rh, Ru)$: Ab initio study. *Journal of superconductivity and novel magnetism, 36*(3), 995–1001.
29. Jamal, M., Asadabadi, S. J., Ahmad, I., & Aliabad, H. R. (2014). Elastic constants of cubic crystals. *Computational materials science, 95,* 592–599.
30. Nabi, M., & Gupta, D. C. (2021). Potential lead-free small band gap halide double perovskites $Cs_2CuMCl_6$ (M = Sb, Bi) for green technology. *Scientific reports, 11*(1), 12945.
31. Rehman, Z. U., Rehman, M. A., Khan, M. R., Rehman, B., Sikiru, S., Rizwan, M., …, & Usman, M. (2024). First principle study of $X_2GaAgCl_6$ (X = Cs, Rb) double perovskites: structural, mechanical, vibrational, electronic, optical, SLME, thermoelectric, and thermodynamic properties for solar cell applications. *Environmental science and pollution research, 31,* 1–15.

CHAPTER 4

# Modeling Structural, Electronic, and Optical Properties of Monolayer MXenes $M_4C_3$ (M = Nb, Ta) Based on DFT

El Mokhtar Darkaoui, Hmad Fatihi, Abderrahman Abbassi, Souad Taj, and Bouzid Manaut

## 4.1 INTRODUCTION

In 2011, it surfaced as a form of new two-dimensional (2D) material. A layer of carbon or nitrogen atoms sits between the metallic layers of MXenes, connecting a few stacked arrays of early transition metal atoms, also known as carbonitrides or nitrides, and transition metal carbides. The chemical formula for MXenes is $M_{n+1}X_n$ (n = 1–3) [1], where $X$ is carbon or nitrogen and M is for transition metal carbides (such as Ti, Nb, Zr, Ta, Hf, V, Sc, Cr, and Mo) [2]. Other significant monolayers, besides monocarbides, have the chemical formula $M_4C_3$ (M = Nb and Ta). Based on ab-initio calculations, the stabilities of monolayers $M_4C_3$ (M = Nb and Ta) were examined, and it was found that they were stable based on their negative formation energies [3]. MXenes often exhibit lateral dimensions between nanometers and micrometers, with thicknesses up to nanometers [4]. MXenes' dual characteristics of hydrophilicity and metallic conductivity have shown them to have a wide range of uses in electrocatalysis, biosensing, water purification, and energy storage [5–8]. Particularly, MXenes' hydrophilicity expands

DOI: 10.1201/9781003499732-4

their potential for numerous biocompatible molecular grafting applications on the MXene surface [9]. The optical characteristics of MXenes, including light absorption, emission, and scattering, have also drawn attention from researchers. Research on MXenes and light interactions, such as scattering, emission, and absorption, shed light on how 2D materials interact with photons and offer viable options for a variety of common uses, including the use of solar energy, energy storage, catalysis, and bioimaging [10]. MXenes' potent light-harvesting ability makes them ideal for applications involving the conversion of light to heat. In the context of photothermal therapy (PTT) for tumor ablation combined with photoacoustic imaging (PAI)-guided diagnostics, the photothermal property of MXenes has been thoroughly investigated. More positively, MXenes have demonstrated desirable absorption ranges at the second biological window (biowindow), indicating a bigger maximum allowable exposure and maybe deeper tissue penetration [11]. As of right now, a number of alluring MXene-based synergistic phototherapeutic treatments that combine PDT, chemotherapy, and photodynamic therapy (PTT) collectively exhibit considerable promise in PTT [1]. Due to their ability to overcome the limitations of tests, first-principles calculations are thought to be helpful for predicting the physical properties of compounds and have been performed for a variety of compounds [12–17]. In this work, we investigated the characteristics of $M_4C_3$ by first-principles calculations. We also recalculated the structural, electronic, and optical properties of the $M_4C_3$MXene system, where M = Nb and Ta. Our results may provide theoretical guidance for future experimental investigation of $M_4C_3$ (M=Nb and Ta).

## 4.2 COMPUTATIONAL METHODOLOGY

Density functional theory (DFT) calculations were conducted using the Quantum ESPRESSO package (QE) [18], with a cut-off energy of 85 *Ry* (1197 *eV*). Norm-conserving pseudopotentials were employed to describe the electron–core interactions [19]. The Broyden–Fletcher–Goldfarb–Shanno (BFGS) algorithm and the Perdew–Burke–Ernzerhof (PBE) functional were utilized to optimize the crystal structure of the studied system and to address the exchange–correlation effects, respectively [20, 21]. A vacuum layer of 20 ($A^0$) in the z-direction was applied to prevent interactions between adjacent layers. Additionally, we used the Monkhorst–Pack approach [22] with k-point mesh samplings of $10\times10\times1$, $20\times20\times1$, and $30\times30\times1$ for structural optimizations, and for electronic and optical property calculations in the Brillouin zone. To achieve more accurate electronic properties, the PBE functional tends to underestimate the band gap of metal materials [23]. The parameters used, including the energy cut-off and k-point meshes, were optimized to meet energy convergence criteria of $10^{-5}$ Ry and atomic forces of $10^{-4}$ Ry/Bohr, respectively. For investigating dynamic stability and the vibrational phonon spectrum, the finite displacement method implemented in the Phononpy code was used [24]. The $M_4C_3$ MXenes monolayer structures are visualized using the VESTA software [25].

## 4.3 RESULTS AND DISCUSSIONS

In this section, we present a detailed discussion of the findings related to the $M_4C_3$ ($M = Nb, Ta$) MXenes monolayers.

### 4.3.1 Structural Optimization and Stability

The structural properties of the two-dimensional $M_4C_3$ MXene system, as depicted in Figure 4.1, reveal important details about the material's

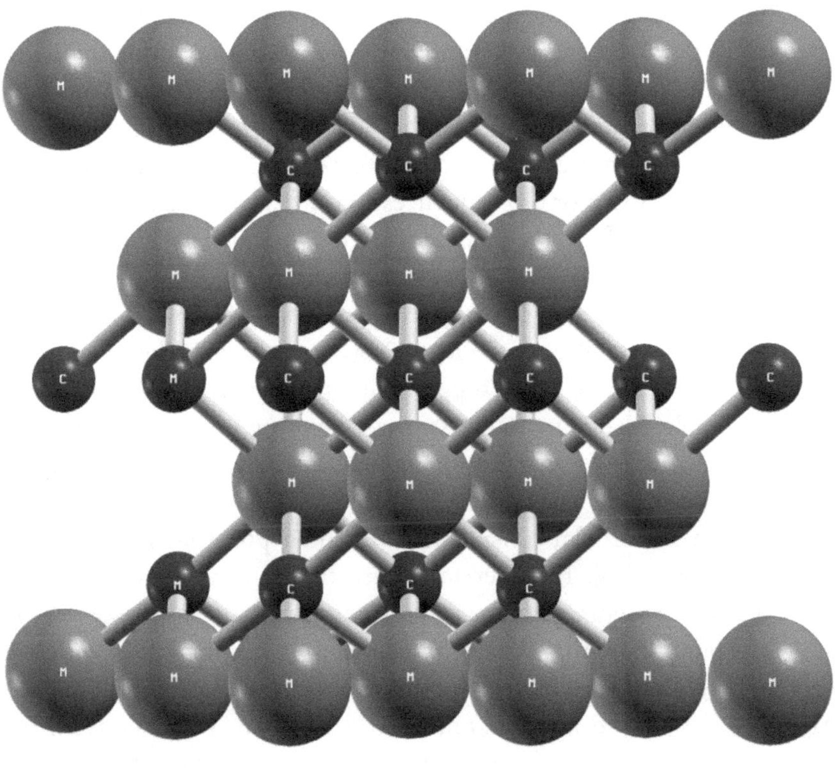

FIGURE 4.1 2D MXene monolayer structure of $M_4C_3$ ($M = Nb, Ta$).

TABLE 4.1 Calculated Lattice Constants ($a = b$) for the $M_4C_3$ ($M = Nb, Ta$) MXenes Monolayers, Bond Lengths ($d_{M-C}$ and $d_{C-C}$), Thickness $h(A^o)$, and Cohesive Energy ($E_{coh}$)

| Compounds | $a = b(A^o)$ | $d_{M-C}(A^o)$ | $d_{C-C}(A^o)$ | $h(A^o)$ | $E_{coh}$ (eV/atom) |
|---|---|---|---|---|---|
| $Nb_4C_3$ | 2.646 | 2.2555 | 3.1385 | 7.6253 | −0.328 |
| Other works | [27] | | | | −0.236 [26] |
| Compounds | $a = b(A^o)$ | $d_{M-C}(A^o)$ | $d_{C-C}(A^o)$ | $h(A^o)$ | $E_{coh}$ (eV/atom) |
| $Ta_4C_3$ | 2,645 | 2.2480 | 3.1176 | 7.6533 | −0.347 |
| Other works | [27] | | | | −0.259 [26] |

atomic-scale organization. The Nb and Ta atoms (M) form a tetragonal arrangement, with the C atoms occupying the center positions. Figure 4.1 illustrates the extended crystal structure of the $M_4C_3$ MXene, highlighting the stacking of the individual MXene layers. The interlayer spacing between the $M_4C_3$ sheets is an important structural parameter that can influence the overall properties of the material. These intricate structural details are crucial in understanding the electronic and optical characteristics of the $M_4C_3$ MXene system, which can be tuned by varying the *M* elements.

Table 4.1 shows the calculated lattice constants ($a = b$), bond lengths ($d_{M-C}$ and $d_{C-C}$), thickness ($h$), and cohesive energy ($E_{coh}$) for the $M_4C_3$ MXene monolayers, where *M* represents either *Nb* or *Ta*. The data reveals that the lattice constants for the $Nb_4C_3$ and $Ta_4C_3$ MXenes are close to those reported in previous studies [26, 27], indicating that the choice of the *M* element (*Nb* or *Ta*) influences the lattice parameters. Similarly, the bond lengths within the MXene structures are also dependent on the *M* and *C* elements.

The cohesive energies ($E_{coh}$) reported in Table 4.1 provide important insights into the stability of the $M_4C_3$ MXene materials. The more negative cohesive energy of $M_4C_3$ suggests that the structure is energetically more favorable, likely due to the stronger bonding interactions between the *M* and *C* atoms [28, 29]. These high cohesive energies imply that the $M_4C_3$ MXenes have sufficient driving force to form stable layered structures, which is a crucial consideration for their potential synthesis and characterization [28]. In order to ensure the dynamic stability of the $M_4C_3$ MXene systems, phonon dispersions were calculated (see Figure 4.2). The plots show the phonon frequencies as a function of the wave vector along the high-symmetry points in the Brillouin zone. For all four $M_4C_3$ MXenes, including $Nb_4C_3$ and $Ta_4C_3$, the phonon dispersions exhibit no imaginary (negative) frequencies, indicating that the structures are dynamically stable.

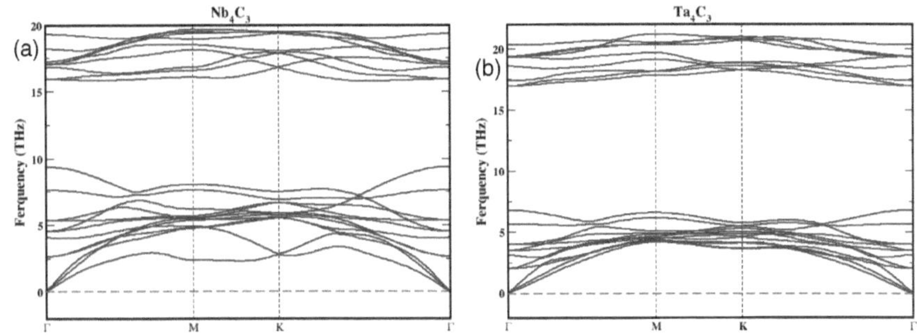


FIGURE 4.2 Phonon dispersions of (a) $Nb_4C_3$ and (b) $Ta_4C_3$ MXenes monolayers.

### 4.3.2 Electronic Properties

In continuation of our investigation, we computed the band gap energy with two different exchange–correlation functionals (PBE-GGA and DFT+U) for the four $M_4C_3$ MXene compounds, as shown in Table 4.2. The results reveal distinct differences in the electronic properties of the MXenes, depending on the choice of the $M$ element ($Nb$ or $Ta$). This indicates that these C-containing MXenes, $Nb_4C_3$ and $Ta_4C_3$, are reported as metallic, with no discernible band gap for either the PBE-GGA or DFT+U [30]. Therefore, the band structure and total density of states (TDOS) will be calculated using DFT+U approximation.

The band structure was sampled along the $\Gamma \rightarrow M \rightarrow K \rightarrow \Gamma$ path, encompassing high-symmetry points in the Brillouin zone. As depicted in Figure 4.3, the compound $M_4C_3$ exhibits metallic properties due to the overlapping orbitals, highlighting its distinct electronic properties compared to the monolayer form [30]. In order to obtain a comprehensive understanding of the electronic structure of $M_4C_3$ ($M = Nb$ and $Ta$), we also calculated the total density of states (*TDOS*).

As shown in Figure 4.4, the TDOS of $Nb_4C_3$ and $Ta_4C_3$ is computed within energy range of $-4\ eV \rightarrow 4\ eV$ and demonstrates the effect of transition

TABLE 4.2 Band Gaps of $M_4C_3$ ($M = Nb$, $Ta$) MXenes Monolayers Calculated Using Different Exchange–Correlation Functionals

| **Compounds** | **Exchange Functionals** | **Band gap (eV)** |
|---|---|---|
| $Nb_4C_3$ | PBE-GGA | Metal |
| | DFT + U | Metal |
| | Other works | [30] |
| $Ta_4C_3$ | PBE-GGA | Metal |
| | DFT + U | Metal |
| | Other works | [30] |

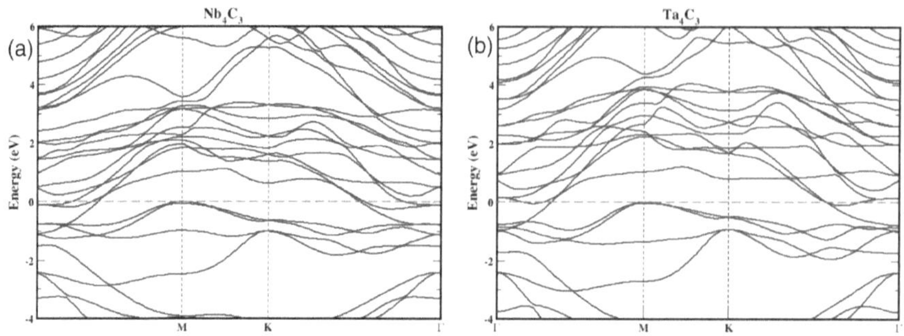


FIGURE 4.3 Band structure of (a) $Nb_4C_3$ and (b) $Ta_4C_3$ MXenes monolayers.

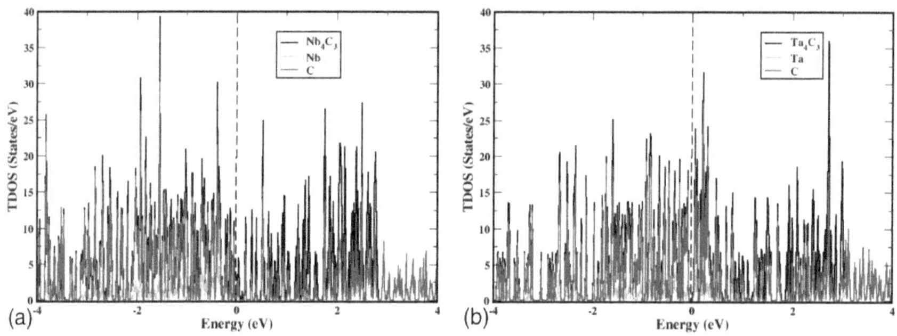


FIGURE 4.4 Total density of States (TDOS) of (a) $Nb_4C_3$ and (b) $Ta_4C_3$ MXenes monolayers.

metals (*M)* on the compounds and their band gap values. The two metals $Nb_4C_3$ and $Ta_4C_3$ have their valence and conduction bands overlaped, resulting in a high density of states at the Fermi level.

### 4.3.3 Optical Properties

The optical properties of MXenes depend on their structure, chemical composition, and surface state. Using DFT, we can model these properties by simulating the interactions between electrons and photons in the material. The results obtained can help us to understand light absorption, dielectric function, refractive index, and optical conductivity, which are crucial factors for potential applications of MXenes in optics.

The complex dielectric function, which has two components, the real and imaginary parts, can be written as follows:

$$\varepsilon(\omega) = \varepsilon_1(\omega) + i\varepsilon_2(\omega) \tag{4.1}$$

where $\varepsilon_1(\omega)$ is the real part, which describes the energy stored in the material, and $\varepsilon_2(\omega)$ represents the energy that is absorbed and dissipated as heat [31].

The computed optical characteristics for the polycrystalline monolayers $M_4C_3$ ($M = Nb$ and $Ta$) carbides are plotted in Figure 4.5. Figure 4.5 shows that for these $M_4C_3$ carbides, the $\varepsilon_1(\omega)$ curves drop with increasing energy before the values of $\varepsilon_1(\omega)$ go negative in the energy range of less than 5eV. According to the imaginary wave vector $K(\omega)$, it prevents light from traveling through

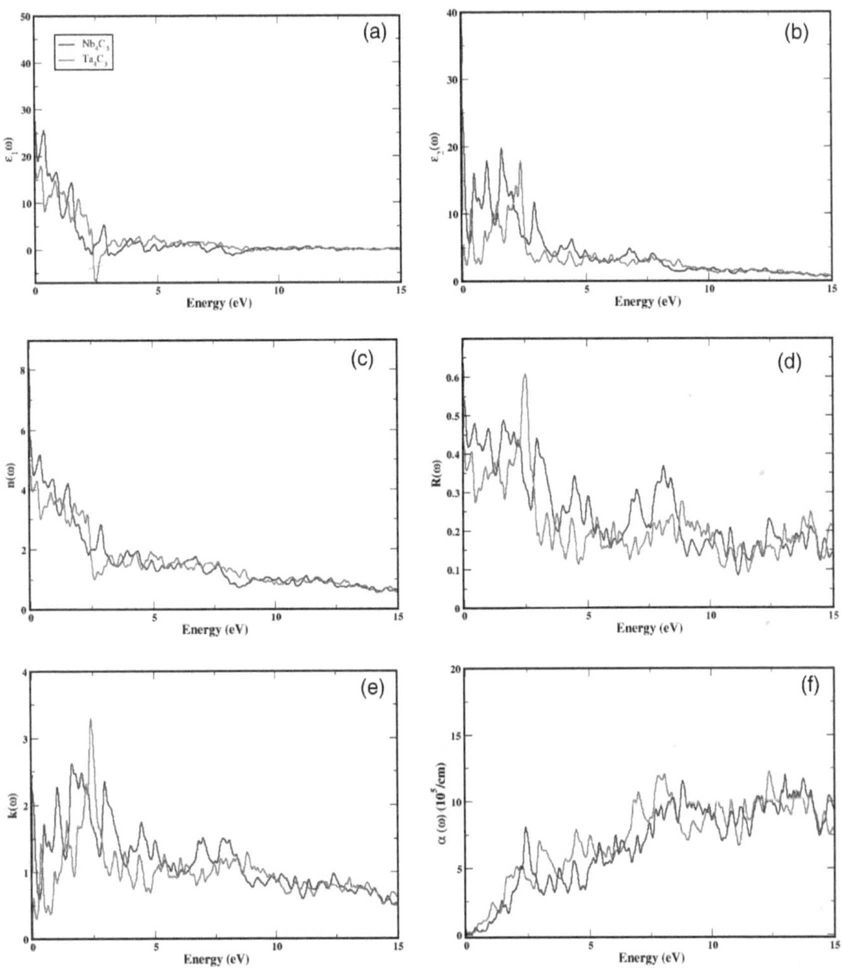


FIGURE 4.5 Real part $\varepsilon_1(\omega)$ (a), and imaginary part $\varepsilon_2(\omega)$) (b), of polycrystalline dielectric function, refractive index $n(\omega)$ (c), reflectivity $R(\omega)$ (d), extinction coefficient $K(\omega)$ (e), and absorption coefficient $\alpha(\omega)$ (f) of $M_4C_3$ ($M = Nb$ and $Ta$).

this range $w^2\varepsilon = C^2(K.K)$, and this also indicates that these $M_4C_3$ carbides are metallic. The values of $\varepsilon_1(0)$ for $Nb_4C_3$ and $Ta_4C_3$ are 25.293 and 23.597, respectively. When the energy grows, the $\varepsilon_2(\omega)$ curves show the same trend as the $\varepsilon_1(\omega)$ curves, but the $\varepsilon_2(\omega)$ curves don't have any negative values. Small peaks are also formed by some minor oscillations in the $\varepsilon_2(\omega)$ curves. These carbides' band topologies show that they don't have any band gaps. Consequently, electronic transitions between M and C can be responsible for the tiny peaks in the $\varepsilon_2(\omega)$ curves. The localization of electrons between M and C is the cause of these transitions, as confirmed by the charge density discrepancies. The refractive index $n(\omega)$, reflectivity $R(\omega)$ and the extinction coefficient $K(\omega)$ make up the complex refractive index $n(\omega)$. According to Figure 4.5(a) and 4.5(b), as the energy increases, the variations in $n(\omega)$ and $K(\omega)$ are the same as those in $\varepsilon_1(\omega)$ and $\varepsilon_2(\omega)$, respectively, because $\varepsilon_1(\omega)$ is directly derived from $\varepsilon_2(\omega)$ and $\varepsilon_1(\omega)^{0.5} = n(\omega)+iK(\omega)$. The values of the static extinction coefficient $K(0)$ for $Nb_4C_3$ and $Ta_4C_3$ are 1.016 and 0, respectively. The values of the static refractive index $n(0)$ for $Nb_4C_3$ and $Ta_4C_3$ are 5.131 and 4.858, respectively. The values of the static refractive index $R(0)$ for $Nb_4C_3$ and $Ta_4C_3$ are 0.469 and 0.434, respectively. According to Figure 4.5(f), $\alpha(\omega)$ exhibits fairly similar variations as the energy increases, where each $\alpha(\omega)$ peak corresponds at the same energy.

## 4.4 CONCLUSION

In this study, we employed DFT within the Quantum ESPRESSO package (QE) to investigate the structural, electronic, and optical properties of $M_4C_3$. Our structural optimizations were conducted using PBE-GGA approximations. Moreover, our investigation confirmed the stability of $Nb_4C_3$ and $Ta_4C_3$, which exhibit a stable behavior despite their negative cohesive energies. This intriguing observation indicates the possible presence of kinetic barriers or phase transitions, requiring further investigation. Our electronic properties analysis revealed distinct characteristics for the tow materials. $Nb_4C_3$ and $Ta_4C_3$ displayed metallic behavior. Furthermore, we observed high absorption coefficients in the ultraviolet (UV) and visible light ranges, indicating potential for optical applications. The optical properties follow the order: $Ta_4C_3 > Nb_4C_3$.

## REFERENCES

1. Naguib M, Kurtoglu M, Presser V, et al. Two-Dimensional Nanocrystals Produced by Exfoliation of $Ti_3AlC_2$. Adv Mater 2011;23:4248–4253.
2. Anasori B, Lukatskaya MR, Gogotsi Y. 2D Metal Carbides and Nitrides (MXenes) for Energy Storage. Nat Rev Mater 2017;2:16098.

3. Yu X-X, Weinberger CR, Thompson GB. Ab Initio Investigations of the Phase Stability in Group IVB and VB Transition Metal Carbides. Comput Mater Sci 2016;112:318–326.
4. Gogotsi Y, Anasori B. The Rise of MXenes. ACS Nano 2019;13:8491–8494.
5. Ghidiu M, Lukatskaya MR, Zhao MQ, Gogotsi Y, Barsoum MW. Conductive Two- Dimensional Titanium Carbide 'clay' With High Volumetric Capacitance. Nature 2014;516:78.
6. Peng J, Chen X, Ong WJ, Zhao X, Li N. Surface and Heterointerface Engineering of 2D MXenes and Their Nanocomposites: Insights into Electro- and Photocatalysis. Chem 2019;5:18–50.
7. Zhao J, Yang Y, Yang C, et al. A Hydrophobic Surface Enabled Salt-Blocking 2D Ti3C2 MXene Membrane for Efficient and Stable Solar Desalination. J Mater Chem A 2018;6:16196–16204.
8. Liu H, Duan C, Yang C, Shen W, Wang F, Zhu Z. A Novel Nitrite Biosensor Based on the Direct Electrochemistry of Hemoglobin Immobilized on MXene-Ti3C2. Sensors Actuat B Chem 2015;218:60–66.
9. Mayerberger EA, Street RM, McDaniel RM, Barsoum MW, Schauer CL. Antibacterial Properties of Electrospun Ti3C2Tz (MXene)/Chitosan Nanofibers, RSC Adv 2018;8:35386–35394.
10. Lin H, Chen Y, Shi J. Insights into 2D MXenes for Versatile Bio-Medical Applications: Current Advances and Challenges Ahead, Adv Sci 2018;5: 1800518.
11. Liu G, Zou J, Tang Q, et al. Surface Modified Ti3C2 MXene Nanosheets for Tumor Targeting Photothermal/Photodynamic/Chemo Synergistic Therapy. ACS App Mater Inter 2017;9:40077–40086.
12. Aouni A, Weisbecker P, Loi TH, Bauer-Grosse E. Search for New Materials in Sputtered V 1 x C x Films, Thin Solid Films 2004;469:315–321.
13. Ma L, Duan Y, Li R. Phase Stability Anisotropic Elastic Properties and Electronic Structures of C15 Type Laves Phases $ZrM_2$ (M =Cr, Mo, and W) from First Principles, Philos Mag 2017;97:2406–2424.
14. Darkaoui E, et al. "First-Principles Investigation of MXenes M4C3 (M= Sc, Cr, and Mn) for Clean Energy." Chinese J Phys 2024;89:884–895. https://doi.org/10.1016/j.cjph.2024.03.041
15. Quan S, Zhang X, Liu C, Jiang W. First-Principles Investigations on Structural Stability, Mechanical, and Thermodynamic Properties of LaT2Al20 (T= Ti, V, Cr, Nb, and Ta) Intermetallic Cage Compounds, Chin Phys B 2018;27:126201.
16. Bao W, Liu D, Li P, Duan Y. Elastic Anisotropies and Thermal Properties of Cubic TMIr (TM = Sc, Y, Lu, Ti, Zr and Hf): A DFT Calculation. Mater Res Exp 2019;6:086574.
17. Pan Y, Guan WM. Exploring the Novel Structure, Elastic and Thermodynamic Properties of W 3 Si Silicides from First-Principles Calculations, Ceram Int 2019;45:15649–15653.
18. Giannozzi P, Baron S, Bonini N, et al., Quantum Espresso: A Modular and Open-Source Software Project for Quantum Simulations of Materials, J. Phys. Condens. Matter, 2009;21:395502. https://doi.org/10.1088/0953-8984/21/39/395502.

19. Troullier N, Martins JL. Efficient Pseudopotentials for Plane-Wave Calculations. Phys Rev B: Condens Matter Mater Phys 1990;43:8861. https://doi.org/10.1103/PhysRevB.43.1993.
20. Paier J, Hirschl R, Marsman M, Kresse G. The Perdew-Burke-Ernzerhof Exchange- Correlation Functional Applied to the G2-1 Test Set Using a Plane-Wave Basis Set, J Chem Phys 2005;122:234102. https://doi.org/10.1063/1.1926272.
21. Dai YH. Convergence Properties of the BFGS Algorithm, SIAM J Optim 2002;13(3):693–701. https://doi.org/10.1137/S1052623401383455.
22. Monkhorst HJ, Pack JD. Specials Points for Brillouin-Zone Integration, Phys Rev B: Solid State 1976;13(12):5188. https://doi.org/10.1103/PhysRevB.13.5188
23. Perdew JP. Density Functional Theory and the Band Gap Problem. Int J Quantum Chem 1985;28:497–523. https://doi.org/10.1002/qua.560280846
24. Togo A, Tanaka I. First Principles Phonon Calculations in Materials Science, Scr Mater 2015;108:1–5. https://doi.org/10.1016/j.scriptamat,2015.07.021.
25. Wolfson R, VESTA Manual, Investig, Y Cienc., Scientific Am. ABRIL, 1983, pp. 1–5.
26. Parajuli D, Samatha K. MXene as Topological Insulator, JETIR 2019; 6(4):689–706.
27. Chibani H, Yahi H, Arshad N, Ouettar C. Revealing Structural, Mechanical, and Electronic Properties of M4C3 (M= Sc, Ti, Zr, Mo, Hf, and W) MXene Monolayers Based on First-Principle Calculations, Comput Condens Matter 2022;32:e00713.
28. Farid B, Godby RW. Cohesive Energies of Crystals, Phys Rev B 1991;43(17). https://doi.org/10.1103/PhysRevB.43.14248.
29. Lou L, et al. "Cohesive Energy Measurement of van der Waals Heterostructures by the Shaft Loaded Blister Test." Extreme Mech Lett 41 (2020) https://doi.org/10.1016/j.eml.2020.100987.
30. Guo Z, Jiang X. Mechanical Properties, Minimum Thermal Conductivities, and Optical Properties of Cubic TM4C3 (TM= V, Nb, Ta) Carbides. J Phys Chem Solids 2020;139:109330.
31. Petzelt J, Rychetský I, in Encyclopedia of Condensed Matter Physics, (2005).

CHAPTER 5

# Accelerated Discovery of High-Efficiency Solar-Cell Materials

## *Integrating Density Functional Theory and Machine Learning*

Hmad Fatihi, El Mokhtar Darkaoui, Abderrahman Abbassi, and Moha El Idrissi

## 5.1 INTRODUCTION

One of the most important answers to the world's energy and environmental problems is solar energy [1]. With the global shift toward more environmentally friendly energy sources [2], solar cell efficiency is becoming increasingly critical [3]. Finding high-efficiency solar-cell materials is still a difficult task, but it is essential to the advancement of solar technology. Innovative techniques are needed to speed up the discovery process since traditional trial-and-error procedures are time-consuming and frequently ineffective [4].

Density functional theory (DFT) has long been a cornerstone in the theoretical evaluation of material properties [5]. By calculating the electronic structure of materials, DFT provides crucial insights into their potential performance in various applications, including photovoltaics [6]. However, while DFT is powerful, its computational intensity limits the scale at which it can be applied [4].

In recent years, machine learning (ML) has revolutionized numerous fields by enabling the analysis of large datasets and the prediction of

DOI: 10.1201/9781003499732-5

complex patterns [7]. When integrated with DFT, ML offers a transformative approach to material discovery. ML models, trained on extensive DFT-computed datasets, can rapidly screen vast libraries of materials, identifying those with the most promising properties for solar-cell applications [4].

This chapter examines how DFT and ML work together to speed the search for effective solar-cell materials. It describes the processes for creating predictive ML models, choosing high-efficiency candidates, and conducting high-throughput screening. Researchers can greatly improve the speed and precision of the material discovery process by combining the advantages of DFT and ML, opening the door for the development of next-generation solar-cell technologies.

## 5.2 DENSITY FUNCTIONAL THEORY (DFT)

In computational physics and chemistry, DFT has emerged as one of the most important and popular techniques for analyzing the electronic structure of many-body systems. DFT's strength comes from its capacity to capture crucial quantum mechanical effects while simplifying the intricate many-electron issue into a more understandable form.

### 5.2.1 Principles of DFT

#### *5.2.1.1 Hohenberg-Kohn Theorems*

Two fundamental theorems put out by Hohenberg and Kohn in 1964 served as the foundation of DFT. According to the first theorem, a many-electron system's ground state features are exclusively dictated by its electron density, or E(r). This implies that all system observables may be obtained directly from the ground state electron density, bypassing the requirement to solve the many-body Schrödinger equation. The second theorem states that a universal functional F[n(r)] exists and that this functional may be reduced to get the system's precise ground state energy. Kinetic and potential energies are only two examples of the many interactions that are included in this functional.

#### *5.2.1.2 Kohn-Sham Equations*

Kohn and Sham presented a practical method to DFT in 1965, building on these principles. They suggested that a system of non-interacting electrons traveling in an effective potential may be used to map the many-body issue. The system's overall energy is represented as a function of the electron density as follows:

$$E[n] = T_s[n] + E_{ext}[n] + E_H[n] + E_{xc}[n]$$

where $T_s[n]$ is the kinetic energy of the non-interacting electrons, $E_{ext}[n]$ is the external potential energy, $E_H[n]$ is the Hartree energy representing electron-electron Coulomb interactions, and $E_{xc}[n]$ is the exchange-correlation energy accounting for the many-body interactions. This energy functional is varied in relation to the electron density to obtain the Kohn-Sham equations, which are a set of iteratively solvable self-consistent field equations.

## 5.3 SPECTROSCOPIC LIMITED MAXIMUM EFFICIENCY (SLME)

The Spectroscopic Limited Maximum Efficiency (SLME) is a crucial statistic for assessing photovoltaic materials' theoretical efficiency. The SLME takes into account actual material characteristics, such as absorption spectra and electronic band structure, in contrast to conventional metrics like the Shockley-Queisser limit, which only takes into account hypothetical circumstances. As a result, SLME becomes a more useful and realistic tool for estimating the possible performance of solar cell materials. The calculation of the SLME involves several steps: First, using DFT calculations, the absorption coefficient ($\alpha(\hbar\omega)$) of the material is determined over a range of photon energies ($\hbar\omega$). Next, the absorption coefficient is integrated over the solar spectrum to calculate the number of absorbed photons by integrating the product of the solar spectral irradiance ($I_{sun}(\hbar\omega)$) and the absorption coefficient. The maximum current ($J_{max}$) that can be generated by the material is then calculated based on the absorbed photons, converting the absorbed photon flux into an electrical current. The open circuit voltage ($V_{oc}$) is determined by considering the material's bandgap and recombination losses, typically involving detailed electronic structure calculations. Finally, the SLME is computed by combining the maximum current and the open circuit voltage, along with other factors such as the fill factor, to estimate the material's overall efficiency. The formula for SLME can be represented as $\eta = Vco*Jmax*FF$, where $J_{max}$ is the maximum current density, Vco is the open circuit voltage, FF is the fill factor, and Pin is the incident power of the solar spectrum.

## 5.4 MACHINE LEARNING (ML)

ML has become a novel technology with applications in a wide range of fields, such as finance, healthcare, materials science, and more. Fundamentally, ML is the use of statistical models and algorithms that allow computers to carry out particular tasks in the absence of explicit instructions by using patterns and inference. A comprehensive review of the basic ideas and varieties of ML is given in this section.

### 5.4.1 Key Concepts in Machine Learning

**Supervised Learning:** One of the most common types of ML is supervised learning. A labeled dataset is used to train the model, meaning that every training sample has an output label associated with it. Learning a mapping from inputs to outputs that can be applied to previously unknown data is the aim of the model. Supervised learning techniques include classification algorithms like decision trees, support vector machines (SVMs), and neural networks that are used to classify data into discrete classes and linear regression that is used to predict continuous outcomes. Applications including picture identification, spam detection, and medical diagnostics frequently make use of supervised learning.

**Unsupervised Learning:** In contrast to supervised learning, unsupervised learning deals with unlabeled data. The objective is to uncover hidden patterns or intrinsic structures within the data without predefined labels. Clustering and dimensionality reduction are two main types of unsupervised learning. Clustering algorithms, such as k-means and hierarchical clustering, group data points into clusters based on similarity, making them useful for market segmentation and anomaly detection. Dimensionality reduction techniques, such as Principal Component Analysis (PCA) and t-Distributed Stochastic Neighbor Embedding (t-SNE), reduce the number of features in a dataset while preserving important information, facilitating data visualization and noise reduction.

**Reinforcement Learning (RL):** A type of ML called RL teaches an agent to make decisions by interacting with its surroundings. Based on its behaviors, the agent is rewarded or penalized, and its goal is to maximize cumulative benefits over time. Inspired by behavioral psychology, RL works especially well in situations requiring sequential decision-making, such as autonomous driving, gaming, and robots. The exploration-exploitation trade-off, in which the agent must weigh the pros and cons of both discovering new actions and utilizing known activities to maximize rewards, and policy optimization, which entails enhancing the agent's decision-making process, are important ideas in RL.

**Deep Learning:** A branch of ML called "deep learning" is centered on deep neural networks, which are multi-layered neural networks.

By using several levels of abstraction, these networks are able to learn intricate data representations. Deep learning has attained cutting-edge results in a number of domains, including audio and picture identification, natural language processing, and gaming. Because they can automatically identify spatial hierarchies of features, Convolutional Neural Networks (CNNs) are a particular kind of deep learning model that performs well for tasks involving images. Language modeling and time series forecasting are two applications that benefit greatly from the sequential data handling capabilities of Recurrent Neural Networks (RNNs) and their derivatives, such as Long Short-Term Memory (LSTM) networks.

**Feature Engineering and Selection:** Feature engineering is the process of using domain knowledge to create new input features from raw data, enhancing the predictive power of ML models. This step is crucial because the quality and relevance of features directly impact the model's performance. Feature selection involves choosing a subset of the most important features for training, reducing the dimensionality of the data, and improving computational efficiency. Techniques for feature selection include filter methods (e.g., correlation coefficients), wrapper methods (e.g., recursive feature elimination), and embedded methods (e.g., Lasso regression).

**Model Evaluation and Validation:** Evaluating and validating ML models are essential steps to ensure their generalizability and robustness. Common evaluation metrics for regression tasks include Mean Absolute Error (MAE), Mean Squared Error (MSE), and R-squared. For classification tasks, metrics such as accuracy, precision, recall, F1-score, and area under the receiver operating characteristic (ROC) curve are used. Cross-validation is a widely used technique for model validation, where the data is split into several folds, and the model is trained and tested multiple times on different folds to ensure it performs well on unseen data. This helps to assess the model's stability and mitigate overfitting, where the model performs well on training data but poorly on new data.

## 5.5 METHODOLOGY FOR COMBINING ML WITH DFT

Finding effective solar-cell materials necessitates a solid methodology that combines advanced computational methods with theoretical predictions. Combining ML with DFT provides a thorough method for predicting and screening high-efficiency materials. This section describes the methods taken from many key papers, emphasizing the integrated approach's process and criteria [4, 8–11].

### 5.5.1 High-Throughput DFT Calculations

DFT, which assesses a material's electrical characteristics and offers vital information about how well it might function in solar cells, is the foundation of the material screening process. The high-throughput DFT methodology consists of the following crucial steps: First, DFT calculations are performed on a wide range of non-metallic materials to ascertain their optical characteristics and electronic band structures. The Tran-Blaha modified Becke-Johnson (mBJ) potential is employed in the computation of many material characteristics, including bandgaps. Determining the optical characteristics of the materials depends heavily on these electrical properties. The SLME metric is then computed utilizing the optical characteristics that were derived from the electronic structure data that was computed using the mBJ potential [12, 13]. By integrating the absorption coefficient throughout the solar spectrum and taking radiative recombination into account, the SLME metric calculates a material's maximal efficiency for use in solar cell applications., which aids in the identification of materials with high theoretical efficiency. Subsequently, the materials undergo screening according to their effective carrier mass, which must be minimal to guarantee effective charge transport. The thermodynamic stability of the materials is evaluated using convex hull stability calculations [14]; only those materials with an energy above the convex hull of less than 0.1 eV/atom are seen to be good candidates [4]. To make sure the materials can effectively absorb sunlight, the process also entails screening for 2D-layered materials, which may offer special benefits in solar-cell applications and computing the absorption coefficient. This all-inclusive method simplifies the process of identifying potential solar-cell materials by means of a thorough theoretical assessment. The methodologies employed for screening and predicting efficient materials are summarized in Figure 5.1.

### 5.5.2 Development of Machine Learning Models

By using DFT-computed data, ML models are created to improve the screening process and enable quick material property prediction. There are several intricate processes in the ML workflow:

First, a large dataset containing the electrical structure, bandgap, dielectric function, and SLME values of materials with DFT calculations

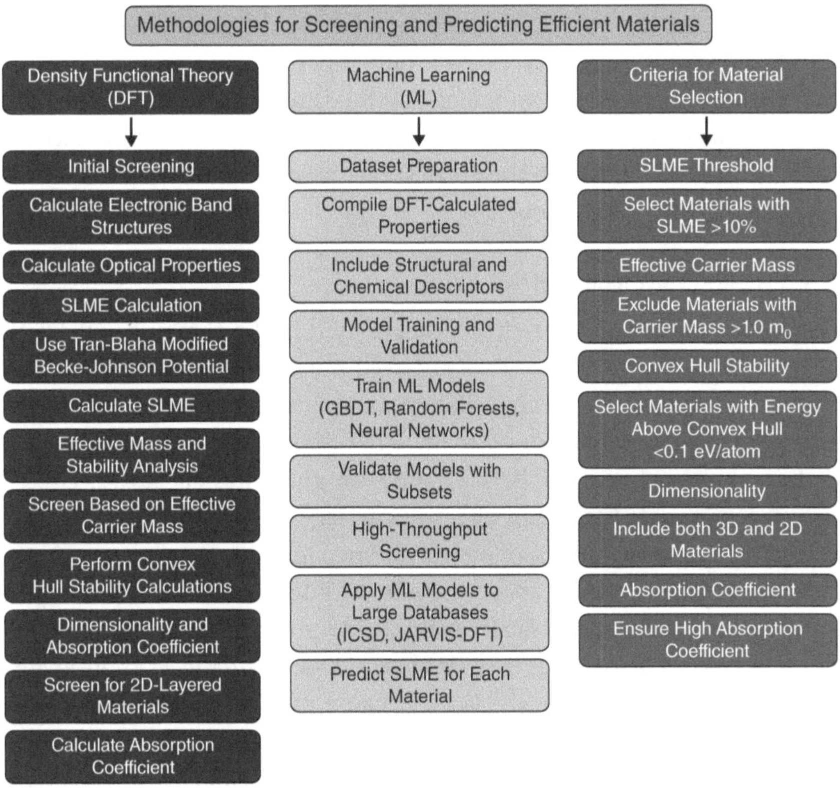


FIGURE 5.1 Methodologies for screening and predicting efficient materials.

is assembled. To increase the accuracy of the ML models, this dataset is enhanced with extra characteristics including structural and chemical descriptors. To find the best-performing model, a variety of ML techniques are explored, such as neural networks, random forests, and gradient-boosting decision trees (GBDTs). To guarantee robust performance, the models are tested on the remaining fraction (10%) after being trained on a portion of the data (usually 90%). Methods such as k-fold cross-validation are employed to evaluate the generalizability of the model and avoid over-fitting. Model performance is assessed using performance indicators like the area under the ROC, AUC, and MSE, using a high Area Under the Curve (AUC) indicating strong predictive capability.

After being taught and verified, the ML models are used on extensive datasets, including prospective materials, such as the JARVIS-DFT database and the Inorganic Crystal Structure Database (ICSD). The materials with the best potential efficiency are quickly identified by the models, which forecast the SLME for each material. By drastically reducing the number of candidates, our high-throughput screening procedure eases the burden of future DFT computations. Finding high-efficiency solar-cell materials will happen faster thanks to the ML-driven pre-screening, which helps rank candidates with the greatest qualities for additional theoretical and experimental validation.

### 5.5.3 Criteria for Material Selection

As shown in Figure 5.1, strict criteria are used in the combined DFT and ML technique to guarantee that only the most promising materials are chosen for additional research. Before materials are taken into consideration for additional screening, their SLME score must be higher than a certain threshold, usually 10%. In order to guarantee successful charge transfer, materials having an effective carrier mass larger than 1.0 $m_0$—where $m_0$ is the mass of a free electron are omitted. Furthermore, only substances whose energy above the convex hull is less than 0.1 eV/atom are deemed thermodynamically stable and appropriate for use in solar-cell technology. The search includes both 2D-layered and 3D bulk materials, with an emphasis on the latter because of their possible benefits in thin-film solar cells. The materials' ability to efficiently absorb sunlight and transform it into electrical energy is lastly assessed by measuring their absorption coefficient. By using an extensive set of criteria, it is ensured that materials chosen for additional research have the highest potential for use in high-efficiency solar cell applications

## 5.6 INITIAL FINDINGS FROM DFT AND ML SCREENING

### 5.6.1 Comprehensive Analysis from Multiple Studies

Finding high-efficiency solar-cell materials has shown to be a successful method when DFT and ML are combined. This method has been used in a number of research, which has produced a large number of potential candidates. The main conclusions from several study publications are summarized below.

### 5.6.2 High-Throughput Screening Process

Tandem organic solar cells were the subject of Greenstein et al.'s (2023) research [8]. They used ML approaches in addition to DFT for material screening, optimizing for both tandem and single-junction designs. Their approach found a number of more stable and effective candidates than traditional methods did. Choudhary et al. (2020) employed a high-throughput screening process [4], starting with a substantial library of non-metallic chemicals. Using DFT simulations, they evaluated the electrical properties and SLME metrics of each material and discovered that a sizable fraction of them had SLME values over the 10% limit, indicating high theoretical efficiency. Other studies, including Yang et al. (2024) [9], have used comparable techniques to use DFT and ML to filter huge datasets of materials. These studies demonstrate how ML may effectively identify workable solutions from hundreds of materials, reducing the time and computational cost of DFT calculations significantly.

### 5.6.3 Identification of Top-Performing Materials

Several top-performing materials have been consistently discovered across several experiments using the combined DFT and ML techniques. The screening procedure was shown to be robust as evidenced by the repeated validation of well-known high-efficiency materials such as CdTe, GaAs, and CuInSe2. These substances continue to be excellent choices for solar cell applications due to their well-established photovoltaic qualities.

Novel materials have also been discovered, including perovskites (CH3NH3PbI3, for example), which exhibit exceptional efficiency and simplicity of synthesis. Furthermore, the identification of 2D-layered materials with excellent efficiency, such as MoS2 and SnSe2, emphasizes their potential for use in flexible and lightweight solar cells.

### 5.6.4 Trends and Characteristics of Discovered Materials

Several significant themes were identified by the examination of the high-efficiency materials that were found. The bandgaps and SLME values are highly correlated. In general, the greatest SLME values were found in materials having direct bandgaps of about 1.1 eV. Compared to the conventional Shockley-Queisser limit, the SLME metric offered a stronger selection criterion that highlighted materials with better absorption and efficiency properties.

A large number of two-dimensional layered materials were found, which are particularly interesting due to their possible benefits in thin-film solar cells. The screening strategy was validated by the minerals that were found, which included chalcopyrites, perovskites, and transition metal chalcogenides, among other classes.

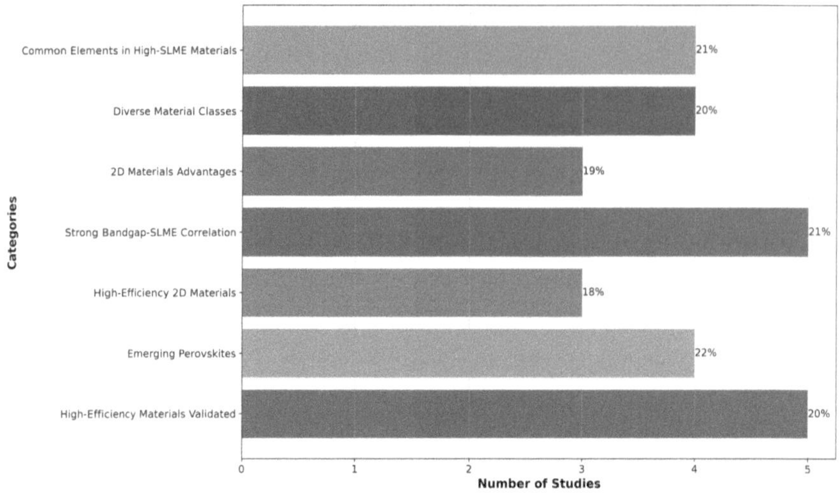


FIGURE 5.2 Summary of findings from DFT and ML screening studies [4, 8–11].

The elemental composition of high-SLME materials was analyzed, and common components including chalcogenides and transition metals were found. Cu, Ag, Mo, W, and Ga were often found in high-performing materials.

The combination of results from several investigations highlights how effective it is to combine DFT and ML approaches to uncover and optimize high-efficiency solar-cell materials more quickly, thereby opening the door to more sophisticated photovoltaic technology.

### 5.6.5 Summary of Findings from DFT and ML Screening Studies

The graph effectively highlights the key trends and findings from multiple studies on high-efficiency solar-cell materials using DFT and ML methodologies. It shows the emphasis placed on different material categories and the significant advancements made in the field through comprehensive research and innovative screening methods. A comprehensive summary of findings across studies is presented in Figure 5.2.

## 5.7 CONCLUSION

In conclusion, the integration of ML and DFT is a revolutionary approach to the discovery of high-efficiency solar-cell materials. By fusing the predictive power of ML with the theoretical rigor of DFT, researchers can expedite the screening process, identify promising materials more effectively, and lessen the computational load associated with traditional methods. The studies reviewed in this chapter demonstrate the efficacy of this combined approach, consistently revealing materials with superior photovoltaic properties. The identification of both novel and well-established high-efficiency materials emphasizes the resilience of this methodology. As the field develops, more research will be conducted to further refine ML models and expand them. The combination of these approaches should be further investigated in the future, with an emphasis on streamlining the process of discovery and broadening the range of useful uses for the materials that have been found.

## REFERENCES


1. Lewis, Nathan S. "Research opportunities to advance solar energy utilization." *Science* 351.6271 (2016): aad1920.
2. Adelekan, Odunayo Adewunmi, et al. "Energy transition policies: a global review of shifts towards renewable sources." *Engineering Science & Technology Journal* 5.2 (2024): 272–287.
3. Kim, Jin Young, et al. "High-efficiency perovskite solar cells." *Chemical Reviews* 120.15 (2020): 7867–7918.
4. Choudhary, Kamal, et al. "Accelerated discovery of efficient solar cell materials using quantum and machine-learning methods." *Chemistry of Materials* 31.15 (2019): 5900–5908.
5. Geerlings, Paul, Frank De Proft, and Wilfried Langenaeker. "Conceptual density functional theory." *Chemical Reviews* 103.5 (2003): 1793–1874.
6. Le Bahers, Tangui, Michel Rerat, and Philippe Sautet. "Semiconductors used in photovoltaic and photocatalytic devices: assessing fundamental properties from DFT." *The Journal of Physical Chemistry C* 118.12 (2014): 5997–6008.
7. Sarker, Iqbal H. "Machine learning: algorithms, real-world applications and research directions." *SN Computer Science* 2.3 (2021): 160.
8. Greenstein, Brianna L., and Geoffrey R. Hutchison. "Screening efficient tandem organic solar cells with machine learning and genetic algorithms." *The Journal of Physical Chemistry C* 127.13 (2023): 6179–6191.
9. Yang, Jiaqi, Panayotis Manganaris, and Arun Mannodi-Kanakkithodi. "Discovering novel halide perovskite alloys using multi-fidelity machine learning and genetic algorithm." *The Journal of Chemical Physics* 160.6 (2024): 064114.
10. Yang, Jiaqi, Panayotis Manganaris, and Arun Mannodi-Kanakkithodi. "A high-throughput computational dataset of halide perovskite alloys." *Digital Discovery* 2.3 (2023): 856–870.
11. Mannodi-Kanakkithodi, Arun. "A guide to discovering next-generation semiconductor materials using atomistic simulations and machine learning." *Computational Materials Science* 243 (2024): 113108.

12. Koller, David, Fabien Tran, and Peter Blaha. "Improving the modified Becke-Johnson exchange potential." *Physical Review B—Condensed Matter and Materials Physics* 85.15 (2012): 155109.
13. Yu, Liping, and Alex Zunger. "Identification of potential photovoltaic absorbers based on first-principles spectroscopic screening of materials." *Physical Review Letters* 108.6 (2012): 068701.
14. Yang, Jinjin, et al. "High-throughput screening of stable sulfide semiconductors for solar cell conversion." *Materials Advances* 5.9 (2024): 3904–3914.

# Integrating Nanomaterials and Advanced Nanoscale Approaches for Boosting the Performance of Next-Generation Photovoltaic Devices Based on III-Nitrides Materials

Redouane En-Nadir, Ayoub El Baraka, Mohamed A. Basyooni-M. Kabatas, Mohamed Tihtih, Ilyass Ez-Zejjari, Hassan Abboudi, Haddou El Ghazi, and Izeddine Zorkani

## 6.1 INTRODUCTION

The advantages of III-Nitride semiconductors, combined with the tunable band-gap energy of (In,Ga)N ternary compounds, make them exceptionally versatile for a wide range of applications in both power and opto-electronic devices. In the context of photovoltaics (PVs), these materials

DOI: 10.1201/9781003499732-6

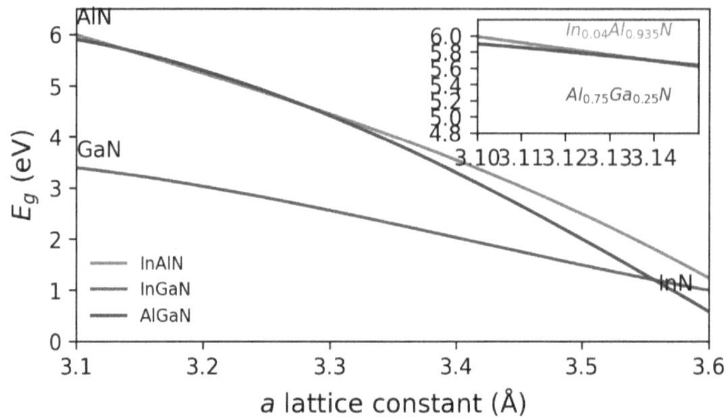


FIGURE 6.1 Relationship between energy gap and in-plane lattice constant in the wurtzite III-Nitride materials system. (Reproduced from [1].)

present an opportunity to enhance the efficiency and performance of solar cells, making III-Nitride-based solar cells a promising area for research and development in the field of next-generation PV technologies.

The graph in Figure 6.1 illustrates the relationship between the energy gap and in-plane lattice constant in the wurtzite III-Nitride materials system. Binary compounds within this system exhibit room-temperature band gaps spanning from 0.61 eV (InN) to 3.44 eV (GaN) to 6.00 eV (AlN). Data on binary band gaps, lattice constants, and ternary bowing parameters are sourced from [1]. The quasi-triangular area enclosed by the binary compounds signifies the Al-In-Ga-N quaternary alloys, while the intersection line denotes lattice-matched quaternary and ternary alloys with GaN. Notably, widely used alloys for optical devices, such as $In_xGa_{1-x}N$ and $Al_xGa_{1-x}N$, deviate from lattice matching between AlGaN and InGaN. For instance, an $In_xGa_{1-x}N$ film with $x = 0.25$ exhibits a lattice mismatch of approximately 3% with GaN. The extent of this mismatch amplifies with increasing indium mole fraction, posing challenges in the growth of emitters for longer wavelengths. However, recently, in 2023, researchers from the Institute of Experimental Physics and Center for Materials Research in Germany have been able to grow high quality of InGaN/GaN heterostructures with indium fraction up to 50% using MOCVD [2].

III-Nitride semiconductor materials and their alloys exhibit a direct bandgap, covering the entire visible solar spectrum. In particular, InGaN ternary material is a promising candidate for high-efficiency solar cells due to its adjustable indium composition, allowing for comprehensive coverage of the solar spectrum [3].

## 6.2 OBJECTIVES: NANOMATERIAL INTEGRATION AND EFFICIENCY ENHANCEMENT

### 6.2.1 Nanomaterial Integration

In the year 2000, Antonio Martí and his research team at the Political University of Madrid-Instituto de Energies Solaire, Madrid, Spain, initiated discussions on a groundbreaking concept—the fabrication of Intermediate Band Solar Cells (IBSCs). This solar cell held the potential to achieve an impressive 63.2% efficiency under concentrated sunlight, utilizing quantum dot (QD) technology. Their proposed method for achieving this feat involved the application of the Stranski and Krastanow technique [4]. Building on this, in 2005, the same research team extended their exploration by proposing a device with a conversion efficiency surpassing the limiting value of 40.7% established for InAs-based QD single-gap cells. This proposal was not just theoretical but also supported by experimental evidence through electroluminescence and quantum efficiency measurements [5]. In the last decade, the study and development of IBSCs have seen limited exploration, both theoretically and experimentally. Various investigations have delved into single-intermediate band solar cells (SIBSCs), employing diverse nanostructures, such as InAsN/AlPSb [6], InAs/InGaAs QDs [7], InN/GaN quantum wells (QWs) [8], and based on InGaN/GaN [9–11]. Moreover, researchers [12] achieved an optimal efficiency of 19% with an $In_{0.59}Ga_{0.41}N$-related PIN cell. Feng et al. [8] reported a numerical investigation suggesting a high conversion efficiency of 23% for $In_{0.75}Ga_{0.25}N$ SC with a 4μm i-layer thickness. Experimentally, Fellip et al  explored InGaN homojunctions, demonstrating a peak external quantum efficiency of 14 ± 2% in the blue-to-orange spectral region [13]. Additionally, (In,Ga)N-based homojunctions obtained by metalorganic vapor phase epitaxy (MOVPE) exhibited promising performance with a peak EQE of 33%, albeit with a spectral cut-off limited to around 420 nm [14, 15].

### 6.2.2 Quantum Wells (QWs) in Photovoltaics

QWs, as illustrated in Figure 6.2, is an ultra-thin layer that confines (quasi-) particles, such as electrons or holes, along the dimension perpendicular

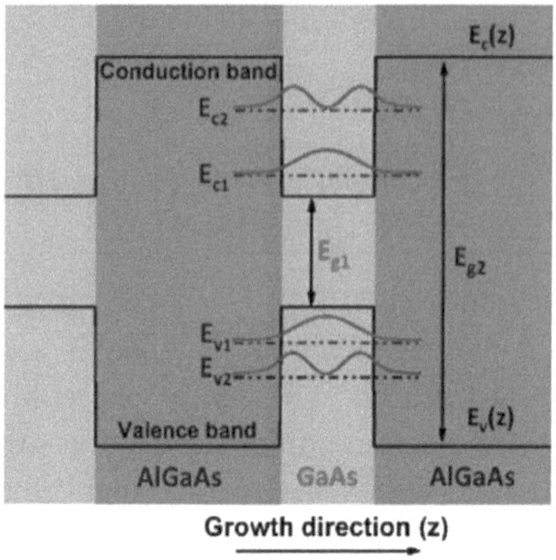


FIGURE 6.2 Illustration of a multiple quantum well structure (MQW): Visualization of the band structure of the quantum wells. (Reproduced from [20].)

to the layer surface, while allowing free movement in other dimensions. This confinement, a quantum effect, has notable impacts on the density of states for the confined particles. In the instance of a QW with a rectangular profile, the density of states remains consistent within defined energy intervals. These nanostructured materials are vital for advancing optoelectronic devices, offering precise control over material band structures through quantum-confined states [16]. They can be found in commercial devices like gallium nitride-based LEDs [17], quantum cascade lasers [18], and infrared photodetectors [19], QWs play a pivotal role.

In PVs, QWs engineer the effective band edge of bulk absorbers, creating a customized bandgap that surpasses lattice matching constraints [21]. An individual QW comprises a thin region of material with a lower bandgap inserted into a higher-bandgap host material, forming a potential well with quantum-confined states. These states enable the absorption of sub-bandgap photons that would typically be lost. Enhanced carrier collection is achieved in the space charge region (SCR) of the diode. Strategically incorporating QWs extends the bulk absorption edge, increasing current density under solar illumination. The energy levels of confined states depend on material composition and well width. In a GaN absorber, $In_xGa_{1-x}N$ wells extend the bandgap beyond 2.3 eV into the infrared. However, as indium composition increases, deeper wells enable further infrared absorption, albeit with a higher barrier for carrier extraction from confined states.

QWs present promising avenues for enhancing current state-of-the-art (SOA) PV designs. Current high-efficiency space PVs typically utilize multiple junctions grown monolithically to optimize solar spectrum utilization [22]. However, challenges arise, particularly in current matching under the AM0 solar spectrum, limiting overall production. The addition of QWs to the GaAs subcell addresses this issue, enhancing current generation and overall efficiency [23]. QWs prove advantageous in this design, facilitating improved current matching without the need for a thick and costly metamorphic grade. Additionally, QWs can enhance radiation tolerance, allowing for thinning of bulk absorber material. In a simplified two- or dual-junction design based on InGaP and GaAs, QWs contribute to improved efficiency, offering a more straightforward architecture compared to 3J counterparts with reduced spectral sensitivity [24].

The investigation of InGaN-based QWs has predominantly focused on the reverse biased regime, while the forward bias reverse current, crucial for PV applications, remains less explored [16]. Advancing beyond conventional theories, there is a pressing need for a comprehensive theory capable of explaining the transport and optical characteristics within an open and interacting non-equilibrium system on a microscopic and unified level [25]. To enhance the efficiency of InGaN/GaN multiple quantum wells (MQWs) solar cells, several critical considerations should be addressed:

a. *Adjusting the Active Region:*
    - Control absorption properties through variations in the thickness and number of QWs and barriers.
    - Modify alloy compositions (In mole fraction) and doping levels to tailor the active region.
b. *Improving Carrier Collection:*
    - Enhance carrier collection by mitigating the impact of the existing piezoelectric field.
    - Adjust doping levels on either side of the active region to optimize carrier screening.
c. *Optimizing Design for Crystal Quality:*
    - Overgrow defects at the GaN/Sapphire interface.
    - Adjust the doping level of p-GaN to enhance p-contact without compromising crystal quality.

## 6.3 LINEAR AND NON-LINEAR OPTICAL ABSORPTION COEFFICIENTS IN INGAN-BASED QWs

In the realm of optical physics, distinct terms delineate various facets of optical absorption. The Linear Optical Absorption Coefficient (LOAC) evaluates a material's linear light absorption concerning intensity, unveiling insights into transparency and absorption traits under low-intensity illumination. Nevertheless, the Non-linear Optical Absorption Coefficient (NLOAC) delves into the non-linear absorption of light at heightened intensities, where the interaction involves multiple photons concurrently, and its manifestation hinges on the incident electromagnetic intensity. NLOAC elucidates the non-linear tendencies of materials when exposed to intense illumination. In addition, the Total Optical Absorption Coefficient (TOAC) sums up both linear and non-linear absorption processes, delivering a comprehensive judgment of a material's absorption behavior across varying light intensities. By the determination of energies and corresponding wave functions with a nanomaterial such as QWs, the linear, non-linear, and TOACs for transitions between any two electronic states are computed utilizing the perturbation expansion and density matrix method. They are expressed analytically as follows:

$$\alpha^{(1)}(\omega)=\hbar\omega\sqrt{\frac{\mu}{\varepsilon_r^*\varepsilon_0}}\frac{\Gamma_{if}\left|M_{fi}\right|^2\sigma_v}{\left(E_{fi}-\hbar\omega\right)^2+\left(\hbar\Gamma_{if}\right)2}, \tag{6.1}$$

$$\begin{aligned}\alpha^{(3)}(\omega,I)=-\sqrt{\frac{\mu}{\varepsilon_r^*}}\left(\frac{I}{2n_r^*\varepsilon_0 c}\right)\frac{4\Gamma_{if}\left|M_{fi}\right|^4\sigma_v\hbar\omega}{\left[\left(E_{fi}-\hbar\omega\right)^2+\left(\hbar\Gamma_{if}\right)^2\right]^2}\\ \times\left[1-\frac{\left|M_{ff}-M_{ii}\right|^2}{|2M_{ff}|^2}\frac{\left(E_{fi}-h\omega\right)^2-\left(\hbar\Gamma_{if}\right)^2+2E_{fi}\left(E_{fi}-h\omega\right)}{E_{fi}^2-\left(\hbar\Gamma_{if}\right)^2}\right]\end{aligned} \tag{6.2}$$

$$\alpha^{(TOAC)}(\omega,I)=\alpha^{(1)}(\omega)+\alpha^{(3)}(\omega,I), \tag{6.3}$$

In the presented equations, $\varepsilon_r^*$ is employed to denote the real component of permittivity, specifically expressed as $n_r^{*2}\varepsilon_0$, where $n_r^*$ represents the refractive index and $\varepsilon_0$ signifies the vacuum permittivity. The variable $\sigma_v$ is indicative of the carrier density, capturing the concentration of charge carriers within the material. $\Delta E_{fi}$ delineates the energy difference between distinct electronic states, with the subscripts '$f$' and '$i$' denoting final and initial states, respectively. Simultaneously, $M_{fi}$ represents the dipole matrix element, characterizing the allowed eigenvalues pertinent to incident radiation polarized along the Z-axis. Furthermore, within this formulation, the symbol $\Gamma_{fi}$ designates the relaxation rate, $c$ stands for the speed of light in free space, and ($I$) represents the intensity of the incident photon beam driving intersubband (ISB) optical transitions. It is crucial to emphasize that both $M_{ff}$ and $M_{ii}$ equate to zero owing to the inherent even symmetry of the confinement potential. This symmetry imparts distinctive characteristics to the dipole matrix elements associated with final and initial electronic states, exerting a discernible influence on their behavior in the context of the considered optical transitions.

### 6.3.1 Effect of Thickness of the QW's Layers: Wells and Barriers

To explore the impact of size variations and indium concentration on the Optical Absorption Coefficients (OACs), we focused on a single QW structure formed by interleaving GaN/InGaN/GaN layers. In this configuration, GaN serves as the barrier material with a length denoted as ($L$), while InGaN acts as the well material with a length denoted as ($l$). The confinement of particles (electrons) occurs due to the energy confinement resulting from the band gap difference between GaN and InGaN. To maintain consistency, we adopted atomic effective units, considering the effective Rydberg ($R^* = 27.60\ meV$) as the unit of energy and the effective Bohr radius ($a^* = 2.69\ nm$) as the unit of length.

#### *6.3.1.1 Intrasuband Optical Transitions*

Intrasuband optical transitions (IBT) refer to electronic transitions within the same subband of energy levels within a material's conduction (CB) or valence band (VB) (see Figure 6.3). These transitions involve the absorption

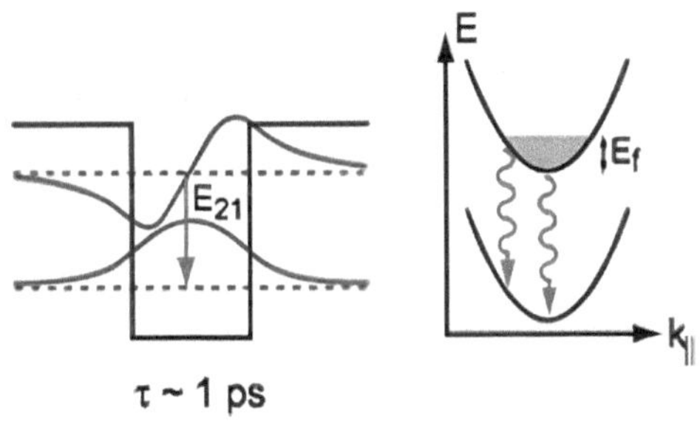


FIGURE 6.3 Illustration of intra-subband (ISB) optical transitions within a quantum well (QW) versus bulk material. Transitions occur within a few picoseconds (ps).

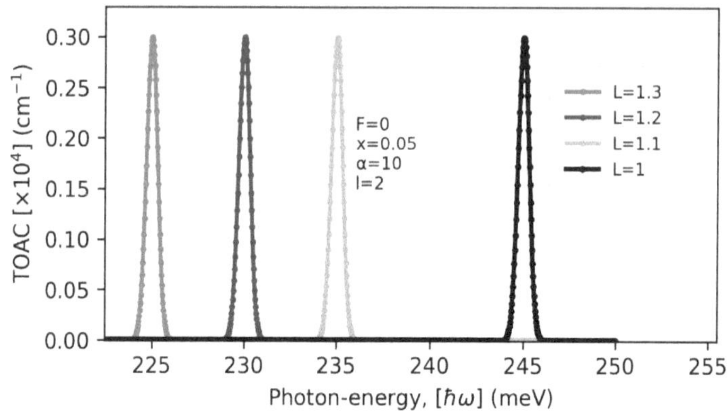


FIGURE 6.4 The variation of the TOAC related to ISB optical transitions versus the incident photon energy for various well and barrier's thicknesses: F=0 (without electric field), $\alpha = 10$ (potential deformation parameter), and $x$ (the indium composition within ternary alloy, InGaN). (Reproduced from [26, 27].)

or emission of photons, typically in the infrared region, and are significant in the study of semiconductor nanostructures and QWs, where confined electron energy levels play a crucial role. Figure 6.4 illustrates how the TOAC varies with incident photon energy for different barrier thicknesses.

The sensitivity of the OACs to layer thickness is intricately associated with the nuanced dynamics of transition energy and the likelihood of transitions within the CB, particularly with respect to intraband transitions (sub-transitions).

As the layer thickness varies, the interplay between these factors manifests a pronounced impact on the absorption characteristics of the material. The observed red-shift in the TOAC peak provides valuable insights into the underlying physics. This shift is a consequence of the diminishing transition energy, a phenomenon accentuated by the enlargement of the barrier's size. The rationale behind this lies in the increased penetration of particles into the well material as the barrier size grows. This penetration effect diminishes the spatial confinement, altering the energy levels and consequently leading to a red-shift in the TOAC peak. Furthermore, the slight drop in the amplitudes of the OACs introduces an additional layer of complexity. This phenomenon can be elucidated by considering the reduced transition probability between the specific states involved, namely the 1s-2p transitions. The larger barrier size, while influencing spatial confinement, also alters the likelihood of electron transitions between these states. The result is a nuanced interplay of quantum mechanical effects that collectively contribute to the observed changes in the optical absorption characteristics of the material. A noteworthy observation emerges, revealing a non-monotonic behavior of the OAC with a peak value at a critical incident photon energy, governed by the specified parameters. The impact of structure size, particularly the well width ($l$=0.5, 1, and 1.5), is elucidated in Figure 6.4 (middle panel), where a substantial blue-shift is evident. This shift is attributed to the increased difference in energy, specifically the transition energy within the CB of the QW. Concurrently, a marked elevation in OAC amplitude is observed. This complex behavior is explicated by the interplay of two competing parameters. On one hand, the absolute value of the dipole matrix element (probability of transition) increases, augmenting the amplitude. On the other hand, the energy difference (transition energy within CB) also increases, resulting in a diminishing effect on the amplitude. Despite these opposing influences, our findings suggest that the impact of transition energy fi is overshadowed by that of osciallator strenght. Furthermore, it is asserted that the third non-linear absorption coefficient significantly contributes to this intricate behavior. This coefficient emerges as a key player in modulating the OAC, emphasizing the intricate relationship between structural dimensions and absorption characteristics within the QW system.

### *6.3.1.2 Intersuband Optical Transitions*

ISB (band-to-band (BTB)) optical transitions refer to electronic transitions between different subbands of energy levels within a material's CB or VB. In these transitions, electrons absorb or emit photons as they move between distinct energy states (Figure 6.5).

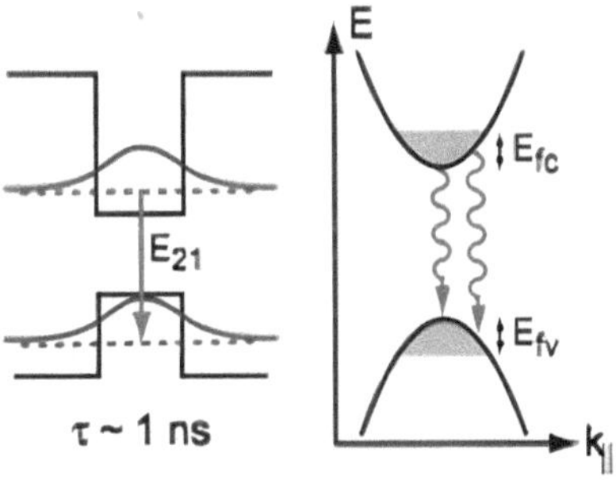


FIGURE 6.5 Illustration of intra-subband (ISB) optical transitions within a quantum well (QW) versus bulk material. Transitions occur within a few nanoseconds (ns) [28].

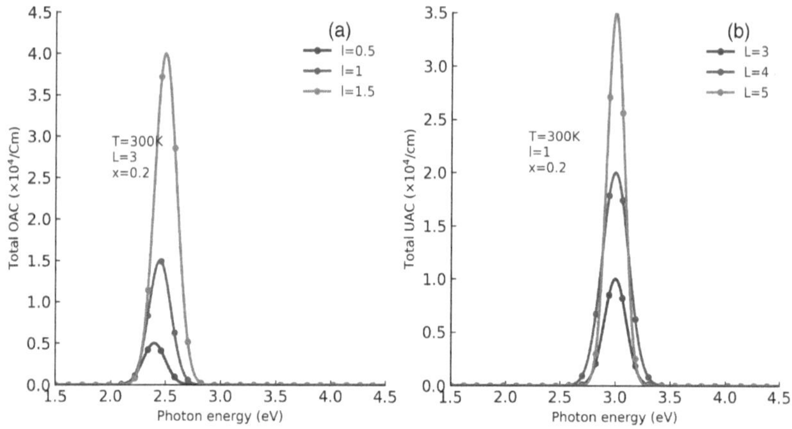


FIGURE 6.6 The variation of the TOAC related to ISB optical transitions versus the incident photon energy for various well and barrier's thicknesses: T=300K (temperature) and $x$ (the indium composition within ternary alloy, InGaN). (Reproduced from [27].)

BTB optical transitions play a crucial role in semiconductor physics, especially in nanostructured materials (QWs), and are relevant to the development and optimization of various optoelectronic devices such as quantum well based solar cells devices and infrared detectors. Figure 6.6 illustrates that, in contrast to the OAC associated with ISB transitions, the OAC related to BTB transitions is less influenced by variations in the well and barrier thicknesses. We shift our focus to inter-valence-CBs recombination. Upon initial examination, it becomes apparent that the effects of well and barrier sizes are less pronounced compared to intraband transitions. The marginal changes observed suggest that the influence of these parameters is overshadowed by other factors.

The energy displacement is notably masked by the band gap energy, a pivotal parameter in interband recombination. Consequently, the absorption spectra exhibit a slight shift for all structure sizes. Therefore, the primary discernible effect is an increase in amplitude corresponding to variations in the well and barrier thicknesses. This information about OAC associate with ISB and BTB optical transitions contribute to the complex interplay between optical absorption and the performance of solar cell conversion.

### 6.3.2 Effect of Indium-Mole-Fraction in the Ternary Alloy $In_xGa_{1-x}N$

It is essential to note that the changes in OACs are contingent upon the transition energy, a parameter influenced by factors such as band gaps,

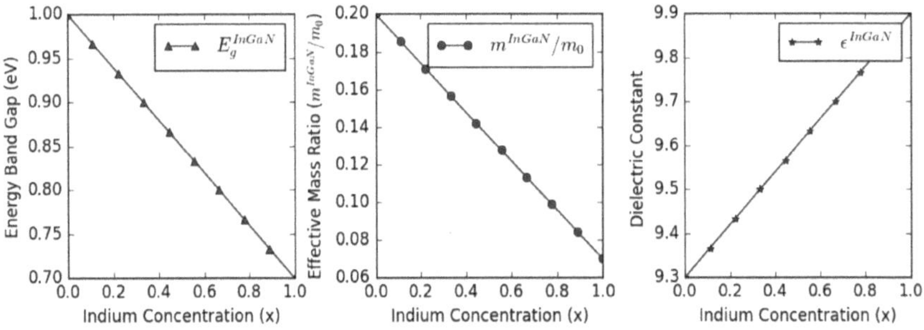


FIGURE 6.7 The variation of the energy band gap (left panel), the effective mass ratio (middle panel), and the dielectric constant (right panel) of InGaN.

dielectric constants, and the effective masses of materials (InN, GaN, InGaN). Therefore, to explore the impact of indium mole fraction in the ternary alloy $In_xGa_{1-x}N$ on OAC, a detailed examination of these parameters becomes imperative. The dependence of indium concentration on various parameters is outlined by the following analytical expression and it is represented in Figure 6.7. The energy band gap of the ternary alloy (InGaN) in the well region is dictated by the indium concentration and is expressed through Varshni's law as follows [29, 30]:

$$\Delta E_g^{InGaN}(x) = xE_g^{InN} + (1-x)E_g^{GaN} - b(1-x) \tag{6.4}$$

where (b = 2.4 eV) is the experimentally obtained bowing parameter between InN and GaN [2].

The dielectric constants of GaN and InGaN versus the temperature are given by the following approximation [31]:

$$\varepsilon_{InGaN}^*(x) = \varepsilon_{GaN}^*(x) + 0.6x \tag{6.5}$$

Based on the $\vec{K}.\vec{P}$ theory the indium-dependent effective mass is given by the expression below:

$$\frac{m_0}{m_i^*(x)} = 1 + \frac{C_i}{E_g^i(x)} \quad i \equiv InN, GaN \tag{6.6}$$

$C_i$ is the energy related to the momentum matrix element $(P = 0)$, they are equal to 14.7 eV (15.50 eV) for GaN, and InN, respectively [32, 33].

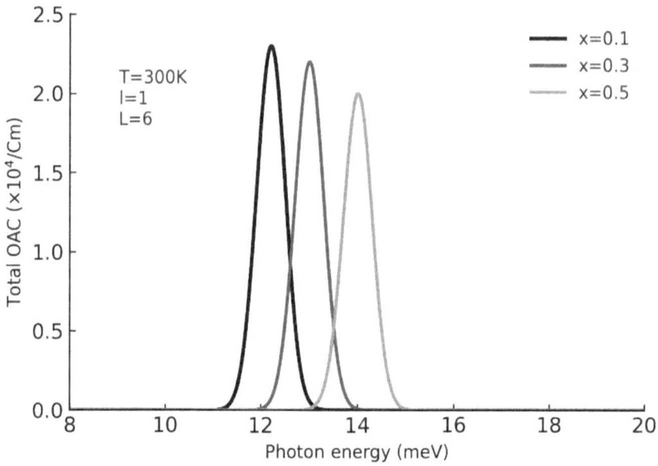


FIGURE 6.8 The variation of the TOAC related to ISB (left-panel) and BTB (right-panel) optical transitions versus the incident photon energy for various well and barrier's thicknesses: T=300K (temperature) and $x$ (the indium composition within ternary alloy, InGaN). (Reproduced from [27, 29].)

The effective mass of InGaN alloy is determined by the following expression:

$$m_e^{InGaN}(x) = x m_e^{InN} + (1-x) m_e^{GaN} \tag{6.7}$$

Figure 6.8 demonstrates the variation of the optical absorption associated to both ISB and BTB optical transitions versus the incident photon energy for various well and barrier's thicknesses at room temperature.

Furthermore, for ISB transitions (left panel), an intriguing observation emerges. With an escalation in the In-fraction, there is a slight reduction in the amplitude of TOAC. This reduction is attributed to the decreased probability of transition, elucidated by the dipole matrix element or oscillator strength. However, when considering TOAC related to BTB transitions (right panel), a distinct trend unfolds. The TOAC peaks exhibit a shift toward lower energy ranges due to the reduction in the energy band gap of InGaN material. Additionally, the maximum amplitudes of these peaks experience a significant drop, primarily influenced by the decrease in oscillator strength, concealed behind the transition probability between the CB and VB. This comprehensive analysis sheds light on the nuanced interplay of well and barrier dimensions, indium concentration, and the intricate dynamics of optical transitions within the ternary alloy.

Optimizing parameters such as well and barrier dimensions, along with controlling the indium concentration in indium gallium nitride (InGaN) QWs, holds the potential to significantly enhance light absorption characteristics. This optimization aims to fine-tune the material's optical properties, particularly the energy band gap and oscillator strengths associated with various transitions. By strategically adjusting these parameters, the absorption spectrum of the QW material can be tailored to better match the solar spectrum. The reduction in energy band gap with an increase in In-fraction, especially for BTB transitions, allows for improved absorption of lower-energy photons in the longer-wavelength range. Additionally, the careful manipulation of oscillator strengths influences the efficiency of transitions between energy states, impacting the material's ability to absorb light. As a result, the optimization of these parameters creates a QW structure that is more attuned to the solar radiation spectrum. This alignment enhances the absorption of a broader range of photons, contributing to improved efficiency in QW-based solar cells. Ultimately, these tailored QW structures have the potential to boost the overall performance and energy conversion efficiency of solar cells, making them more effective in harnessing solar energy for sustainable power generation.

## 6.4 INTERMEDIATE BAND SOLAR CELL CHARACTERISTICS

### 6.4.1 Intermediate Bands: Mechanism, and Operating Principal

The core principle of IBSCs involves the incorporation of intermediate bands (IBs) achieved through QWs or QDs within the material's band gap. This strategic placement enables the absorption of sub-bandgap energies. The inclusion of IBs significantly enhances photon absorption and electron generation processes, resulting in a remarkable improvement in power conversion efficiency (PCE). Figure 6.9 illustrates the operating principle of IBSCs and the physical mechanisms of this novel approach. The incorporation of IBs notably increases the likelihood of absorbing more photons, leading to a corresponding rise in the output current. These nanostructured mini-bands offer a potential solution to overcome the Shockley-Queisser limit, which sets an upper bound for the efficiency of conventional solar cells by balancing photo-generation and radiative recombination of thermalized carriers.

IBSCs, with a radiative limit efficiency of 63.2% [35], surpass the efficiencies of single-gap (40.7%) and two-junction (55.4%) solar cells operated at their radiative limits [36]. Experimentally, the National Renewable

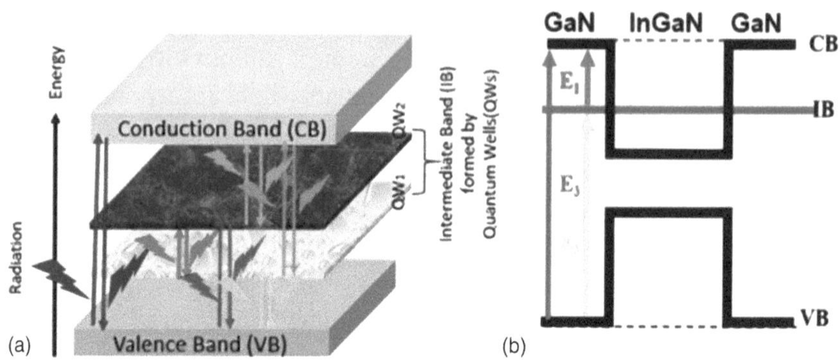


FIGURE 6.9 (a) Visualization of radiative potential optical transitions occurring within an intermediate band (IB). (b) Diagram depicting the band structure of the IB formed by QWs. (Reproduced form [34].)

Energy Laboratory (NREL) estimates that IBSC PV cells have the highest efficiency under experimental conditions (47.1%) [37]. In the operation of IBSCs, the energy of a photon is utilized to promote an electron from the IB to the CB, showcasing the unique and efficient energy conversion mechanism of these advanced solar cells.

As we mentioned above, the IBSC's active region is based mainly on QWs. The crucial parameter governing optical absorption is the band gap energy of the ternary (In,Ga)N, which exhibits non-linear dependence on the In-content, as depicted in Figure 6.10. The design of the IBSC is illustrated in Figure 6.11 demonstrating different parts of the active region of the solar cells.

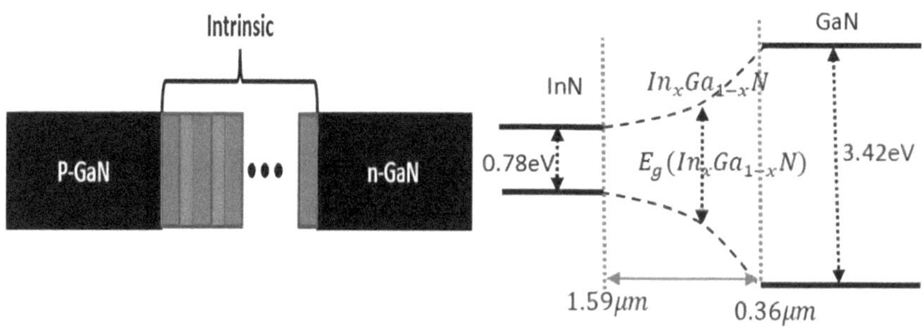


FIGURE 6.10 The energy band gap of the (In,Ga)N structure (left) and P-i-N multiple quantum wells (MQWs) based devices (right). The energy band gap of the (In,Ga)N structure (left) and P-i-N (MQWs) based devices (right). (Reproduced from [38].)

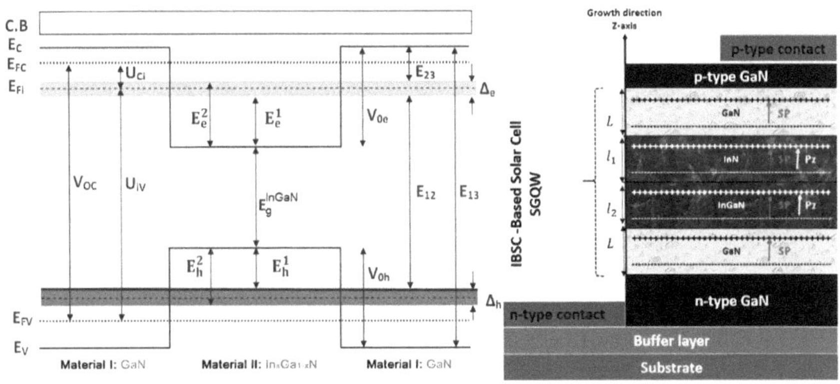


FIGURE 6.11 The energy band diagram illustrates a single intermediate band solar cell (SIBSC) based on InGaN, where $E_C$and $E_V$ represent the energies of the conduction and valence bands of the host material, and $E_{13}$ denotes its bandgap. Additionally, $E^{InGaN}$ represents the band gap of InGaN. $V_{0e}$ and $V_{0h}$ signify the band offsets. $E_{12}$ and $E_{23}$correspond to the energies of the bottom and top sub-gaps. $E_{FC}$, $E_{FV}$, *and* $E_{FI}$ denote the quasi-Fermi levels of the CB, VB, and intermediate band (IB). Finally, $U_{iv}$ and $U_{ci}$ represent the chemical potentials between the VB and IB, and the IB and CB, respectively [9, 10]. The left panel illustrates the intricate design of GaN/InN/InGaN/GaN IBSC Solar Cells, encompassing various components such as contacts, substrate, and buffer layer: $l_1$, $l_2$, and $L$ represent the left/right well and barrier thicknesses, respectively. *SP* and *Pz* are the spontaneous and piezoelectric polarization due to the polar nature of III-Nitride materials. (Reproduced from [35].)

## 6.4.2 Theory and Formalism for Nanoscale Calculations

To numerically calculate the efficiency in the photoelectric conversion process for studying IBSC a detailed theoretical framework is necessary. This framework can be found in detail in Ref. [10, 35, 36, 39]. The key parameters in solar cell investigations encompass photon current density, open circuit voltage, and ultimately efficiency. In the context of full-concentration sunlight, both the number of photons absorbed by the solar cells and those emitted from it determine the density of photons generated. One of the primary physical parameters of a solar cell is the short-circuit (SC) current ($j_{sc}$), formulated as follows:

$$\frac{j_{sc}}{q}=\left[F\left(E_{13},\infty,T_s,0,x\right)-F\left(E_{13},\infty,T_c,\mu_{cV},x\right)\right]+\left[F\left(E_{23},E_{12},T_s,0,x\right)-F\left(E_{23},E_{12},T_c,\mu_{cI},x\right)\right] \tag{6.8}$$

Here, $T_s$ and $T_c$ represent the surface temperature of the sun and the solar cell, respectively. $q$ denotes the elementary charge, while $\mu_{CV}$ and $\mu_{CI}$ stand for the chemical potential differences between the CB and VB, and between the IB and CB, respectively. $E_{13}$, $E_{12}$, and $E_{23}$ are determined by solving the Schrödinger equation for the system. Based on the Roosbroeck-Shockley formula, the flux of photons (F) leaving an object at temperature $T$ can be expressed as:

$$\mathrm{F}(u,\mathrm{v},\mathrm{x},\mu)=\frac{2\pi}{\mathrm{h}^3\mathrm{c}^2}\int_{\mathrm{u}}^{\mathrm{v}}\frac{\mathrm{E}^2\mathrm{dE}}{e^{\frac{(\mathrm{E}-\mu)}{K_B\mathrm{T}}}-1} \tag{6.9}$$

Here, $u$ and $v$ represent the lower and upper energy limits of the photon flux for the respective transitions. $h$ stands for Planck's constant, $c$ is the speed of light in a vacuum, $k_B$ is Boltzmann's constant, and $\mu$ signifies the chemical potential of the transition.

For a P-I-N solar cell, the output voltage $V_{oc}$ of an IBSC can be expressed as:

$$V_{oc}=\mu_{cV}=\mu_{cI}+\mu_{IV} \tag{6.10}$$

Here, $\mu_{CV}$ and $\mu_{IV}$ are given by the following expressions:

$$\mu_{cV}=E_{23}+0.5\Delta e-E_c+E_{FC} \tag{6.11}$$

$$\mu_{IV}=E_{12}+0.5\Delta e-E_{FV}+E_V+V_{0h}+E_h^1 \tag{6.12}$$

where $\Delta e$ is the width of IB for electrons. The quasi-Fermi levels $E_{FC}$ and $E_{EV}$ of CB and VB respectively, can be expressed as:

$$E_c-E_{Fc}=kT\ln\left(\frac{N_c}{n}\right) \tag{6.13}$$

$$E_{FV}-E_V=kT\ln\left(\frac{N_V}{p}\right) \tag{6.14}$$

where $N_C$ and $N_V$ represent the effective densities of states in the CB and VB, respectively. Meanwhile, $n$ and $p$ denote the electron and hole concentrations, respectively, given by:

$$n=N_c\ exp\left[-\frac{Q\Delta E_g^\Gamma(x)}{K_BT}\right] \tag{6.15}$$

$$p = N_V\ exp\left[-\frac{(1-Q)\Delta E_g^{\Gamma}(x)}{K_B T}\right] \tag{6.16}$$

The effective densities of states in the CB and VB, $N_C$ and $N_V$, are expressed as:

$$N_c = N_c^*\ T^{3/2} \tag{6.17}$$

$$N_V = N_V^*\ T^{3/2} \tag{6.18}$$

In the realm to enhance PV conversion efficiency prediction, it is crucial to consider an optimal ($FF \neq 1$). Instead, researchers recognized that the fill factor (FF) is typically a function of the open-circuit voltage, $V_{OC}$, expressed as $\left[V_{OC}/\left(\frac{kT}{q}\right)\right]$, and is formulated as follows:

$$FF = \frac{V_{OC} - Ln\left(V_{OC} + 0.72\right)}{1 + V_{OC}} \tag{6.19}$$

As a result, the efficiency of the IBSC can be derived in the general case by utilizing the output voltage and photo-current density, as follows:

$$\eta = \frac{V_{oc} \cdot J_{sc} \cdot FF}{P_{in}} \tag{6.20}$$

Note that ($P_{in} = \sigma T_s^4$) is the incident power coming from the sun per unit of area and $\sigma = 5.67 \cdot 10^{-8}\ Wm^{-2}k^{-4}$ is the Stefan-Boltzman constant.

### 6.4.3 Open-Circuit Current (Voc)

The performance of a solar cell is assessed through a thorough examination of its current density-versus-voltage (J–V) relationship (as depicted in Figure 6.12). This characterization provides critical insights into the device's performance, offering valuable data on its electrical behavior under varying operating conditions.

The J–V relationship is a key metric used to analyze and optimize the efficiency and functionality of solar cells, contributing to advancements in PV technology.

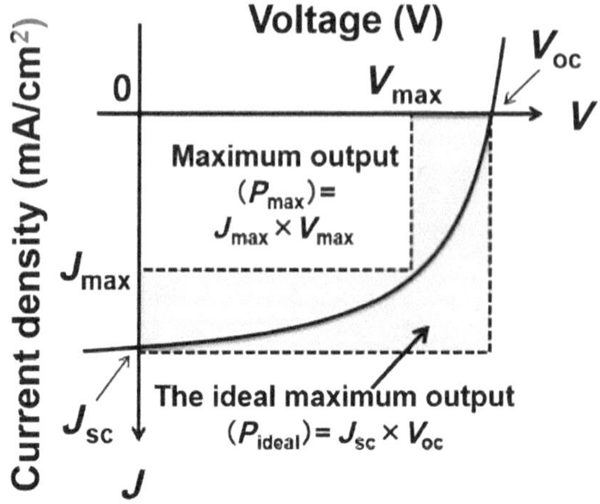


FIGURE 6.12 Schematic of the current-versus-voltage relationship. (Reproduced from [40].)

#### *6.4.3.1 Effect of Thickness of the Layers and Indium Mole-Fraction*

The dimensions of nanostructured materials, particularly the thickness of layers such as well and barrier in structures like QWs and QDs, alongside the indium mole-fraction, play a crucial role in shaping the characteristics of the open-circuit current ($V_{oc}$) in a photodiode. The intricate influence of these parameters on the $V_{oc}$ is meticulously examined and visualized in Figure 6.13. In Left-panel (a), it is evident that, irrespective of the indium composition and pressure value, the $V_{oc}$ undergoes a quasi-linear decrease with an increase in the thickness of both the well (QW2) and the barriers.

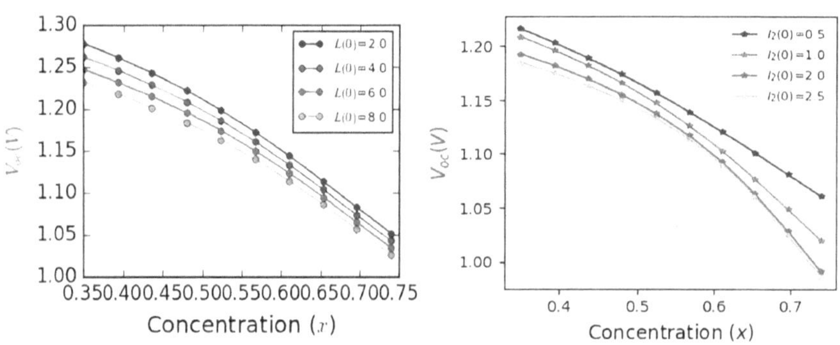


FIGURE 6.13 Changes in the open-circuit current (Voc) of the designed IBSC in relation to the indium concentration ($x$) across various values of barrier width (left-panel) and InGaN layer width (right-panel). (Reproduced from [34].)

This trend signifies that $V_{oc}$ experiences a consistent reduction with the enlargement of these dimensions. The observed decline in $V_{oc}$ is associated with the growth in indium fraction, attributed to an increase in defect density and the consequential impact of spontaneous and piezoelectric polarization. A critical indium fraction ($x_c$= 52.34%) is notable as it distinguishes two distinct behaviors. For $x$ values less than $x_c$, $V_{oc}$ linearly declines, showcasing a consistent trend. Conversely, for $x$ values greater than $x_c$, the influence of the InGaN layer becomes substantial, leading to a non-linear decay in $V_{oc}$.

Upon reducing the thicknesses of both the barrier and well, a pronounced and positive impact on the open-circuit voltage ($V_{oc}$) becomes evident. This observable effect can be attributed to the enhanced spatial confinement, facilitating a more precise control over particles such as electrons and holes. The refined spatial confinement contributes to an overall improvement in $V_{oc}$, signifying a notable enhancement in the photodiode's performance. These results are in good agreement with those reported by El Aouami et al. [39].

### 6.4.4 Short-Circuit Current Density ($J_{sc}$)

The SC current density ($J_{sc}$) is the current flowing through the external circuit when the solar cell electrodes are short-circuited. It relies on the incident light's photon flux, standardized to the AM1.5 spectrum for typical measurements. $J_{sc}$ is influenced by the solar cell area; hence, the $J_{sc}$ is commonly employed to express the maximum current delivery without the solar cell area dependency. This maximum current is significantly influenced by optical properties like absorption and reflection in the absorber layer.

#### *6.4.4.1 Effect of Thickness of the Layers and Indium Mole-Fraction*

For an IBSC, the $J_{sc}$ is given by the analytical expression provided in the Equation (6.8). Examining Figure 6.14, a clear trend emerges wherein the SC current density ($J_{sc}$) exhibits improvement with an increase in indium concentration. Simultaneously, a noteworthy enhancement in $J_{sc}$ is observed with a reduction in the dimensions of the active region, specifically the well and barrier thicknesses. This behavior can be attributed to the intricate interplay between indium concentration and structural dimensions and their ability on leading to enhanced light absorption and improved charge carrier generation within the active region of the solar cell. The positive correlation between indium concentration, reduced

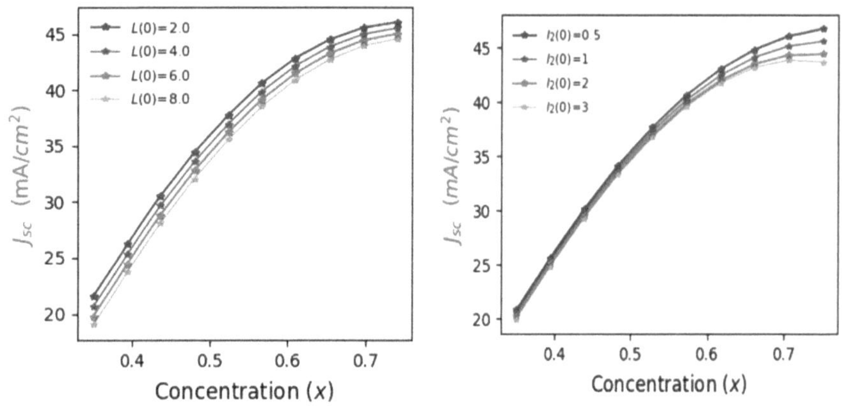


FIGURE 6.14 Changes in the short-circuit current ($J_{sc}$) of the designed IBSC in relation to the indium concentration (x) across various values of barrier width (left-panel) and InGaN layer width (right-panel). (Reproduced from [34].)

active region dimensions, and the resulting boost in $J_{sc}$ underscores the significance of these factors in optimizing the solar cell's performance.

### 6.4.5 Efficiency of Intermediate Band Solar Cell

The efficiency of an IBSC is a critical parameter that gauges the effectiveness of this innovative PV technology. This efficiency metric encapsulates the ability of the IBSC to convert incident sunlight into usable electrical power, considering factors such as the IB's impact on the energy conversion process. Therefore, understanding the efficiency of IBSCs is pivotal for assessing their performance and potential applications in the field of solar energy.

#### *6.4.5.1 Effect of Thickness of the Layers and Indium Mole-Fraction*

Figure 6.15 reveals a comprehensive insight into the performance of IBSCs across diverse parameters, with a particular focus on indium fraction and well/barrier thicknesses. Notably, the efficiency of IBSC is intricately linked to both its design and alloy compositions. As depicted, the performance of IBSC exhibits a consistent improvement until it attains a peak efficiency at a critical indium concentration of approximately 56%. Beyond this point, the efficiency experiences a decline, suggesting a crucial balance in material composition. The decrease in efficiency post the critical indium concentration is attributed to the degradation of material quality, resulting in delocalization phenomena that adversely impact performance. This underscores the delicate equilibrium required for optimizing IBSC efficiency.

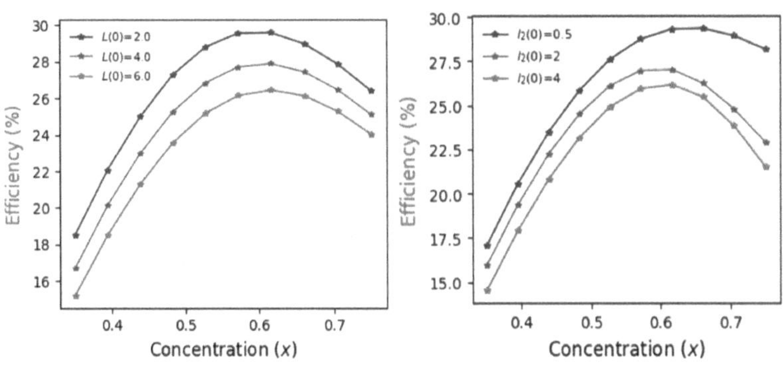


FIGURE 6.15 Changes in the performance of the designed IBSC in relation to the indium concentration (x) across various values of barrier width (left-panel) and InGaN layer width (right-panel). (Reproduced from [34].)

Interestingly, irrespective of the indium composition, a noteworthy observation is the enhancement in PV conversion performance with the reduction of well and barrier thicknesses. This improvement is attributed to the enhanced confinement of carriers within the QW. The correlation between performance and structural dimensions highlights the significance of design parameters in achieving superior efficiency in IBSCs. These findings agreed largely with those reported previously in the literature [39].

## 6.5 CONCENTRATED SUNLIGHT INFLUENCE OF IBSC PERFORMANCE

Before delving into the assessment of InGaN-based IBSC (as designed in Figure 6.16) performance under solar conditions, it is imperative to acknowledge that concentrated sunlight holds the potential to significantly

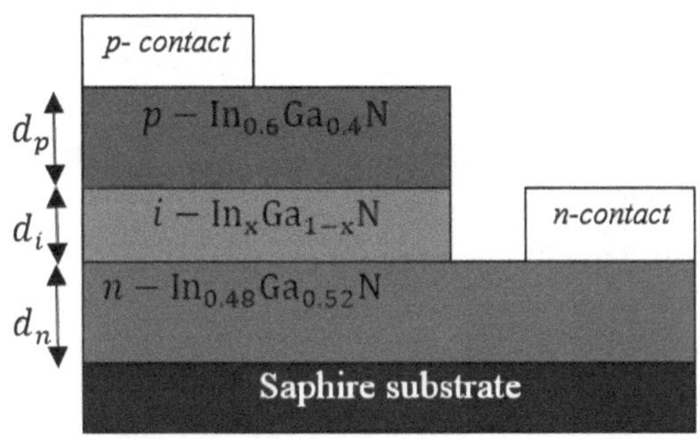


FIGURE 6.16 Design of an InGaN-based P-I-N IBSC showing different parts of the photo-diode (solar cell) subjected to light radiation (sun) [40].

augment solar cell efficiency. This enhancement is achieved by intensifying light, thereby improving conversion efficiency and enabling the use of smaller solar cells, reducing material requirements, and potentially mitigating additional costs related to concentrating optics. However, it is essential to exercise meticulous thermal management to address potential temperature rises. Challenges such as the need for precise tracking systems, which can add complexity and cost, as well as the introduction of optical components leading to possible losses, should be carefully navigated. While concentration presents clear advantages, the effective resolution of these challenges is paramount to optimizing solar cell performance in concentrated solar power systems. Innovative approaches have been developed to tackle some of these challenges.

### 6.5.1 Effect of Concentrated Sunlight on the Optical Absorption Spectra

Generating electron-hole pairs via sunlight absorption is crucial for solar cell functionality. Fundamentally, absorption entails the direct excitation of an electron from the VB to the CB, leaving a hole behind. Modeling InGaN-based solar cells necessitates an accurate model for light absorption across the solar spectrum and varied x-indium contents. To achieve this, we employed a previously suggested phenomenological model for InGaN [41].

$$\alpha(E,x) = 2.2\ 10^5 \sqrt{a(x)[E - Eg(x)] + b(x)[E - Eg(x)]^2}\ \left(cm^{-1}\right) \quad (6.21)$$

where $E$ is the photon energy, and $(x)$, $(x)$ are coefficients related to indium-mole concentration. These coefficients are derived from experimental results in Ref [42]. Their In-mole dependency is approximated using a fourth-degree polynomial for $(x)$ and a quadratic polynomial for $(x)$. InGaN boasts a high absorption coefficient of $10^5\ cm^{-1}$ [43], substantial radiation resistance [44], suitability for harsh environments, increased carrier mobility, excellent thermal conductivity, and low effective mass. Furthermore, the narrowing of the band gap (BGN), attributed to high doping levels, is accounted for using the Klaassen model, outlined as follows [45]:

$$\Delta E_g = \alpha\left[\frac{\ln(N)}{\beta} + \sqrt{\left(\frac{\ln(N)}{\beta}\right)^2 + \gamma}\right] \quad (6.22)$$

where $\alpha$, $\beta$, and $\gamma$ are the constants.

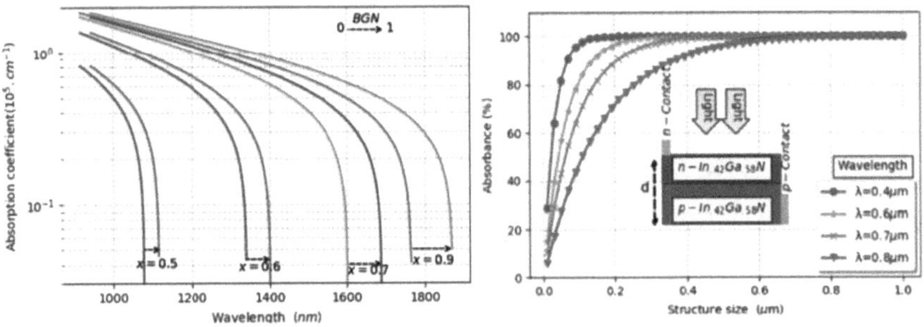


FIGURE 6.17 Semilogarithmic plot depicting the absorption coefficient (a) and absorbance (b) of the unintentionally doped layer ($In_xGa_{1-x}N$) with $5\times10^{17}\,cm^{-3}$ versus the wavelength. The graph represents various indium-mole compositions ($x \geq 0.43$) With the inclusion of the Burstein-Moss shift impact (0 ≡ Without, 1 ≡ With) [11, 40].

Figure 6.17 (left panel, right panel) illustrates its variation absorption versus the wavelength for different In-content values and absorbance as a function of the system (IBSC) dimension for different wavelengths. With increasing In-content, the intrinsic layer demonstrates enhanced absorption across a broad wavelength spectrum from 900 nm to 1900 nm. The absorption coefficient reaches a magnitude of $10^5\ cm^{-1}$ and rapidly decreases below the band edge. Moreover, the rise in In-composition enhances the absorption of low-energy photons, advantageous for PIN-based solar cells. The influence of Band Gap Narrowing (BGN), resulting from unintentional and/or intentional doping, further broadens the absorption spectrum toward higher wavelengths. Notably, the BGN effect is more pronounced than the impact of defect concentration and in-mole fraction augmentation [46]. These attributes position InGaN as a highly promising material for solar PV applications [47, 48].

Moreover, to achieve a nonpolar solar cell (SC) and counteract polarization charge, perpendicular growth to the c-axis is essential. The investigation focuses on a nonpolar n-(In,Ga)N/p-(In,Ga)N structure, assuming flat surroundings. High-quality $In_{0.41}Ga_{0.59}N$ SC with 2-μm thickness and 10% reflection can attain efficiencies of 28%, 32%, and 32.6% under 1-sun AM0, AM1.5D, and AM1.5G illuminations at 42% In-composition. The study centers on exploring light concentration effects on PV conversion in a nitride-based $In_{0.42}Ga_{0.58}N$ solar cell (1.42eV band gap). Absorption coefficient analysis reveals thickness-dependent behaviors with saturation for thickness above a critical value ($d_c$) and an abrupt drop below $d_c$.

Critical value $d_c$ varies with wavelength (e.g., 180 nm for λ = 400 nm and 760 nm for λ = 800 nm). Absorbance increases with thickness, reaching saturation, indicating improved photon absorption and electron–hole pair generation. Absorbance decreases with longer wavelengths, necessitating a thinner layer for effective absorption. For instance, a 0.2 μm thickness is sufficient for λ = 400 nm, compared to 0.8 μm for λ = 800 nm.

### 6.5.2 Solar Spectral Irradiance in InGaN Layer

Figure 6.18 presents the spectral irradiance under 1-sun conditions, specifically AM0, AM1.5D, and AM1.5G, in units of $W.m^{-2}nm^{-1}$ as a function of wavelength (nm). The spectrum encompasses various spectral regions, including the ultraviolet (UV), visible, and infrared (IR), along with the range corresponding to the $In_xGa_{1-x}N$ ternary alloy. To assess solar cell (SC) performance, the ASTM G173-03 reference spectra are employed, generated from the Simple Model of Atmospheric Radiative Transmission of Sunlight (SMARTS v.2.9.2). These power density profiles versus wavelength serve as a universal reference for evaluating spectrally selective PV materials, considering their effectiveness under different manufactured and natural light sources with varying spectral data patterns [49].

Two spectra are distinguished for terrestrial applications. The first is the direct normal spectrum, denoted as AM1.5D, with a power density of

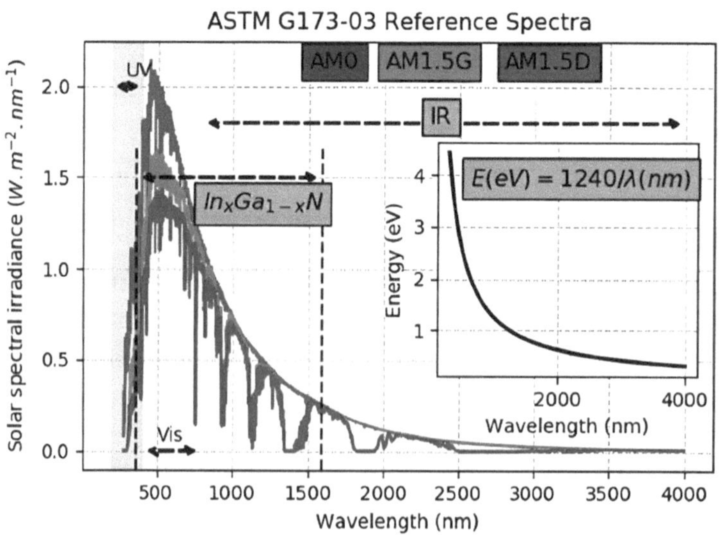


FIGURE 6.18 The solar spectral irradiance under 1-sun conditions, specifically AM0, AM1.5G, and AM1.5D, is depicted based on wavelength. The inset provides a representation of energy versus wavelength. (Reproduced from [48].)

900 $W.m^{-2}$, representing the direct component of the entire global (hemispherical) spectrum for solar concentrator work. This includes the direct sunlight beam and the circumsolar component within a disk 2.5 degrees in diameter revolving around the sun. The second is the AM1.5G global spectrum designed for flat plate modules, featuring an integrated power of 1000 $W.m^{-2}$ (100 $W.m^{-2}$). The inset within the figure illustrates the well-known changes in the band gap energy within the same wavelength range, providing additional insights into the semiconductor material's optical properties.

### 6.5.3 Solar Cell Characteristics under X-Sun Irradiation

Modeling solar cell performance under X-sun irradiation is possible through simulations using software such as 1D-SCAPS, 1D-AMPS, Silvaco, PC-1D, a widely adopted approach. Also, utilizing the finite difference method implemented in Python (version 3.9) for data calculation and visualization. This section addresses well-known equations to provide a comprehensive understanding of our modeling. When a solar cell is exposed to light, only photons with energies exceeding the band gap energy ($E_g$) of the semiconductor are absorbed, generating electron-hole pairs. The cutoff wavelength for photons conducive to carrier formation is dictated by the material's band gap energy. In this situation, the 1-sun SC current density (as depicted in Figure 6.19), $J_{sc}^{X}$, is determined by the solar emission spectrum and is expressed as follows [50]:

$$J_{sc}^{1} = \frac{q(1-s)}{hc} \int_{0}^{\lambda_g(x)} \left[1-R(\lambda,x)\right]\left[1-e^{(-\alpha(\lambda,x).d)}\right].\mathcal{M}_{\lambda}^{0}.\lambda.d\lambda \qquad (6.23)$$

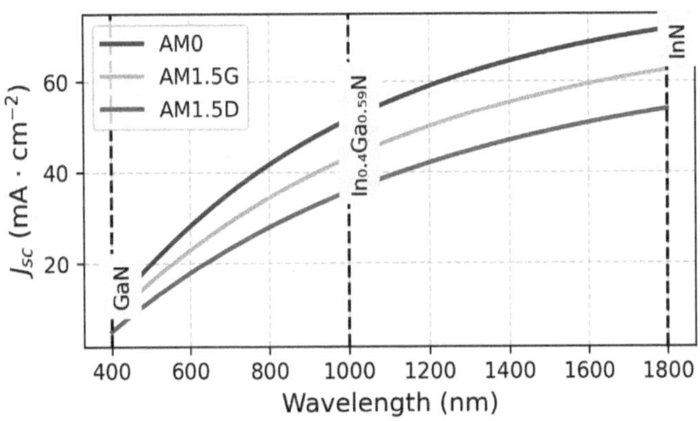


FIGURE 6.19 The integrated maximum short-circuit current with respect to wavelength for 1-sun AM0, AM1.5D, and AM1.5G spectra. (Reproduced from [48].)

where $s$ is the grid-shadowing factor, q is the elementary charge, $\mathcal{M}_{\lambda}^{0}$ is a special intensity $\left(W/m^{-2}.nm^{-1}\right)$, $h$ is the Planck's constant, $\lambda$ is the photon wavelength, $c$ is the light velocity in vacuum, $d$ is the layer's thickness, $\lambda_g(x)$ is the indium-related wavelength cutoff.

The X-Sun SC current density is given as follows:

$$J_{sc}^{X} = X.J_{sc}^{1} \tag{6.24}$$

The absorbance of each InxGa1−x N layer can be defined as follows:

$$A(E,x) = \frac{I(E,x,z) - I(E,x,z+z_1)}{I(E,x,z)} \tag{6.25}$$

The expression for light intensity ($I$) at a position ($z$) within the $In_xGa_{1-x}N$-layer, denoted by ($E$, $x$, $z$), is described by the equation:

$$I(E,x,z) = I_0(E)e^{-\alpha(\lambda,x).z}$$

Here, $I_0(E)$ represents the incident light intensity of solar radiation corresponding to different spectra such as AM1.5G, AM1.5D, and AM0.

The open-circuit voltage related to the X-sun, denoted as $V_{oc}(X,x)$ can be defined as follows [51]:

$$V_{oc}(X,x) = \frac{n_I k_B T}{q} \ln\left(1 + \frac{J_{sc}^{X}}{J_0}\right) \tag{6.26}$$

In this context, $n_I$ represents the ideal factor. Generally, the ideality factor for a real diode based on III-V material falls within the range of 1.1–1.5. In the case of III-N materials, $n_I$ is considered to be 1.2.

The FF, characterized as the ratio of the maximum power output to the product of open-circuit voltage and SC current density, constitutes the third fundamental attribute delineating the solar cell. In accordance with Green, it is defined for X-related illuminations through the following expression [52].

$$FF^{X} = \frac{\frac{V_{oc}^{X}}{V_{th}} - \ln\left[\frac{V_{oc}^{X}}{V_{th}} + 0.72\right]}{1 + \frac{V_{oc}^{X}}{V_{th}}} \tag{6.27}$$

In the end, the pivotal factor is the efficiency of the solar cell, expressed as the ratio of the maximum output to incoming power [53].

$$\eta^X = \frac{V_{oc}^X \times J_{sc}^X \times FF^X}{P_i^X} \tag{6.28}$$

Delving into the PCE based on (In,Ga)N of relevance requires scrutinizing the results obtained from calculations that match the open-circuit voltage and FF concerning the X-sun concentration. This preliminary analysis is essential for acquiring sufficient insights to investigate the behavior of electrical properties, specifically the PCE, in response to X-sun concentration.

The computation of the 1-sun integrated maximum SC current density, J1sc, is conducted utilizing AM0, AM1.5D, and AM1.5G ASTM experimental spectrum data, and it is significantly influenced by the material band gap energy. As depicted in Figure 6.19, the obtained $J_{sc}^1$ results for the AM0 spectrum exhibit the highest values, while those for the AM1.5D spectrum are the lowest, irrespective of the wavelength. For instance, our AM1.5G-related results reveal that GaN, InN, and $In_{0.42}Ga_{0.58}N$ current densities are 0.8, 61.1, and 33.2 $mA{\cdot}cm^{-2}$, respectively.

Figure 6.20 delineates the variations in different solar cell characteristics concerning X-sun concentration for two distinct doping levels. Subplots (a), (b), and (c) depict the changes in X-sun AM1.5G-related open-circuit voltage ($V_{oc}$), FF, and optimum efficiency, while subplot (d) illustrates the alterations in optimum efficiency for AM0, AM1.5G, and AM1.5D illuminations. Notably, all SC characteristics exhibit a robust increase with rising sun concentration, irrespective of doping and illumination. A key observation is the tendency of all SC parameters toward a saturation regime at high X-sun concentrations, regardless of doping and illumination. Panel (d) of Figure 6.20 delineates the changes in optimum efficiency versus X-sun concentration for three spectral illuminations, considering doping at $10^{16}\,cm^{-3}$ and room temperature. It is evident that the curvature of $\eta_{max}$ is similar and strongly influenced by X-sun concentration, irrespective of the illumination. The results indicate that the highest PCE is associated with AM1.5G, while AM0 performances are the least favorable for all X-sun concentrations. These findings align well with reported literature [54], who investigated electrical properties in high-concentration PV systems. Similarly, Abu Kowsar et al. [55] and Outes et al. [56] explored the impact of material band gap energy, X-sun concentration, and surface recombination velocity on multi-junction

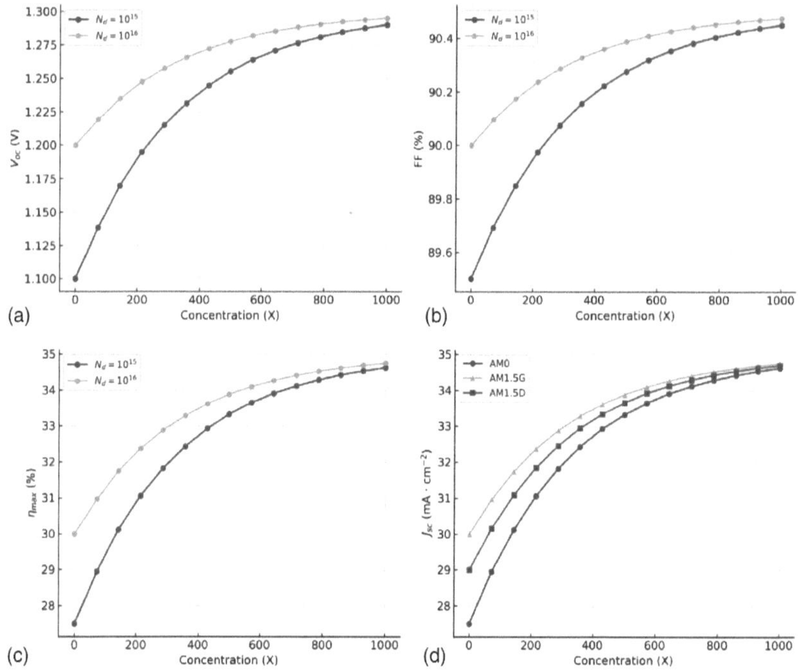


FIGURE 6.20 Outcomes for the open-circuit voltage (Voc) under 1-sun AM1.5G conditions are presented concerning the X-sun concentration at room temperature. The investigation considers two distinct levels of doping. (Reproduced from [49].)

solar cells, showing efficiency improvements with increased concentration. It is noteworthy that the measured Jsc for commercial silicon solar cells typically falls within the range of 28–36 $mA{\cdot}cm^{-2}$, while for laboratory cells, it is around 42 $mA{\cdot}cm^{-2}$. The corresponding outcomes for different materials are delineated by vertical dashed lines [49]. The impact of doping concentration on the performance of the (In,Ga)N solar cell under AM1.5G illumination is evident. With an increase in doping concentration from $1 \times 10^{15}\,cm^{-3}$ to $1 \times 10^{16}\,cm^{-3}$, various key parameters experience notable changes. The open-circuit voltage ($V_{oc}$) exhibits an increase from $1.121\ V$ to $1.275\ V$, indicating a variation of roughly 13.74% at the lower doping concentration and a more significant variation of about 20.27% at the higher concentration. The FF shows a minor improvement from 89.29% to 90.23% at the lower doping level and a slight increase from 89.28% to 90.64% at the higher concentration. However, the maximum efficiency ($\eta_{max}$) demonstrates a substantial improvement from 27.63% to 33.78% (22.26% variation) at the lower doping concentration and from 29.57% to 35.71% (20.76% variation) at the higher concentration. These findings highlight the sensitivity of the (In,Ga)N solar cell to doping concentration, particularly in terms of efficiency enhancement under AM1.5G conditions [49].

## 6.6 FUTURE DIRECTIONS AND RECOMMENDATIONS FOR FURTHER RESEARCH

This chapter unfolds a progression from the initial findings to future directions, offering a chronological perspective on the exploration of factors influencing the efficiency of intermediate solar cells (IBSCs). The findings, indicating potentially higher values, prompt a nuanced interpretation considering inherent simplifications in numerical calculations. Acknowledging certain assumptions and idealized conditions becomes crucial, setting the stage for subsequent investigations. Subsequent studies will undertake a more intricate analysis, delving into a comprehensive exploration of factors impacting IBSCs' efficiency. The spectrum of these factors encompasses temperature effects, electric fields, laser irradiation, pressure, confinement of energy profiles, impurities, defects, and magnetic fields. Understanding their influence draws from insights gained in previous studies, particularly on the optical and electronic properties of nanostructures, such as InGaN/GaN QWs [57–63]. This extended investigation aims to provide a deeper understanding of how these factors uniquely contribute to modifying energy levels within the IB formed by QWs in IBSCs. Each factor plays a distinct role; temperature effects optimize thermal energy for carriers, electric fields influence carrier movement, laser irradiation selectively excites carriers, pressure alters band structure, confinement of energy profiles creates discrete energy states, impurities and defects induce additional energy states, and magnetic fields alter electronic structure and carrier dynamics. Collectively, these modifications aim to create a tailored energy landscape within the IB, ensuring efficient photon utilization across a broader spectrum. This intricate engineering of the landscape represents a significant advancement in renewable energy technologies, contributing to the evolution of sustainable and efficient solar energy.

The subsequent discussion further details the impact of each factor:

- Temperature effects optimize thermal energy for carriers, enhancing their contribution to the photoelectric effect. Electric fields influence carrier movement and enhance the separation of photoexcited carriers, contributing to overall efficiency.

- Laser irradiation selectively excites carriers, optimizing photon absorption and utilization.
- Pressure alters the band structure, impacting energy levels within the IB.
- Confining energy profiles within QWs creates discrete energy states, allowing tailored optimization of energy levels.
- Controlled introduction of impurities or defects induces additional energy states, enhancing carrier multiplication and overall efficiency.
- Magnetic and electric fields alter the electronic structure and carrier dynamics, improving the separation and collection of charge carriers.

Emphasizing the insufficiency of relying solely on numerical and theoretical calculations, experimental measurements play a pivotal role in advancing IBSC technology. Given the polar nature of III-Nitrides, particularly InGaN/GaN QWs, this consideration becomes paramount during the growth and characterization of solar cells. A comprehensive approach integrating theoretical insights with experimental validation is essential, propelling the development of IBSCs as an emerging and efficient PV technology. This holistic strategy aims to establish a robust foundation for unraveling the intricacies of the studied phenomena and accelerating the progress of IBSCs in the realm of renewable energy.

## 6.7 CONCLUSIONS

In conclusion, this chapter serves as a comprehensive exploration of Integrating Nanomaterials and Advanced Nanoscale Approaches for Enhancing the Performance of Next-Generation PV Devices, with a specific focus on III-Nitrides nanomaterials in IBSCs. The study delved into a thorough examination of the electrical characteristics of this PV technology, encompassing open-circuit current, SC current density, FF, and overall performance. Significantly, we underscored the importance and advantages of integrating nanomaterials, particularly emphasizing the potential benefits derived from low-dimensional systems like QWs and QDs. These nanomaterials play a pivotal role in shaping the novel and promising landscape of IBSCs. Theoretical frameworks essential for determining the characteristics of IBSCs were presented, providing a foundational understanding of the technology. Moreover, we elucidated the influence of size, specifically the well and barrier thicknesses, on the optical absorption and energy transitions within InGaN/GaN-based QWs that form the IB of IBSCs. The examination of these size-dependent factors contributes to a nuanced understanding of the device's performance. A detailed theoretical framework was also established to assess the performance of this third-generation solar cell technology under X-sun concentration, a critical parameter for investigating PV performance. This investigation sheds light on the technology's capabilities under concentrated sunlight, offering insights into its potential efficacy. As we conclude this chapter, we turn our attention to future directions and recommendations. These considerations are vital for propelling the advancement of this promising PV technology, ultimately contributing to the reduction of energy costs and achieving sustainability goals. The outlined future directions provide a roadmap for further research and development, ensuring that IBSCs continue to evolve as an efficient and sustainable solution in the realm of renewable energy.

## REFERENCES


1. S. N. Alam, V. Z. Zubialevich, B. Ghafary, and P. J. Parbrook, 'Bandgap and refractive index estimates of InAlN and related nitrides across their full composition ranges', *Sci. Rep.*, vol. 10, no. 1, Art. no. 1, Oct. 2020, doi: 10.1038/s41598-020-73160-7.
2. M. F. Zscherp *et al.*, 'Overcoming the Miscibility Gap of GaN/InN in MBE Growth of Cubic In x Ga1−x N', *ACS Appl. Mater. Interfaces*, vol. 15, 2023.
3. J. Wu, 'When group-III nitrides go infrared: New properties and perspectives', *J. Appl. Phys.*, vol. 106, no. 1, p. 011101, Jul. 2009, doi: 10.1063/1.3155798.
4. A. Martí, L. Cuadra, and A. Luque, 'Quantum dot intermediate band solar cell', in *Conference Record of the Twenty-Eighth IEEE Photovoltaic Specialists Conference-2000 (Cat. No. 00CH37036)*, IEEE, 2000, pp. 940–943.
5. A. Luque *et al.*, 'Experimental analysis of the quasi-Fermi level split in quantum dot intermediate-band solar cells', *Appl. Phys. Lett.*, vol. 87, no. 8, 2005.
6. F. Blaabjerg, Y. Yang, K. A. Kim, and J. Rodriguez, 'Power Electronics Technology for Large-Scale Renewable Energy Generation', *Proc. IEEE*, vol. 111, no. 4, pp. 335–355, Apr. 2023, doi: 10.1109/JPROC.2023.3253165.
7. Y. Wang, F. Hu, and K. Hane, 'Patterned growth of InGaN/GaN quantum wells on freestanding GaN grating by molecular beam epitaxy', *Nanoscale Res. Lett.*, vol. 6, no. 1, p. 117, Feb. 2011, doi: 10.1186/1556-276X-6-117.
8. S.-W. Feng, C.-M. Lai, C.-H. Chen, W.-C. Sun, and L.-W. Tu, 'Theoretical simulations of the effects of the indium content, thickness, and defect density of the i-layer on the performance of p-i-n InGaN single homojunction solar cells', *J. Appl. Phys.*, vol. 108, no. 9, p. 093118, Nov. 2010, doi: 10.1063/1.3484040.

9. H. Abboudi, H. El Ghazi, F. Benhaddou, R. En-nadir, A. Jorio, and I. Zorkani, 'Temperature-related photovoltaic characteristics of (In,Ga)N single-intermediate band quantum well solar cells for different shapes', *Phys. B Condens. Matter*, p. 413495, Nov. 2021, doi: 10.1016/j.physb.2021.413495.
10. H. Abboudi, H. El Ghazi, A. Jorio, and I. Zorkani, 'Impurity-related photovoltaic efficiency of (In,Ga)N/GaN quantum well-single intermediate band solar cell considering heavy hole impact', *Superlattices Microstruct.*, vol. 150, p. 106756, 2021.
11. H. El Ghazi, 'Numerical investigation of one-intermediate band InN/GaN QW solar cell under electric field, impurity and size effects', *Phys. B Condens. Matter*, vol. 602, p. 412427, 2021.
12. A. Adaine, S. Ould Saad Hamady, and N. Fressengeas, 'Simulation study of a new InGaN p-layer free Schottky based solar cell', *Superlattices Microstruct.*, vol. 96, pp. 121–133, Aug. 2016, doi: 10.1016/j.spmi.2016.05.020.
13. J. Ajayan, D. Nirmal, P. Mohankumar, M. Saravanan, M. Jagadesh, and L. Arivazhagan, 'A review of photovoltaic performance of organic/inorganic solar cells for future renewable and sustainable energy technologies', *Superlattices Microstruct.*, vol. 143, p. 106549, Jul. 2020, doi: 10.1016/j.spmi.2020.106549.
14. M. R. Islam, M. S. Hasan, and M. R. Islam, 'Estimation of exciton confinement in III-nitride quantum wires', in *2016 9th International Conference on Electrical and Computer Engineering (ICECE)*, IEEE, 2016, pp. 311–313.
15. X. Cai *et al.*, 'Investigation of InGaN p-i-n Homojunction and Heterojunction Solar Cells', *IEEE Photonics Technol. Lett.*, vol. 25, no. 1, pp. 59–62, Jan. 2013, doi: 10.1109/LPT.2012.2227702.
16. K. W. J. Barnham *et al.*, 'Quantum well solar cells', *Phys. E Low-Dimens. Syst. Nanostructures*, vol. 14, no. 1, pp. 27–36, Apr. 2002, doi: 10.1016/S1386-9477(02)00356-9.
17. S. Nakamura, 'The roles of structural imperfections in InGaN-based blue light-emitting diodes and laser diodes', *Science*, vol. 281, no. 5379, pp. 956–961, Aug. 1998, doi: 10.1126/science.281.5379.956.
18. C. Gmachl, F. Capasso, D. L. Sivco, and A. Y. Cho, 'Recent progress in quantum cascade lasers and applications', *Rep. Prog. Phys.*, vol. 64, no. 11, p. 1533, Oct. 2001, doi: 10.1088/0034-4885/64/11/204.
19. C. L. Tan and H. Mohseni, 'Emerging technologies for high performance infrared detectors', *Nanophotonics*, vol. 7, no. 1, pp. 169–197, Jan. 2018, doi: 10.1515/nanoph-2017-0061.
20. N. Tran,' Mid-Infrared intersubband polaritonic devices'. Condensed Matter [cond-mat]. Université Paris-Saclay, 2020. https://hal.science/tel-03122850/
21. K. W. J. Barnham and G. Duggan, 'A new approach to high-efficiency multi-band-gap solar cells', *J. Appl. Phys.*, vol. 67, no. 7, pp. 3490–3493, Apr. 1990, doi: 10.1063/1.345339.
22. T. Takamoto, M. Kaneiwa, M. Imaizumi, and M. Yamaguchi, 'InGaP/GaAs-based multijunction solar cells', *Prog. Photovolt. Res. Appl.*, vol. 13, no. 6, pp. 495–511, 2005, doi: 10.1002/pip.642.

23. D. Aiken *et al.*, 'Lattice-matched solar cells with 40% average efficiency in pilot production and a roadmap to 50%', in *2012 IEEE 38th Photovoltaic Specialists Conference (PVSC) PART 2*, Jun. 2012, pp. 1–7. doi: 10.1109/PVSC-Vol2.2012.6656724.
24. R. Hoheisel *et al.*, 'Quantum-well solar cells for space: The impact of carrier removal on end-of-life device performance', *IEEE J. Photovolt.*, vol. 4, no. 1, pp. 253–259, Jan. 2014, doi: 10.1109/JPHOTOV.2013.2289935.
25. M. A. Omar, 'Development of InxGai-xN alloys based multi-quantum well solar cells: An overview', *2018 International Conference on Computing, Mathematics and Engineering Technologies (ICoMET)*, Jan. 2018, Accessed: Feb. 25, 2024. [Online]. Available: https://www.academia.edu/103900744/Development_of_InxGai_xN_alloys_based_multi_quantum_well_solar_cells_An_overview
26. R. En-nadir *et al.*, 'Tailoring optoelectronic properties of InGaN-based quantum wells through electric field, indium content, and confinement shape: A theoretical investigation', *Phys. B Condens. Matter*, vol. 663, p. 414976, Aug. 2023, doi: 10.1016/j.physb.2023.414976.
27. R. En-nadir, H. El Ghazi, A. Jorio, and I. Zorkani, 'Inter and intra band impurity-related absorption in (In,Ga) N/GaN QW under composition, size and impurity effects', in *MATEC Web of Conferences*, EDP Sciences, 2020.
28. R. En-nadir *et al.*, 'Enhancing Emission via Radiative Lifetime Manipulation in Ultrathin InGaN/GaN Quantum Wells: The Effects of Simultaneous Electric and Magnetic Fields, Thickness, and Impurity', *Nanomaterials*, vol. 13, no. 21, Art. no. 21, Jan. 2023, doi: 10.3390/nano13212817.
29. R. En-nadir *et al.*, 'Intrasubband-related linear and nonlinear optical absorption in single, double and triple QW: The compositions, temperature and QW's number effects', *Philos. Mag.*, vol. 103, pp. 1–14, 2022.
30. I. Maouhoubi, R. En-nadir, I. Zorkani, A. O. T. Hassani, and A. Jorio, 'The effects of the dielectric screening, temperature, magnetic field, and the structure dimension on the diamagnetic susceptibility and the binding energy of a donor-impurity in quantum disk', *Phys. B Condens. Matter*, vol. 646, p. 414371, 2022.
31. R. En-nadir, H. El Ghazi, W. Belaid, A. Jorio, and I. Zorkani, 'Intraconduction band-related optical absorption in coupled (In,Ga)N/GaN double parabolic quantum wells under temperature, coupling and composition effects', *Results Opt.*, vol. 5, p. 100154, Dec. 2021, doi: 10.1016/j.rio.2021.100154.
32. R. En-nadir *et al.*, 'Analyzing the combined influences of external electric field, impurity-location, in-content, and QW's number on donor-impurity binding energy in multiple quantum wells with finite squared potential', *Opt. Quantum Electron.*, vol. 55, no. 7, p. 597, May 2023, doi: 10.1007/s11082-023-04893-8.
33. R. En-nadir, H. El Ghazi, A. Jorio, I. Zorkani, H. Abboudi, and F. A. Jabouti, 'Numerical Study of Temperature and Electric Field Effects on the Total Optical Absorption Coefficient Related-Conduction-Subband Optical Transitions in InGaN/GaN Single Parabolic QW', *Fluid Dyn. Mater. Process.*, Mar. 2022, doi: 10.32604/fdmp.2022.021759.

34. H. Abboudi *et al.*, 'Enhancing the photovoltaic efficiency of In0.2Ga0.8N/GaN quantum well intermediate band solar cells using combined electric and magnetic fields', *Materials*, vol. 17, no. 21, Art. no. 21, Jan. 2024, doi: 10.3390/ma17215219.
35. H. Abboudi, H. EL Ghazi, R. En-nadir, M. A. Basyooni-M. Kabatas, A. Jorio, and I. Zorkani, 'Efficiency of InN/InGaN/GaN intermediate-band solar cell under the effects of hydrostatic pressure, in-compositions, built-in-electric field, confinement, and thickness', *Nanomaterials*, vol. 14, no. 1, Art. no. 1, Jan. 2024, doi: 10.3390/nano14010104.
36. A. Martí, C. R. Stanley, and A. Luque, 'Intermediate band solar cells (IBSC) using nanotechnology', in *Nanostructured Materials for Solar Energy Conversion*, Elsevier, 2006, pp. 539–566.
37. 'Best Research-Cell Efficiency Chart | Photovoltaic Research | NREL'. Accessed: Feb. 25, 2024. [Online]. Available: https://www.nrel.gov/pv/cell-efficiency.html
38. H. El Ghazi, 'Nanomedicine & Nanotechnology Open Access Committed to Create Value for Researchers Nanomaterials-Based Multiple Quantum Wells for High Photovoltaic Conversion Solar Cells Nanomed Nanotechnol Nanomaterials-Based Multiple Quantum Wells for High Photovoltaic Conversion Solar Cells', vol. 9, Feb. 2024, doi: 10.23880/nnoa-16000288.
39. A. El Aouami *et al.*, 'Internal polarization electric field effects on the efficiency of InN/InxGa1-xN multiple quantum dot solar cells', *Sol. Energy*, vol. 201, pp. 339–347, May 2020, doi: 10.1016/j.solener.2020.03.011.
40. Md. Shahiduzzaman *et al.*, 'Subchapter 4.3 – Organometal halide perovskite photovoltaics', in *Comprehensive Guide on Organic and Inorganic Solar Cells*, Md. Akhtaruzzaman and V. Selvanathan, Eds., in Solar Cell Engineering., Academic Press, 2022, pp. 273–317. doi: 10.1016/B978-0-323-85529-7.00007-4.
41. A. Luque and A. Martí, 'Theoretical Limits of Photovoltaic Conversion', in *Handbook of Photovoltaic Science and Engineering*, John Wiley & Sons, Ltd, 2003, pp. 113–151. doi: 10.1002/0470014008.ch4.
42. G. F. Brown, J. W. Ager III, W. Walukiewicz, and J. Wu, 'Finite element simulations of compositionally graded InGaN solar cells', *Sol. Energy Mater. Sol. Cells*, vol. 94, no. 3, pp. 478–483, 2010.
43. R. M. Farrell *et al.*, 'High quantum efficiency InGaN/GaN multiple quantum well solar cells with spectral response extending out to 520 nm', *Appl. Phys. Lett.*, vol. 98, no. 20, p. 201107, May 2011, doi: 10.1063/1.3591976.
44. E. N. Afshar, R. Rouhi, and N. E. Gorji, 'Review on the degradation and device physics of quantum dot solar cells', *Mod. Phys. Lett. B*, vol. 29, no. Supplement 1, p. 1530008, Dec. 2015, doi: 10.1142/S0217984915300082.
45. J. M. López-González, 'Numerical simulations of rear point-contacted solar cells on 2.2 Ωcm p-type c-Si substrates', *Prog. Photovolt. Res. Appl.*, Jan. 2013, Accessed: Feb. 24, 2024. Online]. Available: https://www.academia.edu/53873649/Numerical_simulations_of_rear_point_contacted_solar_cells_on_2_2_%CE%A9cm_p_type_c_Si_substrates

46. H. El Ghazi, Y. R. Eker, R. En-nadir, S. E. Zaki, and M. A. Basyooni-M. Kabatas, 'Enhancing performance of polar InGaN-based thin film solar cells through intrinsic layer impact optimization: Numerical modeling', *Results Eng.*, vol. 21, p. 101909, Mar. 2024, doi: 10.1016/j.rineng.2024.101909.
47. T. Selmi and R. Belghouthi, 'A novel widespread Matlab/Simulink based modeling of InGaN double hetero-junction p-i-n solar cell', *Int. J. Energy Environ. Eng.*, vol. 8, no. 4, pp. 273–281, Dec. 2017, doi: 10.1007/s40095-017-0243-7.
48. R. Belghouthi, J. P. Salvestrini, M. H. Gazzeh, and C. Chevallier, 'Analytical modeling of polarization effects in InGaN double hetero-junction p-i-n solar cells', *Superlattices Microstruct.*, vol. 100, pp. 168–178, Dec. 2016, doi: 10.1016/j.spmi.2016.09.016.
49. H. E. Ghazi, Y. E. Ramazan, and R. En-nadir, 'Numerical analysis of InGaN/GaN intermediate band solar cells under X-sun concentration, in-compositions, and doping: Unlocking the potential of concentrated photovoltaics', *Arab. J. Sci. Eng.*, Jan. 2024, doi: 10.1007/s13369-023-08645-4.
50. T. Sogabe *et al.*, 'High-efficiency InAs/GaAs quantum dot intermediate band solar cell achieved through current constraint engineering', *Mater.*, vol. 1, no. 2, p. 100013, Jun. 2023, doi: 10.1016/j.nxmate.2023.100013.
51. 'Reference Air Mass 1.5 Spectra'. Accessed: Nov. 26, 2023. [Online]. Available: https://www.nrel.gov/grid/solar-resource/spectra-am1.5.html
52. J. R. Sites, 'Calculation of impact ionization enhanced photovoltaic efficiency', *Sol. Cells*, vol. 25, no. 2, pp. 163–168, Nov. 1988, doi: 10.1016/0379-6787(88)90020-8.
53. D. Qingwen *et al.*, 'Theoretical study on In {sub x} Ga {sub 1-x} N/GaN quantum dots solar cell', *Phys. B Condens. Matter*, vol. 406, 2011.
54. P. Megantoro, M. A. Syahbani, I. H. Sukmawan, S. D. Perkasa, and P. Vigneshwaran, 'Effect of peak sun hour on energy productivity of solar photovoltaic power system', *Bull. Electr. Eng. Inform.*, vol. 11, no. 5, Art. no. 5, Oct. 2022, doi: 10.11591/eei.v11i5.3962.
55. A. Kowsar, S. F. U. Farhad, and S. Sakib, 'Effect of the bandgap, sun concentration and surface recombination velocity on the performance of a III-V bismide multijunction solar cells', *Int. J. Renew. Energy Res.*, vol. 8, pp. 2218–2227, Dec. 2018.
56. C. Outes, E. F. Fernández, N. Seoane, F. Almonacid, and A. J. García-Loureiro, 'Numerical optimisation and recombination effects on the vertical-tunnel-junction (VTJ) GaAs solar cell up to 10,000 suns', *Sol. Energy*, vol. 203, pp. 136–144, Jun. 2020, doi: 10.1016/j.solener.2020.04.029.
57. R. En-nadir, 'Theoretical study of the non-parabolicity and size effects on the diamagnetic susceptibility of donor impurity in Si, HgS and GaAs cylindrical quantum dot and quantum disk: applied magnetic field influence is considered', *Philos. Mag.*, pp. 1–14, Dec. 2022, doi: 10.1080/14786435.2022.2158384.
58. R. En-nadir *et al.*, 'The confinement profile effect on the optical properties in different inverse-shaped single InGaN/GaN quantum wells', *Iraqi J. Phys.*, vol. 20, no. 1, pp. 1–13, 2022.

59. R. En-nadir *et al.*, 'The electric and magnetic field effects on the optical absorption in double QWs with squared, U-shaped and V-shaped confinement potentials', *Philos. Mag.*, vol. 0, no. 0, pp. 1–14, Nov. 2022, doi: 10.1080/14786435.2022.2145027.
60. R. En-nadir *et al.*, 'Effects of laser and electric fields on ultrathin InGaN/GaN heterostructure: Exploring binding energy, electron distribution, and diamagnetic susceptibility with harmonic and inharmonic potential', *Phys. E Low-Dimens. Syst. Nanostructures*, p. 115933, Feb. 2024, doi: 10.1016/j.physe.2024.115933.
61. R. En-nadir, H. E. Ghazi, W. Belaid, A. Jorio, I. Zorkani, and H. Ş. Kiliç, 'Ground and first five low-lying excited states related optical absorption in In.1Ga.9N/GaN double quantum wells: Temperature and coupling impacts', *Solid State Commun.*, vol. 338, p. 114464, 2021, doi: https://doi.org/10.1016/j.ssc.2021.114464.
62. W. Belaid, H. El Ghazi, I. Zorkani, and A. Jorio, 'Pressure-related binding energy in (In,Ga)N/GaN double quantum wells under internal composition effects', *Solid State Commun.*, vol. 327, p. 114193, Mar. 2021, doi: 10.1016/j.ssc.2021.114193.
63. W. Belaid, H. El Ghazi, R. En-Nadir, H. Ş. Kılıç, I. Zorkani, and A. Jorio, 'Temperature-related electronic low-lying states in different shapes in. 1Ga. 9N/GaN double quantum wells under size effects', *Trends Sci.*, vol. 19, no. 17, p. 5777, 2022.

CHAPTER 7

# New Approaches for Developing Efficient Eco-Friendly Lead-Free Perovskite Thin-Film Solar Cells

## *Modeling and Design Methodology*

Abdelhak Maoucha, Faycal Djeffal, and Hichem Ferhati

## 7.1 INTRODUCTION

Thin-film solar cells (TFSCs), including amorphous silicon (Si), perovskite, CZTS, CdTe, and CIGS technologies, have attained significant interest owing to their cost-effectiveness and reduced material consumption [1–7]. Recently, intensive research efforts have been directed toward developing novel emerging materials to enhance the efficiency-to-cost ratio of TFSCs [1, 2, 7–10]. Lead-based perovskites have emerged as a promising candidate due to their exceptional electrical and optical properties, including a high absorption coefficient, low exciton binding energy, extended charge diffusion lengths, tunable bandgap, high carrier mobility, and low-cost fabrication processes. Notably, lead-based perovskite solar cells (PSCs) have already achieved a power conversion efficiency (PCE) of 26.1% [10]. Even with these remarkable properties, lead-based PSCs are hindered by two major limitations: degradation issues and lead toxicity [10–13]. To

DOI: 10.1201/9781003499732-7

address these challenges, research has shifted toward exploring alternative lead-free perovskite materials, such as those incorporating Ge, Cs, Bi, and Sn halides. These materials exhibit promising potential as substitutes for lead-based perovskites due to their tunable bandgaps, enhanced optical absorption behavior, and excellent optical conductivity [14]. Notably, the successful development of lead-free perovskites like $MAGeI_3$ and $CsSnGeI_3$ has opened exciting avenues for achieving high-performance TFSCs [14, 15]. Their captivating electrical, optical, and processing characteristics, including bandgap tunability, high carrier mobility, strong visible-to-infrared (Vis-IR) absorption coefficient, and improved stability compared to $MAPbI_3$, contribute significantly to this potential [15]. Extensive numerical and experimental investigations of lead-free PSCs have been conducted, exploring various electronic configurations with different electron transport layers (ETLs) and hole transport layers (HTLs) [11–18]. These studies demonstrate the significant potential of lead-free perovskites for achieving improved PCEs. However, the use of suboptimal transport layers can induce detrimental degradation mechanisms associated with energy level misalignment and lattice mismatch effects. Consequently, novel design strategies are crucial to address these challenges and further enhance the performance of lead-free PSCs. To achieve this goal, a comprehensive understanding of device behavior and performance is essential, facilitated by the development of accurate simulation models.

Conventional methods for studying and modeling semiconductor device behavior rely on analytical and numerical approaches [17–23]. Analytical models provide closed-form solutions to the governing physical equations, while numerical methods, such as finite difference, finite element, and Monte Carlo techniques, solve them iteratively. Despite their widespread use and valuable insights into device behavior, these approaches face limitations. They struggle to handle complex physical phenomena and accurately model non-linear behavior, potentially leading to inaccurate performance predictions [23]. To address these limitations, soft computing techniques have emerged as powerful tools for modeling and optimizing complex systems across various domains. Soft computing represents a subfield within computer science that excels at managing imprecision, uncertainty, and partial truth in data [24]. In contrast to traditional, hard computing methods reliant on precise mathematical models and algorithms [24], soft computing techniques draw inspiration from human reasoning and decision-making processes [25]. This field encompasses a diverse array of methodologies, including fuzzy logic, neural networks, genetic algorithms (GAs), and machine learning. These techniques empower the development of systems that can learn and adapt from experience, recognize patterns within data, and make informed decisions even with incomplete or uncertain information. Soft computing approaches are particularly valuable for tackling complex, real-world problems where conventional methods may prove impractical or ineffective [24, 25]. Their applicability extends across a broad spectrum of domains, as evidenced by their successful implementation in robotics, finance, engineering, natural language processing, image and speech recognition, medicine, and optimization problems [25–31].

To enhance device performance while minimizing development time and cost the semiconductor industry is relentlessly driven. In this pursuit, soft computing techniques have become instrumental in enabling the analysis of vast datasets and the optimization of device parameters. Within the domain of optoelectronic devices, it finds applications in modeling the behavior of light-emitting diodes, photodetectors, and solar cells, with the ultimate goal of performance optimization. For instance, photodetector design optimization can be achieved through the synergistic application of soft computing and numerical techniques, such as GAs. Alternatively, a combined approach utilizing semi-analytical modeling and PSO (particle swarm optimization) can be employed to improve sensor performance [30–32]. In the realm of photovoltaics, researchers unceasingly work toward maximizing efficiency through material exploration and optimization techniques. Soft computing proves particularly valuable in this context, as it excels at modeling complex, non-linear relationships between input and output variables. Also, it offers a versatile toolbox for tackling challenges within the realm of optoelectronic devices. For example, fuzzy logic demonstrates competence in modeling solar cell behavior and extracting key parameters through accurate predictions of output variables based on input data [33, 34]. This capability stems from its inherent ability to manage imprecise and uncertain information, allowing for the formulation of accurate control decisions based on real-time operating conditions of photovoltaic systems, a concept often referred to as maximum power point tracking [34–37]. Furthermore, techniques like GAs and PSO can be leveraged to optimize the design of PV systems. This optimization encompasses the selection of optimal material combinations, component sizing, and overall system layout configuration [37–40]. Additionally, soft computing approaches excel at analyzing large datasets and extracting valuable patterns and insights. This proves particularly useful for gaining a deeper understanding of complex optoelectronic device behavior and pinpointing areas for performance enhancement.

## 7.2 MODELING AND OPTIMIZATION OF PHOTOVOLTAIC DEVICES

Modeling photovoltaic devices play a critical role in understanding and predicting their performance under various operating conditions [39, 41]. This is achieved through the development of mathematical equations and computer simulations that represent the behavior of the materials and the complete device [42–44]. Such models empower engineers and scientists to optimize the design of microelectronic devices. At the core of PV device modeling lies a comprehensive understanding of the underlying physical phenomena governing the materials [45]. These phenomena include charge carrier transport, carrier generation and recombination, and the interaction of carriers with impurities, defects, and applied electric fields [18, 46]. The electrical properties of the materials, along with their dependence on temperature and other parameters, are also crucial aspects of the modeling process. Several modeling approaches exist for photovoltaic devices, categorized as analytical, numerical, and physical (empirical) models. Analytical models leverage mathematical equations to describe device behavior and are often employed for simpler devices or to gain fundamental insights into their operation. Accurately modeling device behavior necessitates the application of numerical models that employ sophisticated computational algorithms to address the inherent non-linearities of the governing equations. This capability allows them to model intricate devices and conduct in-depth performance analyses [47, 48]. Once a model is established, optimization techniques like GAs, PSO, or gradient descent come into play. These techniques guide the search for the optimal design and operating conditions that will maximize device performance [47]. This optimization process can encompass various aspects, such as identifying the ideal device dimensions and doping levels or pinpointing the most suitable materials and fabrication processes to achieve peak efficiency in photovoltaic devices.

The field of modeling and optimization techniques constitutes a vibrant area of research and development, driven by its pivotal role in the design and optimization of novel devices and technologies. The relentless pursuit in this domain centers on achieving enhanced device efficiency, cost-effectiveness, and reliability [49–51].

### 7.2.1 Analytical Modeling

Analytical models serve as a cornerstone for photovoltaic device modeling by leveraging mathematical equations and analytical solutions derived from fundamental concepts like calculus, linear algebra, and differential equations [52]. These models enable predictions of system behavior, processes, or phenomena based on defined assumptions and known inputs. The solutions typically take a closed form, expressible through established mathematical functions [53]. Applications of analytical models span diverse fields, including economics, fluid dynamics, and circuit analysis. While advantageous in providing a clear and intuitive understanding of photovoltaic device behavior, analytical models often rely on simplifying assumptions that may not hold true in all scenarios [54]. Consequently, they are frequently employed in conjunction with numerical simulations to yield more accurate performance predictions. Photovoltaic device modeling incorporates a core set of governing equations. However, depending on the specific device application and the underlying physics of the problem being modeled, additional equations may be necessary.

#### *7.2.1.1 Poisson's Equation*

Poisson's equation stands as a cornerstone of physics, governing the distribution of electric potential within a specific region [41, 43, 45]. In the context of semiconductors, this equation describes how the distribution of charge carriers within the material dictates the electric potential. Mathematically, Poisson's equation for a semiconductor can be expressed as:

$$div(\varepsilon \nabla \psi) = -q(N_D + N_A + n + p) \tag{7.1}$$

where $\psi$ represents the electric potential, q signifies the elementary charge, ε denotes the permittivity of the semiconductor material, and p and n symbolize the concentrations of holes and electrons, respectively, ND and NA represent the doping concentrations of donor and acceptor impurities, respectively.

#### *7.2.1.2 The Continuity Equation*

The continuity equation serves as a fundamental principle in semiconductor physics, upholding the law of charge conservation [41, 43, 45].

It dictates that the rate of change in charge density within a semiconductor equates to the net current density entering and exiting a specific, infinitesimal volume of the material.

$$\frac{\partial n}{\partial t} = \frac{1}{q} div\vec{J_n} + G_n - R_n \tag{7.2}$$

$$\frac{\partial p}{\partial t} = -\frac{1}{q} div\vec{J_p} + G_p - R_p \tag{7.3}$$

where n and p are the electron and hole concentration, Jn and Jp denote the electron and hole current densities, respectively, and Gn and Gp represent the generation rates of electrons and holes, respectively. Finally, Rn and Rp designate the recombination rate of electrons and holes, respectively.

#### *7.2.1.3 The Drift-Diffusion Equations*

Drift-diffusion equations constitute a fundamental set of equations governing the transport mechanisms for charge carriers (electrons and holes) within a semiconductor device [41, 43, 45]. These equations are established upon the principles of current density balance for both electrons and holes, alongside the continuity equation.

$$\vec{J_n} = q\mu_n n\vec{E} + qD_n\overrightarrow{\nabla n} \tag{7.4}$$

$$\vec{J_p} = q\mu_p p\vec{E} - qD_p\overrightarrow{\nabla p} \tag{7.5}$$

where Jn and Jp represent the electron and hole current densities respectively, μn and μp signify the electron and hole mobilities, Dn and Dp denote the electron and hole diffusivities, and E stands for the electric potential field.

### 7.2.2 Numerical Modeling

In contrast to analytical models, numerical modeling tackles complex or uncertain systems through intricate computer algorithms and simulations [55]. These models become necessary when the governing equations and relationships defy analytical solutions due to their complexity [55]. The core principle of numerical modeling involves segmenting a system or process into smaller, more tractable components. Subsequently, computer algorithms are employed to solve the equations governing each of these segments. Unlike analytical solutions, numerical solutions are approximations derived from the input data and the inherent assumptions embedded within the model [55, 56]. In the context of photovoltaic devices, numerical modeling leverages computer simulations to predict the behavior of materials and complete devices. This can be achieved through various techniques, such as solving the Schrödinger equation for electrons or employing the drift-diffusion equations to model charge carrier transport. Finite difference, finite element, and boundary element methods represent some of the common numerical techniques used for this purpose [57]. The ultimate objective of numerical modeling in this domain is to predict the performance and characteristics of photovoltaic devices, including solar cells, transistors, and diodes. This predictive capability serves as a valuable tool for the design and optimization of such devices [57].

#### *7.2.2.1 Finite Element Method*

The finite element method (FEM) has garnered significant traction within the engineering domain for addressing challenges in structural mechanics, predating its formal mathematical foundation [58, 59]. This widespread adoption stems from two key advantages: geometric adaptability and relative ease of application in complex scenarios. Mathematically, the FEM approximates solutions for continuous problems using a weighted residual approach. This approach involves formulating the problem in a variational form, often expressed as $\sum w_i \phi_i$ where the $\phi_i$ represent trial functions and $w_i$ are weights calculated to minimize a specific function [60]. The Rayleigh-Ritz-Galerkin technique serves as the cornerstone of these methods. A defining innovation of the FEM lies in its decomposition of trial functions into smaller, piecewise polynomial functions, each applicable within a designated subdomain or element, as opposed to employing functions spanning the entire domain. This unique characteristic contributes to the extensive applicability of FEMs across diverse engineering fields. Standard FEMs are characterized by the absence of additional techniques such as upwinding, fitting, or inverse averaging during their development [61, 62]. Conversely, the inclusion of these techniques classifies the method as "non-standard." The FEM process generally follows a well-defined sequence of steps (Figure 7.1)

Discretization of the Domain: The problem domain, encompassing a structure or a spatial region, is subdivided into smaller, manageable elements.

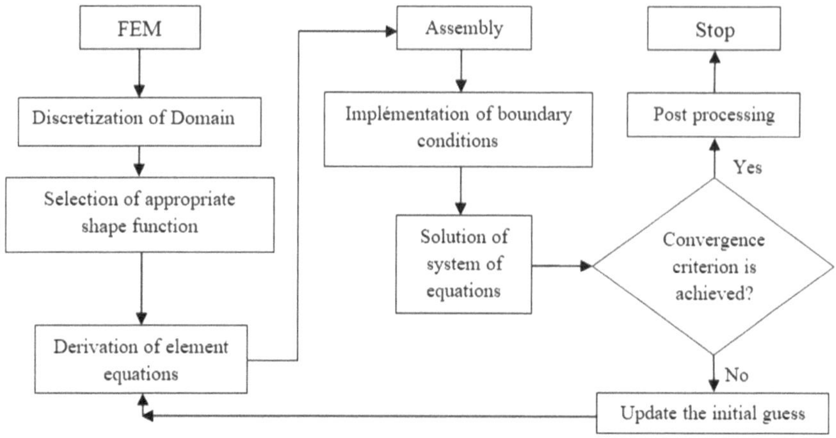


FIGURE 7.1 The general flowchart of the finite element method (FEM).

Selection of Shape Functions: Shape functions are chosen to approximate the solutions within each element. These selections depend on the element type (linear, quadratic, etc.) and the specific problem being addressed.

- Development of Element Equations: Utilizing the chosen shape functions and the governing equations for the problem, element equations are derived for each element.
- Assembly of Global Equations: The element equations are assembled to form a comprehensive set of global equations. This step incorporates the continuity conditions at the nodes shared by multiple elements.
- Incorporation of Boundary Conditions: The global equations are modified to account for any prescribed boundary conditions. These conditions specify the values of unknowns at designated boundary locations.
- Solution of Global Equations: Numerical methods such as Gaussian elimination or iterative methods like the conjugate gradient method are employed to solve the global equations for the unknowns.
- Interpolation of Results: Interpolation techniques are used to approximate the solution at all other points within the elements, leveraging

the values of unknowns at the nodes and the shape functions. This step effectively interpolates the results across the entire domain.

- Post-processing: The final step involves analyzing the obtained solution and generating visual representations for further interpretation.

#### *7.2.2.2 Density Functional Theory Method*

Density functional theory (DFT) stands as a prominent computational method for investigating the electronic structure of diverse materials, encompassing molecules, solids, and surfaces. Unlike traditional methods that focus on individual electron wave functions, DFT adopts theoretical and statistical approaches centered on the electron density within a system [22–69]. This shift in perspective positions DFT as a powerful and widely utilized tool in computational chemistry and materials science. Its ability to accurately predict the properties of a vast array of systems, while maintaining a relatively low computational cost, underpins its extensive applicability [67–69]. The core principle behind DFT lies in the concept that a system's total energy can be expressed as a function of its electron density. In simpler terms, knowledge of the electron distribution within a system allows for the calculation of its total energy. The Kohn-Sham method serves as the most prevalent approach to solving the DFT equations [67, 68]. This method ingeniously transforms the complex problem of interacting electrons into a system of non-interacting electrons governed by an effective potential. Local density approximation (LDA) and generalized gradient approximation (GGA) represent two of the most popular functions employed in DFT calculations.

DFT has emerged as a cornerstone methodology within the modeling and simulation of semiconductor devices [69]. A prominent application of DFT in this domain involves the prediction of electronic band structures and the density of states within semiconductor materials. This information serves as the bedrock for comprehending the electronic properties of the materials, ultimately determining their suitability for various device types, including light-emitting diodes, solar cells, and transistors [70]. The flowchart outlining the DFT modeling process for semiconductor devices is presented in Figure 7.2 [69]. The latter depicts this fundamental flowchart for DFT modeling of semiconductor devices. It is noteworthy that this flowchart can be adapted to accommodate varying levels of accuracy and computational resource constraints. Additionally, the specific system

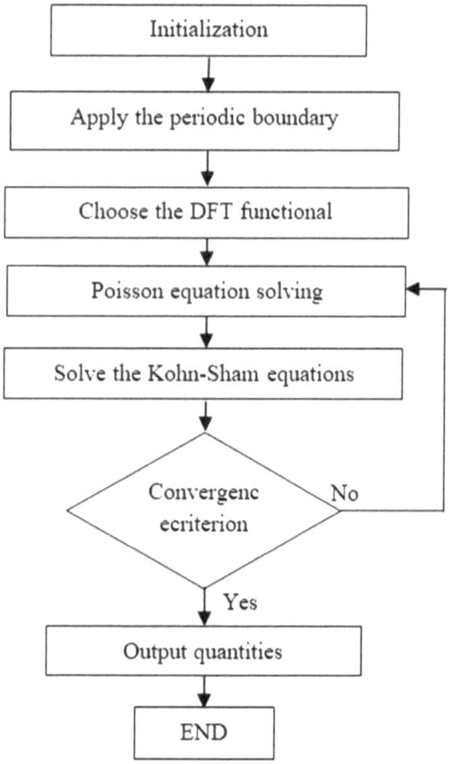


FIGURE 7.2 The general flowchart of density functional theory method.

and application under investigation will influence the choice of functional approximation and the method employed to solve the Kohn-Sham equations. Therefore, selecting the most appropriate methods and parameters for a given system and application is crucial.

## 7.3 METAHEURISTIC TECHNIQUES

In the domain of optimization, metaheuristics stand as a collection of high-level problem-solving methodologies. These approaches serve to guide local heuristic search methods in their exploration of a problem's solution space [71, 72]. Drawing inspiration from natural phenomena, metaheuristic algorithms encompass techniques like PSO, bee colony optimization, ant colony optimization, simulated annealing, and artificial and GAs [73–75]. Their effectiveness in tackling global optimization problems has been well-demonstrated. Unlike problem-specific algorithms, metaheuristics are applicable across diverse domains. They incorporate strategies to prevent stagnation at local optima while also employing mechanisms to balance the exploration of uncharted solution space regions with the exploitation of promising areas to identify high-quality solutions [71, 72].

### 7.3.1 Multi-Objective Genetic Algorithm Optimization (MOGA)

Multi-objective genetic algorithms (MOGA) represent a specific variant of thetraditional GA tailored to handle optimization problems involving multiple objectives [73]. While sharing the core principles of GA, such as population evolution through selection, crossover, and mutation, MOGA tackles the simultaneous optimization of these multiple objective functions, rather than focusing on a single one [73]. The population of candidate solutions is visualized as a set of points within a multidimensional space, where each point corresponds to a specific combination of parameter values for the problem at hand. These points can also be referred to as individuals within the population [76–78]. The MOGA process begins by initializing a population with randomly generated individuals. Subsequently, genetic operators are applied to drive the population's evolution across generations until a satisfactory solution is achieved. The selection process leverages a combination of elitist and non-elitist strategies. Elitism ensures a portion of the best individuals from the current generation are preserved for the next, while non-elitist selection chooses the remaining individuals based on their fitness values for subsequent mating and offspring generation [76–79].

### 7.3.2 Particle Swarm Optimization (PSO)

PSO stands as a metaheuristic optimization algorithm inspired by the collective movement of birds or insects foraging for food [80–82]. This approach utilizes a population of candidate solutions, termed "particles," that navigate the solution space. Particle positions dynamically adapt based on two key factors: their own historical best solution (pbest) and the global best solution discovered by any particle within the population [80]. The core principle underlying PSO revolves around each particle representing a potential solution to the problem. The algorithm leverages principles of social interaction and communication to guide particle movement. Each particle possesses a position within the solution space and a velocity that dictates the direction and magnitude of its movement. Updates to a particle's position consider its current velocity and prior location. Two fundamental concepts govern PSO: pbest and the global best solution (gbest). The pbest of a particle represents the most promising solution it has encountered thus far, while the gbest signifies the best solution identified by any particle in the population [81]. Particle movement updates factor in the current velocity, pbest, and gbest, employing an equation to determine the new velocity.

$$V_i^{k+1} = wV_i^k + c_1 r_1^k \left(p_{li}^k - X_i^k\right) + c_2 r_2^k \left(p_{gi}^k - X_i^k\right) \tag{7.6}$$

$$X_i^{k+1} = X_i^k + V_i^{k+1} \quad \text{for} \quad i = 1 \ldots \text{n} \tag{7.7}$$

where n is the swarm size, $X_i^k$ and $V_i^k$ are the particle position and velocity, $p_{li}^k$ and $p_{gi}^k$ present the particle's best position and particle group's best position in the swarm respectively, c1 and c2 represent the cognitive and social acceleration factors, r1 and r2 reveal the random numbers, which are picked from the range [0, 1], w represents the inertia weight applied to balance the global exploration and $X_i^k$ is the actual position of the particle in the swarm.

Within the realm of semiconductor device optimization, PSO emerges as a valuable tool for optimizing various design parameters. A detailed illustration of the algorithm's flowchart can be found in Figure 7.3.

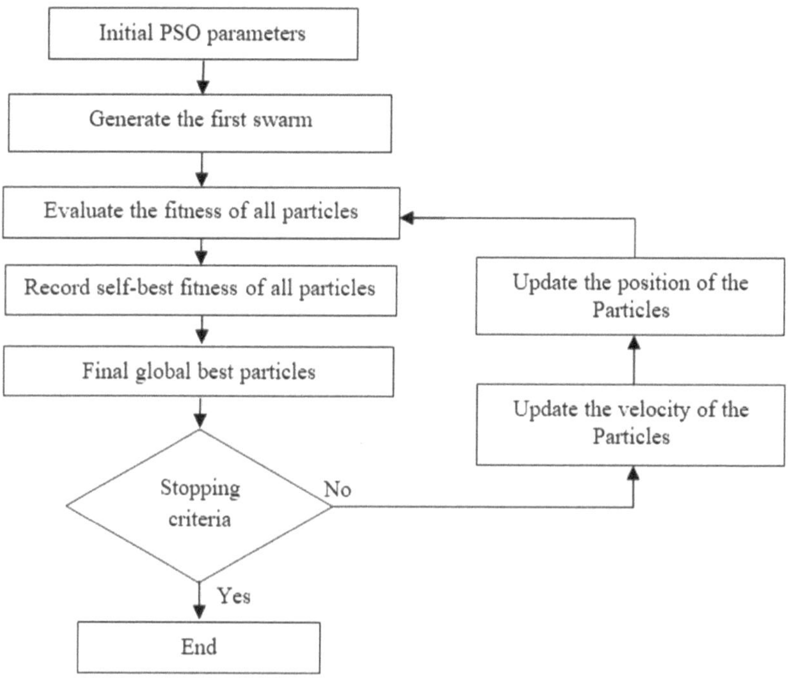


FIGURE 7.3 The general flowchart of particle swarm optimization (PSO).

## 7.4 TANDEM CELL ARCHITECTURE AND SIMULATION METHODOLOGY

Tandem solar cells can achieve higher efficiency than single-layer cells by reducing energy loss through thermalization and increasing the open-circuit voltage. This research introduces a tandem cell design that captures a broader spectrum of sunlight. The top layer, made from a lead-free perovskite material, efficiently absorbs high-energy, short-wavelength sunlight. The light that passes through this layer is then absorbed by the bottom layer, which is made of Si and captures lower-energy light. Figure 7.4a details the top layer's structure, consisting of a transparent conductive base (ITO glass), layers that manage electrical charges, a light-absorbing layer, and a final gold layer for electrical connection (ITO/HTL/perovskite/ETL/FTO/Au). The study tested various material combinations for the top layer to determine the best configuration. Figure 7.4b describes the bottom layer's design, which includes crystalline Si (ITO/n-nc-SiOx/i-a-Si/n-c-Si) with i-a-Si and p-a-Si layers each 5 nm thick. The thickness of the n-c-Si layer is variable, ranging from 30 to 200 μm.

To effectively model this proposed tandem structure, we can treat it as two individual solar cells connected in series. One cell would be made of the perovskite material (FASnI3) and the other of Si. These cells would have different energy gaps (bandgap energies). In this simulation setup, sunlight with the standard AM1.5G spectrum hits the top layer first. The light that passes through the top layer, the unabsorbed spectrum or transmitted spectrum, is then used to simulate the bottom layer's light absorption.

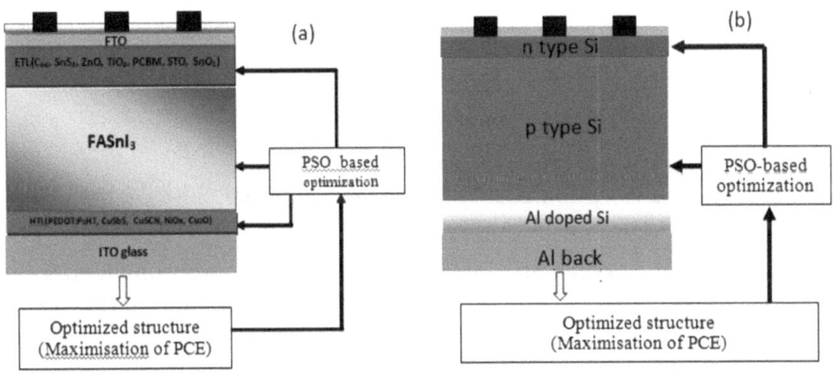


FIGURE 7.4 The investigated solar cells associated with the global optimization approach for (a) perovskite and (b) silicon solar cells.

TABLE 7.1 Set of the Top Sub-Cell Parameters Adopted for the Simulations

| Parameters | $Cu_2O$ | FASnI3 | $SnO_2$ | FTO |
|---|---|---|---|---|
| Thickness (µm) | 0.14 | 1.35 | 0.05 | 0.4 |
| Bandgap Eg(eV) | 2.17 | 1.42 | 3.4 | 3.5 |
| Electron affinity χ (eV) | 3.2 | 4 | 4 | 4.3 |
| Dielectric permittivity | 7.11 | 8.2 | 9 | 9 |
| CB effective density of states ($cm^{-3}$) | $2 \times 10^{17}$ | $1 \times 10^{18}$ | $2.2 \times 10^{18}$ | $2.2 \times 10^{18}$ |
| VB effective density of states ($cm^{-3}$) | $1.1 \times 10^{19}$ | $1 \times 10^{18}$ | $1.8 \times 10^{19}$ | $1.8 \times 10^{19}$ |
| Electron mobility ($cm^2$/V.s) | 20 | 22 | 100 | 20 |
| Hole mobility ($cm^2$/V.s) | 80 | 22 | 25 | 10 |
| Donor Concentration $N_D$ ($cm^{-3}$) | $1 \times 10^7$ | 0 | $1 \times 10^{15}$ | $1 \times 10^{18}$ |
| Acceptor Concentration $N_A$ ($cm^{-3}$) | $1 \times 10^{18}$ | $7 \times 10^{19}$ | 0 | 0 |

TABLE 7.2 Set of the Bottom Sub-Cell Parameters Adopted for the Simulations

| Parameters | n-Si | p-Si | A doped Si |
|---|---|---|---|
| Thickness (µm) | 0.5 | 90 | 10 |
| Bandgap Eg(eV) | 1.12 | 1.12 | 1.12 |
| Electron affinity χ (eV) | 4.05 | 4.05 | 4.05 |
| Dielectric permittivity | 11.9 | 11.9 | 11.9 |
| CB effective density of states ($cm^{-3}$) | $2.8 \times 10^{19}$ | $2.8 \times 10^{19}$ | $2.8 \times 10^{19}$ |
| VB effective density of states ($cm^{-3}$) | $1 \times 10^{19}$ | $1 \times 10^{19}$ | $1 \times 10^{19}$ |
| Electron mobility ($cm^2$/V.s) | $1.5 \times 10^3$ | $1.5 \times 10^3$ | $1.5 \times 10^3$ |
| Hole mobility ($cm^2$/V.s) | 100 | 25 | 100 |
| Donor Concentration $N_D$ ($cm^{-3}$) | 0 | 0 | $1 \times 10^{18}$ |
| Acceptor Concentration $N_A$ ($cm^{-3}$) | $3 \times 10^{18}$ | $1 \times 10^{15}$ | 0 |

To achieve this, we use an optical transmission filter. This filter's properties are calculated using the following equation:

$$S(\lambda) = S0(\lambda).\exp(-\alpha ITO.dITO).\exp(-\alpha SnO2.dSnO2).\exp(-\alpha ETL.dETL)$$
$$.\exp(\alpha perov.dperov).\exp(-\alpha HTL.dHTL)$$

where α is the absorption coefficient of each material, S0(λ) is the global AM1.5G spectrum, and d represents the thickness of different solar cell layers. The basic characteristics of the structures, which were obtained from previously published literature [83, 84] are detailed in Tables 7.1 and 7.2.

## 7.5 MODELING FRAMEWORK

This research employed a combination of numerical simulations and device optimization to identify the optimal design for a high-efficiency solar cell. We utilized the SCAPS-1D tool, a widely recognized software for simulating semiconductor-based devices. This software solves critical equations governing optoelectronic properties in these materials. The main objective was to design an innovative solar cell by carefully selecting materials and fine-tuning the thickness of various layers within the cell. This optimization process took into account both the intrinsic properties of the materials and their interfacial compatibility to maximize the device's performance.

## 7.6 RESULTS AND DISCUSSIONS

After performing numerical simulations using SCAPS software, an optimization process was undertaken to determine the best structure configuration that maximizes device performance. The optimization variables included layer thicknesses, suitable charge transport materials, and ideal doping profiles. Figures 7.5a and b compare the current-voltage (J-V)

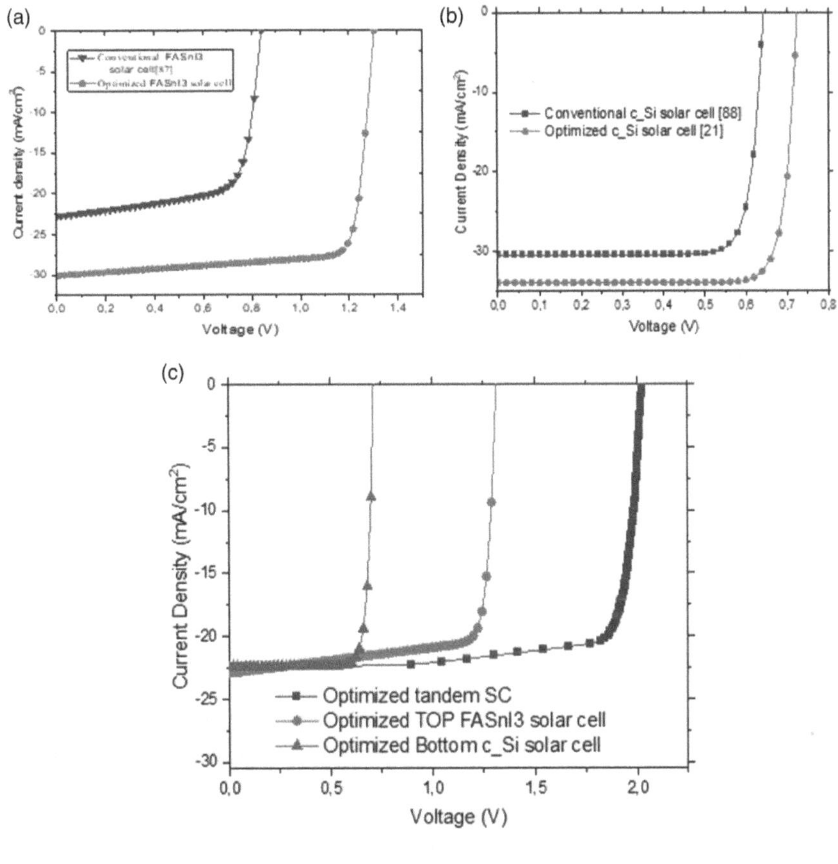


FIGURE 7.5 I-V Characteristics of the proposed lead-free perovskite tandem cell (a) top cell (b) bottom cell, and (c) tandem structure.

characteristics of the optimized design with those of the conventional structure, clearly showing the significant performance enhancement achieved through PSO. Figure 7.5c presents the J–V characteristics of the top sub-cell under the standard AM1.5G solar spectrum and the bottom sub-cell under the filtered spectrum from the top cell simulation. The results indicate a remarkable open-circuit voltage (Voc) of 2.02 V, a short-circuit current density (Jsc) of 22.71 $mA/cm^2$, and a PCE of 37.4% for the tandem solar cell.

## 7.7 SUMMARY

In this chapter, we introduced an innovative modeling and design methodology aimed at enhancing the efficiency of thin-film tandem solar cells, comprising a lead-free perovskite top sub-cell and a Si-based bottom sub-cell. Our proposed design strategy leverages the PSO approach to optimize both the lead-free perovskite absorber material and the Si-based cell, ultimately augmenting the overall performance of the tandem solar cell. The PSO technique systematically identifies the optimal device configuration by analyzing various parameters, including different HTL and ETL materials, along with determining appropriate layer thicknesses and doping profiles. This method holds substantial promise for boosting tandem solar cell performance by accurately adjusting band alignment and mitigating recombination effects. Furthermore, the optimized thickness configuration not only enhances light absorption across the device but also provides better adaptability in adjusting band alignment at the ETL/absorber/HTL interfaces. We believe that our proposed design strategy offers new insights toward the elaboration of eco-friendly and efficient photovoltaic devices.

## REFERENCES

1. M. A. Green, E. D. Dunlop, J. Hohl-Ebinger, M. Yoshita, N. Kopidakis, and X. Hao, “Solar cell efficiency tables (Version 56),” Progress in Photovoltaics, vol. 28, pp. 3–15, 2020.
2. P. K. Nayak, S. Mahesh, H. J. Snaith, and D. Cahen, “Photovoltaic solar cell technologies: Analysing the state of the art,” Nature Review Materials, vol. 4, pp. 269–285, 2019.
3. A. Maoucha, H. Ferhati, F. Djeffal, and F. AbdelMalek, “Highly efficient Cd-Free ZnMgO/CIGS solar cells via effective band-gap tuning strategy,” Journal of Computational Electronics, vol. 22, pp. 887–896, 2023.
4. K. Branker, M. Pathak, and J. M. Pearce, “A review of solar photovoltaic levelized cost of electricity,” Renewable and Sustainable Energy Reviews, vol. 15, pp. 4470–4482, 2011.

5. M. A. Green, "Tracking solar cell conversion efficiency," Nature Reviews Physics, vol. 2, pp. 172–173, 2020.
6. H. Ferhati, F. Djeffal, and L. B. Drissi, "Performance improvement of Perovskite/CZTS tandem solar cell using low-cost ZnS/Ag/ITO multilayer spectrum splitter," Superlattices and Microstructures, vol. 148, pp. 106727, 2020.
7. K. Kacha, F. Djeffal, H. Ferhati, D. Arar, and M. Meguellati, "Numerical investigation of a double-junction a: SiGe thin-film solar cell including the multi-trench region," Journal of Semiconductors, vol. 36, pp. 064004, 2015.
8. H. Ferhati, F. Djeffal, and K. Kacha, "Optimizing the optical performance of ZnO/Si-based solar cell using metallic nanoparticles and interface texturization," Optik, vol. 153, pp. 43–49, 2018.
9. M. C. Beard, J. M. Luther, and A. J. Nozik, "The promise and challenge of nanostructured solar cells," Nature Nanotechnology, vol. 9, pp. 951–954, 2014.
10. Best Research-Cell Efficiencies chart, Available online: https://www.nrel.gov/pv/cell-efficiency.html.
11. T. Zhang, H. Li, H. Ban, Q. Sun, Y. Shen, and M. K. Wang, "Efficient $CsSnI_3$ -based inorganic perovskite solar cells based on a mesoscopic metal oxide framework via incorporating a donor element," Journal of Materials Chemistry, vol. 8, pp. 4118–4124, 2020.
12. F. Giustino and H. J. Snaith, "Toward lead-free perovskite solar cells," ACS Energy Letters, vol. 1, pp. 1233–1240, 2016.
13. F. Baig, Y. H. Khattak, B. Marí, S. Beg, A. Ahmed, and K. Khan, "Efficiency enhancement of $CH_3NH_3SnI_3$ solar cells by device modeling," Journal of Electronic Materials, vol. 47, pp. 5275–5282, 2018.
14. W. Ming, H. Shi, and M. H. Du, "Large dielectric constant, high acceptor density, and deep electron traps in perovskite solar cell material $CsGeI_3$," Journal of Materials Chemistry, vol. 4, pp. 13852–13858, 2016.
15. S. Bhattarai and T. D. Das, "Optimization of carrier transport materials for the performance enhancement of the $MAGeI_3$ based perovskite solar cell," Solar Energy, vol. 217, pp. 200–207, 2021.
16. S. Bhattarai, R. Pandey, J. Madan, F. Ahmed, and S. Shabnam, "Performance improvement approach of all inorganic perovskite solar cell with numerical simulation," Materials Today Communications, vol. 33, p. 104364, 2022.
17. A. Maoucha, F. Djeffal, and H. Ferhati, "Numerical Investigation of a New Double-Absorber Lead-free Perovskite Solar Cell via SCAPS-1D," 2023 20th International Conference on Electrical Engineering, Computing Science and Automatic Control (CCE), Mexico City, Mexico, 2023.
18. C. M. Snowden. The Semiconductor Device Modelling. Springer-Vedag, 1989.
19. D. Báez-Lópe and F. E. Guerrero-Castro, "Semiconductor Device," Circuit Analysis with Multisim, Springer, pp. 117–122, 2011.
20. C. M. Snowden. Introduction to Semiconductor Device Modelling. World Scientific Publishing Co, 1998.

21. S. Prasad, H. Schumacher, and A. Gopinath, "Review of Semiconductor Materials and Physics," High-Speed Electronics and Optoelectronics: Devices and Circuits. Cambridge University Press, pp. 3–45, 2009.
22. M. Shaul. Applications of Monte Carlo Method in Science and Engineering. InTech, 2011.
23. I. Ramezani, K. Moshkbar-Bakhshayesh, N. Vosoughi, and M. B. Ghofrani, "Applications of soft computing in nuclear power plants: A review," Progress in Nuclear Energy, vol. 149, p. 104253, 2022.
24. W. Zhang, Y. Zhang, X. Gu, C. Wu, and L. Han. Application of Soft Computing, Machine Learning, Deep Learning and Optimizations in Geoengineering and Geoscience. Springer, 2021.
25. S. B. Gupta, Shivani, "A Review of Soft Computing Techniques and Applications," International Journal of Engineering Research & Technology (IJERT) ICRADL, vol. 09, pp. 2278–0181, 2021.
26. R. S. Yadav, "Review of academic performance using soft computing techniques," International Journal of Computer Engineering and Applications, vol. 11, 2017.
27. S. K. Das, A. Kumar, B. Das, and A. P. Burnwal, "On soft computing techniques in various areas," Computer Science and Information Technology, vol. 3, pp. 59–68, 2013.
28. S. Gambhir, S. K. Malik, and Y. Kumar, "Role of soft computing approaches in healthcare domain: A mini review," Journal of Medical Systems, vol. 40, p. 287, 2016.
29. M. Khan, A. Manzoor, K. Rohail, S. M. Ali, A. Iftikhar, and M. Alam, "Soft Computing Applications in Education Management—A Review," IEEE International Conference on Innovative Research and Development (ICIRD), Bangkok, Thailand, pp. 1–4, 2018.
30. N. H. M. Shamsuddin, N. A. Ali, and R. Alwee, "An Overview on Crime Prediction Methods," 6th ICT International Student Project Conference, pp. 1–5, 2017.
31. F. Djeffal, N. Boubiche, H. Ferhati, J. Faerber, F. Le Normand, N. Javahiraly, and T. Fix, "Highly efficient and low-cost multispectral photodetector based on RF sputtered a-Si/Ti multilayer structure for Si-photonics applications," Journal of Alloys and Compounds, vol. 876, p. 160176, 2021.
32. H. Ferhati and F. Djeffal, "Role of optimized grooves surface-textured front glass in improving $TiO_2$ thin film UV photodetector performance," IEEE Sensors Journal, vol. 16, pp. 5618–5624, 2016.
33. H. Ferhati and F. Djeffal, "A novel high-performance self-powered ultraviolet photodetector: Concept, analytical modeling and analysis," Superlattices and Microstructures, vol. 112, pp. 480–492, 2017.
34. T. Bendib, F. Djeffal, D. Arar, and M. Meguellati, "Fuzzy-logic-based approach for organic solar cell parameters extraction," Proceedings of the World Congress on Engineering, vol. 2, 2013.
35. G. Dileep and S. N. Singh, "Application of soft computing techniques for maximum power point tracking of SPV system," Solar Energy, vol. 141, pp. 182–202, 2017.

36. V. Indragandhi, V. Subramaniyaswamy, and R. Logesh, "Resources, configurations, and soft computing techniques for power management and control of PV/wind hybrid system," Renewable and Sustainable Energy Reviews, vol. 69, pp. 129–143, 2017.
37. M. Balamurugan, S. K. Sahoo, and S. Sukchai, "Application of soft computing methods for grid connected PV system: A technological and status review," Renewable and Sustainable Energy Reviews, vol. 75, pp. 1493–1508, 2017.
38. A. Maoucha and F. Djeffal, "Multi-objective-optimization-based approach to improve the electrical efficiency for organic solar cells," Journal of Computational Electronics, vol. 11, pp. 336–343, 2012.
39. H. Ferhati and F Djeffal, "Exceeding 30% efficiency for an environment-friendly tandem solar cell based on earth-abundant Se/CZTS materials," Physica E: Low-Dimensional Systems and Nanostructures, vol. 109, pp. 52–58, 2019.
40. F. Z. Ramadan, F. Djeffal, L. B. Drissi, S. Saidi, and H. Ferhati, "Highly efficient ACdTS kesterite solar cell based on a new photovoltaic material," Journal of Physics and Chemistry of Solids, vol. 161, p. 110458, 2022.
41. E. A. B. Cole. Mathematical and Numerical Modelling of Heterostructure Semiconductor Devices: From Theory to Programming. Springer, 2009.
42. F. Brezzi, L. D. Marini, and P. Pietra, "Numerical simulation of semiconductor devices," Computer Methods in Applied Mechanics and Engineering, vol. 75, pp. 493–514, 1989.
43. F. Babarada, "Semiconductor Processes and Devices Modeling," Semiconductor Technologies, edited by Grym Jan. InTech, pp. 1–26, 2010.
44. D. Vasileska, D. Mamaluy, H. R. Khan, K. Raleva, and S. M. Goodnick, "Semiconductor device modeling, "Journal of Computational and Theoretical Nanoscience, vol. 5, pp. 1–32, 2008.
45. S. Selberherr. The Stationary Semiconductor Device Equations. Springer-Verlag Wien GmbH, 1986.
46. S. Shamsir, Md. S. Hasan, O. Hassan, et al, "Semiconductor Device Modeling and Simulation for Electronic Circuit Design," Modeling and Simulation in Engineering – Selected Problems, edited by Jan Valdman and Leszek Marcinkowski. IntechOpen, pp. 207–221, 2020.
47. M. S. Khan, F. Jabeen, S. Ghouzali, Z. Rehman, S. Naz, and W. Abdul, "Metaheuristic algorithms in optimizing deep neural network model for software effort estimation," IEEE Access, vol. 9, pp. 60309–60327, 2021.
48. X.-F. Xie, W.-J. Zhang, and De-C. Bi, "Optimizing semiconductor devices by self-organizing particle swarm," *Proceedings of the 2004 Congress on Evolutionary Computation (IEEE Cat. No.04TH8753)*, Vol. 2, pp. 2017–2022, 2004.
49. D. Jiang, W. Lin, and N. Raghavan, "Semiconductor manufacturing final test yield optimization and wafer acceptance test parameter inverse design using multi-objective optimization algorithms," IEEE Access, vol. 9, pp. 137655–137666, 2021.
50. M. Burger, M. Hinze, and R. Pinnau, "Optimization Models for Semiconductor Dopant Profiling," Semiconductor Technologies, edited by Carlo Cercignani. Ester Gabetta, Birkhauser Boston, pp. 91–116, 2007.

51. S. E. De León-Aldaco, H. Calleja, and J. Aguayo Alquicira, "Metaheuristic optimization methods applied to power converters: A review," IEEE Transactions on Power Electronics, vol. 30, no. 12, pp. 6791–6803, 2015.
52. C. M. Snowden. Introduction to Semiconductor Device Modelling. World Scientific Publishing Co. Pte. Ltd. 1998.
53. H. Ferhati, F. Djeffal, and L. B. Drissi, "A new approach to the modeling and simulation of multi-junction solar cells," Optik, vol. 200, p. 163452, 2020.
54. H. Ferhati and F. Djeffal, "An efficient analytical model for tandem solar cells," Materials Research Express, vol. 6, p. 076424, 2019.
55. J. N. Burghartz, "Technology Computer Aided Design," Guide to State-of-the-Art Electron Devices, edited by Joachim N. Burghartz. Wiley-IEEE Press, pp. 97–106, 2013.
56. H. Ferhati and F. Djeffal, "Novel high-performance SOI junctionless FET-based phototransistor using channel doping engineering: Numerical investigation and sensitivity analysis," Optik, vol. 138, pp. 119–126, 2017.
57. H. Ferhati, F. Djeffal, and T. Bentercia, "Numerical investigation of a new junctionless phototransistor for high-performance and ultra-low power infrared communication applications," Telkomnika, vol. 14, pp. 1213–1216, 2016.
58. J. J. Barnes and R. J. Lomax, "Finite-element methods in semiconductor device simulation," IEEE Transactions on Electron Devices, vol. 24, pp. 1082–1089, 1977.
59. D.-M. Yu, X.-M. Pan, and X.-Q. Sheng, "An Accurate and Stable Finite Element Method for Self-Heating Effects Simulation of Semiconductor Devices," International Conference on Electromagnetics in Advanced Applications (ICEAA), Granada, Spain, pp. 1107–1110, 2019.
60. N. Mohamed and M. Z. Sujod, "Finite Elements in Semiconductor Devices," International Conference on Information Management and Engineering, Kuala Lumpur, Malaysia, pp. 108–110, 2009.
61. N. Mohamed and M. Z. Sujod, "Finite Elements in Semiconductor Devices," International Conference on Information Management and Engineering, 2009.
62. J. J. H. Miller, W. H. A. Schilders, and S. Wang, "Application of Finite Element Methods to the Simulation of Semiconductor Devices," Reports on Progress in Physics, edited by Grym Jan, InTech, vol. 62, pp. 287–300, 1999.
63. C. Moglestue. Monte Carlo Simulation of Semiconductor Devices. Chapman & Hali, 1993.
64. P. Lugli, "The Monte Carlo method for semiconductor device and process modeling," IEEE Transactions on Computer-Aided Design of Integrated Circuits and Systems, vol. 9, pp. 1164–1177, 1990.
65. R. W. Kelsall, "The Monte Carlo Method for Semiconductor Device Simulation," IEE Colloquium on Physical Modelling of Semiconductor Devices, pp. 1–5, 1995.
66. G. U. Jensen, B. Lund, T. A. Fjeldly, and M. Shur, "Monte Carlo simulation of semiconductor devices," Computer Physics Communications, vol. 67, pp. 1–61, 1991.

67. K. Burke, et al. The ABC of DFT. University of California, 2007.
68. M. Freyss. "Density Functional Theory," NEA, State-of-the-Art Report on Multi-scale Modelling of Nuclear Fuels. Nuclear Energy Agency of the OECD (NEA), pp. 225–235, 2015.
69. S. Farah, Z. Dibi, H. Ferhati, and F. Djeffal, "DFT-FDTD modeling of a new broadband mid-infrared IGZO thin-film phototransistor based on black phosphorus capping layer incorporating intermediate metallic film," Journal of Physics and Chemistry of Solids, vol. 162, pp. 110528, 2022.
70. A. Kebabi, A. Bentabet, F. Djeffal, H. Ferhati, N. Benmekideche, A. Benmakhlouf, and A. Chala, "DFT study of X-doped (X= Cu, Ag, Au) boron nitride nanotubes for spintronic and optoelectronic applications,". Optik, vol. 225, p. 165863, 2021.
71. C. Blum and A. Roli, "Metaheuristics in combinatorial optimization: Overview and conceptual comparison," ACM Computing Surveys, vol. 35, pp. 268–308, 2003.
72. H. Ferhati, F. Djeffal, and L. B. Drissi, "Metaheuristic-based decision maker framework for the development of multispectral IGZO thin-film phototransistors," Journal of Science: Advanced Materials and Devices, vol. 7, pp. 100414, 2022.
73. J. H. Holland, Adaptation in Natural and Artificial Systems: An Introductory Analysis with Applications to Biology, Control, and Artificial Intelligence. University of Michigan Press, 1975.
74. S. Kirkpatrick, C. D. Gelatt, and M. P. Vecchi, "Optimization by simulated annealing," Science, vol. 220, pp. 671–680, 1983.
75. I. Boussaıd, J. Lepagnot, and P. Siarry, "A survey on optimization metaheuristics," Information Sciences, vol. 232, pp. 82–117, 2013.
76. F. Djeffal, M. Meguellati, and A. Benhaya, "A two-dimensional analytical analysis of subthreshold behavior to study the scaling capability of nanoscale graded channel gate stack DG MOSFETs," Physica E: Low-Dimensional Systems and Nanostructures, vol. 41, pp. 1872–1877, 2009.
77. P.-C. Chang, J.-C. Hsieh, and C.-Y. Wang, "Adaptive multi-objective genetic algorithms for scheduling of drilling operation in printed circuit board industry," Applied Soft Computing, vol. 7, pp. 800–806, 2007.
78. K. Atashkari, N. Nariman-Zadeh, A. Pilechi, A. Jamali, and X. Yaob, "Thermodynamic Pareto optimization of turbojet engines using multi-objective genetic algorithms," International Journal of Thermal Sciences, vol. 44, pp. 1061–1071, 2005.
79. A. Maoucha and F. Djeffal, "Multi-objective-optimization-based approach to improve the electrical efficiency for organic solar cells," Journal of Computational Electronics, vol. 11, pp. 336–343, 2012.
80. M. Clerc and J. Kennedy, "The particle swarm – explosion, stability, and convergence in a multidimensional complex space," IEEE Transactions on Evolutionary Computation, vol. 73, pp. 57–70, 2002.
81. F. Djeffal, N. Lakhdar, M. Meguellati, and A. Benhaya, "Particle swarm optimization versus genetic algorithms to study the electron mobility in wurtzite GaN-based devices," Solid-State Electronics, vol. 53, pp. 988–992, 2009.

82. F. Djeffal, T. Bendib, R. Benzid, and A. Benhaya, "An approach based on particle swarm computation to study the nanoscale DG MOSFET-based circuits," Turkish Journal of Electrical Engineering and Computer Sciences, vol. 18, pp. 1131–1140, 2010.
83. T. Wu et al., "Heterogeneous $FASnI_3$ absorber with enhanced electric field for high-performance lead-free perovskite solar cells," Nano-Micro Letters, vol. 14, no. 1, Apr. 2022.
84. T. Shawky, M. H. Aly, and M. Fedawy, "Performance analysis and simulation of c-Si/SiGe based solar cell," IEEE Access, vol. 9, pp. 75283–75292, Jan. 2021.

# CHAPTER 8

# Magnetic Spinel for Smart Applications

Ketankumar Rameshbhai Gayakvad
and Ketaki Ketan Patankar

## 8.1 INTRODUCTIONS

Magnetic spinels popularly known as spinel ferrites have a general chemical formula $MFe_2O_4$ where M represents divalent metal cation namely cobalt (Co), zinc (Zn), nickel (Ni), manganese (Mn), cadmium (Cd), magnesium (Mg), and copper (Cu) [1, 2]. These magnetic spinels are called spinel ferrites because of their closely packed cubic structure which resembles the structure naturally found in spinel $MgAl_2O_4$ [3]. The closely packed cubic structure is shown in Figure 8.1 [4].

A closed-packed cubic structure of spinel ferrite comprises 8 units of $MFe_2O_4$ resulting in a total of 56 ions. It comprises 16 $Fe^{3+}$ cations, 32 $O^{2-}$ anions, and 8 $M^{2+}$ metal cations. The net total charge on the spinel ferrite

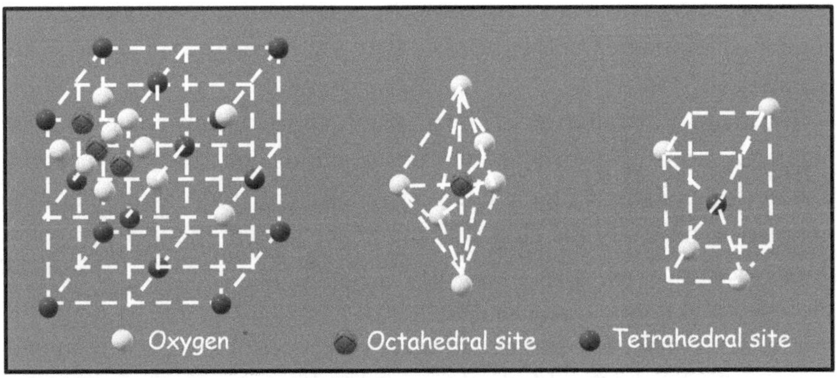


FIGURE 8.1 Spinel ferrite cell structure.

DOI: 10.1201/9781003499732-8

unit cell is zero; hence, the spinel ferrite cell is electrically neutral. Because of the availability of two kinds of cations ($M^{2+}$ and $Fe^{3+}$), two crystallographic sites namely tetrahedral A-site and octahedral B-site are present in spinel structure. Four oxygen ions accompany the tetrahedral A-site whereas six oxygen ions surround the octahedral B-site. $M^{2+}$ cations in the A-site and $Fe^{3+}$ in the B-site result in a normal spinel structure. $ZnFe_2O_4$ and $CdFe_2O_4$ are examples of normal spinel structures. However, there are some magnetic spinels in which $M^{2+}$ cations found on the A-site and $Fe^{3+}$ cations are equally distributed among the A-site and B-site. These types of spinel structures are generally called inverse spinel structures and they are all ferrimagnetic. $CoFe_2O_4$, $NiFe_2O_4$, and $FeFe_2O_4$ are inverse spinel structures. These types of spinel structures endue extraordinary chemical stability, good thermal properties, and excellent electric and magnetic properties [5, 6].

## 8.2 MAGNETISM OF MAGNETIC SPINEL

Magnetic spinel exhibits ferrimagnetism. This spinel follows processes of magnetic saturation and magnetic hysteresis loop. These substances cease the property of spontaneous magnetization above a particular temperature $T_c$ called the Curie temperature and then they transfer into paramagnetic. The magnetic spinel with cubic crystal structure namely $MnFe_2O_4$, $NiFe_2O_4$, $MgFe_2O_4$, $CoFe_2O_4$ and mixed ferrites shows properties of ferrimagnetism. The coercivity is the external applied magnetic field $H_c$ needed to drop down the magnetization or the induction $\vec{B}$ to zero. Magnetic spinels with low coercivity are called soft whereas those spinel ferrites with high coercivity are called hard. Therefore, $CoFe_2O_4$ is magnetically hard compared to others. In other words, $MnFe_2O_4$, $NiFe_2O_4$, and $MgFe_2O_4$ are magnetically soft [7].

The magnetic properties of any magnetic spinel mainly come from the magnetic moment of orbital electrons of atoms constituting the atom. The nuclear magnetic moment is very much smaller as compared to the magnetic moment of the electrons and hence nuclear magnetic moment does not contribute significantly to the magnetic properties. Therefore, the magnetic moment of electrons plays a crucial role in the magnetic properties of magnetic spinel. There are two types of magnetic moments associated with the two kinds of motions (orbital and spin) of electrons. The motion of electrons in the circular orbit of the radius $r$ around the nucleus may considered the current loop and hence the magnetic moment associated with this current loop may be estimated using the following formula,

$$\mu = area\ of\ loop \times current$$

$$\mu_{orbit} = \frac{eh}{4\pi m} \text{ for the orbit } n = 1$$

Here, $h$ is Planck's constant and $m$ is the mass of the electron.

The magnetic moment associated with the spinning motion of the electron can be expressed as,

$$\mu_{spin} = \frac{eh}{4\pi m} \text{ for the orbit } n = 1$$

We could see the spin magnetic moment and orbital magnetic moment both are exactly equal for the n=1 orbit. The magnetic moment corresponding to the n=1 orbit is taken as a fundamental quantity for expressing the magnetic moments. It is denoted by $\mu_B$ and termed as Bohr magneton. Its estimated value turns out to be around $\mu_B = 1.17 \times 10^{-29}$ Wb m

The net magnetic moment of the atom is the vector sum of magnetic moments of all magnetic moments associated with the electrons. If the net magnetic moment is estimated and found null then the substance is called diamagnetic and it exhibits diamagnetism. If the total magnetic moment of the atom results in a finite value (greater than zero), the substance shows paramagnetic, ferromagnetic, antiferromagnetic, or ferrimagnetic behaviors.

## 8.3 SATURATION MAGNETIZATION

The saturation magnetization of a magnetic spinel can be estimated at 0 K with the knowledge of the moment on each ion and the distribution of the ions among A and B sites provided the condition that the exchange interaction between A and B crystallographic sites is negative. The AA, BB, and AB interactions are leading to be negative. However, all these interactions are not all negative simultaneously. In general, the interaction AB is the strongest. The magnetic moment vectors at A sites are parallel to each other and anti-parallel to the magnetic moment vectors at B sites. In the case of magnetically soft ferrites like $MnFe_2O_4$, $NiFe_2O_4$, and $MgFe_2O_4$, the crystallographic directions of magnetic moments are of the nature <111>.

That is the crystallographic directions of these magnetic moments are parallel to a body diagonal of the unit cell in all soft magnetic spinels. These soft magnetic spinels exhibit remarkable magnetic and electric properties like low coercivity and electric conductivity, intermediate saturation magnetization, and good hysteresis characteristics. In the case of cobalt ferrite, these magnetic moments are parallel to the cube edge direction <100>. The saturation magnetization of magnetic spinel per molecule in terms of magnetic moment $\mu_B$ for typical magnetic spinels are as mentioned in the Table 8.1:

TABLE 8.1 Saturation Magnetization of Magnetic Spinels in Terms of $\mu_B$

| **Spinel Ferrite** | **$ZnFe_2O_4$** | **$CuFe_2O_4$** | **$NiFe_2O_4$** | **$CoFe_2O_4$** | **$FeFe_2O_4$** | **$MnFe_2O_4$** |
|---|---|---|---|---|---|---|
| Calculated $\mu_H$ | 0 | 1 | 2 | 1 | 4 | 5 |
| Measured $\mu_H$ | 0 | 1.3 | 2.3 | 1.3 | 4.1 | 4.6 |

The Coercivity (in oersted), magnetization $M_s$ (in gauss), and Curie temperature (in kelvin) for some of the spinel ferrites are as following Table 8.2 [7]

TABLE 8.2 Coercivity, Magnetization, and Curie Temperature of Magnetic Spinel

| **Magnetic Spinels** | **Coercivity (oersted)** | **Room Temperature Magnetization $M_s$ (gauss)** | **Curie Temperature (kelvin)** |
|---|---|---|---|
| $MnFe_2O_4$ | 113 | 410 | 573 |
| $FeFe_2O_4$ | 2150 | 480 | 858 |
| $NiFe_2O_4$ | 10-300 | 270 | 858 |
| $CuFe_2O_4$ | 86.83 | 135 | 758 |
| $ZnFe_2O_4$ | – | $<10$ | $T_N \sim 10$ K |
| $CoFe_2O_4$ | 500–10000 | 4000–5000 | 793 |

## 8.4 TYPES OF MAGNETISM IN MAGNETIC SPINEL

- **Ferrimagnetism:** This type of magnetism is found in magnetic spinel which shows a substantial spontaneous magnetization at room temperature. Such magnetic spinels consist of magnetic domains that are magnetically saturated.
- **Antiferromagnetism:** This kind of magnetism is observed in substances having a small positive susceptibility at all temperatures.
- **Ferromagnetism and paramagnetic:** A ferromagnetic substance exhibits spontaneous magnetic moment. It has a magnetic moment

even in a zero-applied magnetic field. Beyond Curie temperature $T_c$, a ferromagnetic material becomes paramagnetic.

- **Curie temperature ($T_c$):** The Curie temperature $T_c$ is the temperature above which the spontaneous magnetization ceases. Above Curie temperature $T_c$, a ferromagnetic material transforms into paramagnetic and its susceptibility follows the Curie-Wiess law. Curie temperature bifurcates the disordered paramagnetic phase at $T > T_c$ from the ordered ferromagnetic phase at $T < T_c$.
- **Neel temperature ($T_N$):** The Neel temperature $T_N$ is the temperature below which the magnetic spins in antiferromagnet are ordered in an antiparallel distribution resulting in a null net magnetic moment.

## 8.5 ELECTRICAL AND OPTICAL PROPERTIES OF SPRAY-DEPOSITED $CoFe_2O_4$ THIN FILMS

The dependency of the energy bandgap and electrical resistivity of spray-deposited $CoFe_2O_4$ thin film on the concentration of the solution is mentioned in Table 8.3. With the increase in the concentration of the solution, the optical bandgap energy and electrical resistivity decrease. Therefore, we can fabricate the spinel ferrite especially $CoFe_2O_4$ thin film of desired bandgap and electrical resistivity for their application in artificial intelligence and quantum computing.

TABLE 8.3 Electrical and Optical Properties of $CoFe_2O_4$ Thin Films of Various Concentrations [8]

| Sr. No. | Solution Concentration (M) | Optical Bandgap (eV) | Electrical resistivity ($\Omega$ *cm*) At 300 K $(\times 10^3)$ |
|---|---|---|---|
| 1. | 0.05 | 2.45 | 288 |
| 2. | 0.10 | 2.39 | 83.2 |
| 3. | 0.15 | 2.33 | 28.2 |
| 4. | 0.20 | 2.30 | 2.63 |
| 5. | 0.25 | 2.28 | 9.77 |

## 8.6 MAGNETIC SPINEL: SYNTHESIS TECHNIQUE AND SMART APPLICATIONS

Table 8.4 elaborates on the magnetic spinel, synthetic route, scientific finding, and their smart applications. Spinel ferrites for diverse applications are synthesized using synthesis routes like hydrothermal method, dip coating, spray pyrolysis, pulsed laser deposition (PLD), Co-precipitation,

TABLE 8.4 Spinel Ferrite: Synthesis Route, Scientific Findings, and Smart Applications

| Sr. No. | Magnetic Spinel | Synthetic Technique | Scientific Findings | Smart Applications | Ref. |
|---|---|---|---|---|---|
| 1. | Smart spinel ferrite nanoparticles | Hydrothermal | Temperature dependence of magnetic properties | Quantum computing and Qubit | [9] |
| 2. | $CoFe_2O_4$ | Dip coating | Magnetoionic effect | Lithium-ion and sodium-ion batteries | [10] |
| 3. | $CoFe_2O_4$, $NiFe_2O_4$, $Co_{0.5}$ $Ni_{0.5}$ $Fe_2O_4$ | Dip coating | Electrochemical tuning of magnetism | Lithium-ion battery, magnetic actuation, and spintronics applications | [11] |
| 4. | $CoFe_2O_4$ | Hydrothermal | Contrast agent for magnetic resonance imaging (MRI) | MRI | [12] |
| 5. | $CoFe_2O_4$ | Spray pyrolysis | Dependence of concentration of precursor solution on physio-chemical properties of cobalt ferrite | Electrode material for energy storage devices | [8] |
| 6. | $CoFe_2O_4$ | Pulsed laser deposition | An increase in the A-B inter-sub-lattice super-exchange interaction leads to enhancement in magnetization of $CoFe_2O_4$ thin film | Multifunctional magnetoelectric sensor and non-volatile multilevel resistive switching memory | [13] |
| 7. | $CoFe_{2-x}Ce_xO_4$ | Co-precipitation | Ce composition with x=0.25 shows a comparatively enhanced hysteresis loop | Non-volatile resistive random access memory applications | [14] |
| 8. | $Ni_{1-x}Co_xFe_2O_4$ | Spin coating | Gas sensing dependence on cobalt content | Gas sensor | [15] |

*(Continued)*

TABLE 8.4 Spinel Ferrite: Synthesis Route, Scientific Findings, and Smart Applications *(Continued)*

| Sr. No. | Magnetic Spinel | Synthetic Technique | Scientific Findings | Smart Applications | Ref. |
|---|---|---|---|---|---|
| **9.** | $CoFe_2O_4$ | Spin coating | Bipolar multifunctional resistive switching | Multifunctional memory devices | [16] |
| **10.** | $CoFe_2O_4$ | Spin coating | Compliance current dependent properties of memory device | Resistive switching device | [17] |
| **11.** | $CoFe_2O_4$ | Spin coating | Thickness dependence of the resistive switching | RRAM Applications | [18] |
| **12.** | $CoFe_2O_4$ | Spin coating | Temperature dependence of the current in low and high resistance state | Resistive switching memory applications | [19] |
| **13.** | $CuFe_2O_4$ | Spin coating | Multifunctional multistate resistive and magnetic switching | High-density data storage devices | [20] |
| **14.** | $Co_{1-x}Cu_xFe_2O_4$ | Spin coating | Hoping phenomenon between Fe and Cu ions | RRAM device | [21] |
| **15.** | $CoFe_2O_4$ | Spray pyrolysis | Bipolar resistive switching with the coexistence of mem elements | Neuromorphic Applications | [22] |
| **16.** | $NiFe_2O_4$ | Spin coating | Enhancement of the performance of memristor device due to embedded particles | Memristive applications | [23] |
| **17.** | $NiFe_2O_4$ | Hydrothermal | Forming free bipolar resistive switching | Memristive applications | [24] |

*(Continued)*

TABLE 8.4 Spinel Ferrite: Synthesis Route, Scientific Findings, and Smart Applications *(Continued)*

| Sr. No. | Magnetic Spinel | Synthetic Technique | Scientific Findings | Smart Applications | Ref. |
|---|---|---|---|---|---|
| 18. | $Ni_{1-x}Cu_xFe_2O_4$ | Spin coating | $Cu^{2+}$ ions control the conducting filament path | Multifunctional RRAM devices | [25] |
| 19. | $Ni_{0.5}Zn_{0.5}Fe_2O_4$ | Radio frequency magnetron sputtering | Bipolar resistive switching phenomenon | Non-volatile Memory | [26] |
| 20. | $NiFe_{1.95}Cr_{0.05}O_4$ | Spin coating | Unipolar resistive switching | Multistate logic switching | [27] |
| 21. | $Ni_{0.5}Zn_{0.5}Fe_2O_4$ | Pulsed laser deposition | Thickness-dependent resistive switching | Memristive applications | [28] |
| 22. | $NiFe_2O_4$ | Pulsed laser deposition | Coexistence of unipolar and bipolar resistive switching | Resistive memory devices | [29] |
| 23. | $NiFe_2O_4$ | Spin coating | Ag and Cu nanoparticle layers support the quality of resistive switching cycles | Multilayer memory device | [30] |
| 24. | $ZnFe_2O_4$ | Pulsed laser deposition | Improved resistive switching characteristics due to partial pressure of oxygen | Resistive random access memory applications | |
| 25. | $NiFe_2O_4$ | Co-precipitation | Bipolar resistive switching | Resistive switching and electrochemical applications | [31] |
| 26. | $NiFe_2O_4$ | Radio frequency magnetron | Relation of conducting filament formation mechanism and grain microstructure | Resistive switching applications | [32] |
| 27. | $NiFe_2O_4$ | Reactive co-sputtering | Resistive hysteresis and capacitance effect | Non-volatile RRAM devices | [33] |
| 28. | $NiFe_2O_4$ | Spin coating | Unipolar resistive switching | Resistive random-access memory applications | [4] |

*(Continued)*

TABLE 8.4 Spinel Ferrite: Synthesis Route, Scientific Findings, and Smart Applications *(Continued)*

| Sr. No. | Magnetic Spinel | Synthetic Technique | Scientific Findings | Smart Applications | Ref. |
|---|---|---|---|---|---|
| **29.** | $ZnFe_2O_4$ | Spin coating | Bipolar and tri-state unipolar resistive switching | Memristive applications | [34] |
| **30.** | $SnFe_2O_4$ | Spin coating | Thickness effect on the stability of unipolar resistive switching | Resistive random access memory applications | [35] |
| **31.** | $ZnFe_2O_4$ | Radiofrequency magnetron | Bipolar resistive switching | Non-volatile memory applications | [36] |
| **32.** | $NiFe_2O_4$ | Sol-gel-assisted auto combustion | Polaron hopping mechanism | Spintronics applications | [37] |
| **33.** | $NiFe_2O_4$ | Spin coating | Effect of Ag nanoparticles on resistive switching and associated saturation magnetization | Multifunctional electromagnetic devices | [38] |
| **34.** | $NiFe_2O_4$ | Pulsed laser deposition | Bipolar resistive switching characteristics | Non-volatile multilevel resistive switching memory | [39] |
| **35.** | $NiFe_{2-x}Ce_xO_4$ | Spin coating | Ce doping improves the performance of resistive switching device | Resistive Random-Access Memory Device | [40] |
| **36.** | $NiFe_{2-x}Gd_xO_4$ | Spin coating | Doping of Gd into magnetic spinel enhances the resistive switching properties | Resistive random-access memory device | [41] |
| **37.** | $NiFe_2O_4$ | Spin coating | Influence of Au doping on unipolar resistive switching | Resistive random access memory device | [42] |

*(Continued)*

TABLE 8.4 Spinel Ferrite: Synthesis Route, Scientific Findings, and Smart Applications *(Continued)*

| Sr. No. | Magnetic Spinel | Synthetic Technique | Scientific Findings | Smart Applications | Ref. |
|---|---|---|---|---|---|
| **38.** | $NiFe_2O_4$ | Spin coating | Coexistence of unipolar and bipolar resistive switching using Ag nanoparticles | Resistive switching applications | [43] |
| **39.** | $ZnFe_2O_4$ | Spin coating | Coexistence of unipolar and bipolar resistive switching | Non-volatile memory device | [44] |
| **40.** | $ZnFe_2O_4$ | Spin coating | Reversible transitions among four modes of nonpolar resistive switching cycles | Non-volatile memory applications | [45] |

radio frequency magnetron, and Reactive Co-Sputtering technology. We can conclude that most of the spinel ferrite for memristive application is fabricated using a sol-gel-assisted spin coating technique. Memristors find applications in artificial intelligence and quantum computing. Specially synthesized spinel ferrites are used in energy storage devices like lithium-ion batteries and supercapacitors. Spinel ferrites are also employed for their applications in Magnetic actuation, spintronic applications, magnetic resonance imaging (MRI), magnetoelectric sensors, gas sensors, neuromorphic applications, and non-volatile memory applications.

Spinel ferrite and its typical applications are demonstrated in Figure 8.2. Spinel ferrites are employed in diverse applications like quantum technology, memristive applications, spintronics, gas sensors, magnetoelectric sensors, and energy storage devices.

## 8.7 SYNTHETIC TECHNIQUES

Spinel ferrites can be fabricated using various techniques. The two-dimensional spinel ferrite materials are synthesized via PLD, Spin coating, Dip coating, Radio Frequency Magnetron, Reactive Co-sputtering, and Spray pyrolysis.

Figure 8.3 represents various synthesis techniques for synthesizing spinel ferrite materials. In this chapter, let us briefly discuss PLD and spin coating techniques.

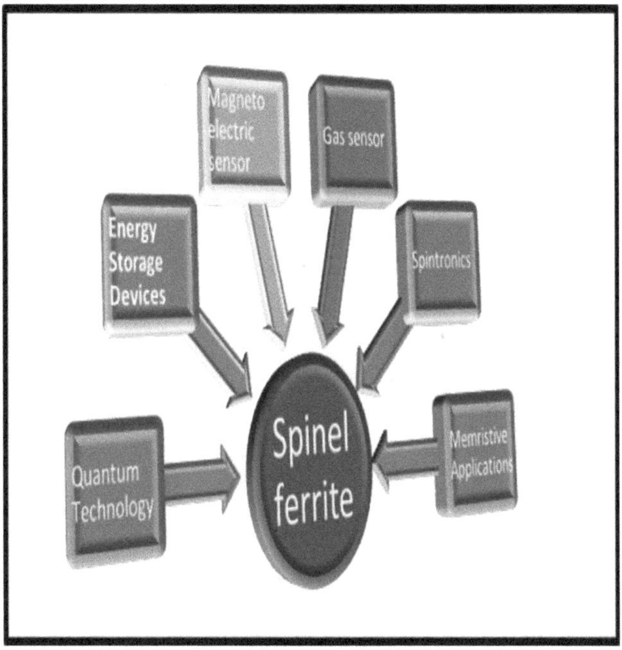


FIGURE 8.2 Spinel ferrite and its applications.

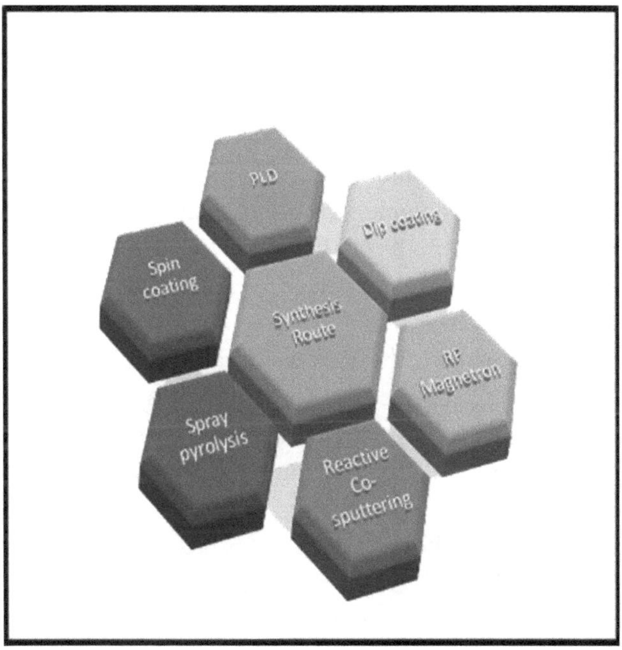


FIGURE 8.3 Synthesis techniques.

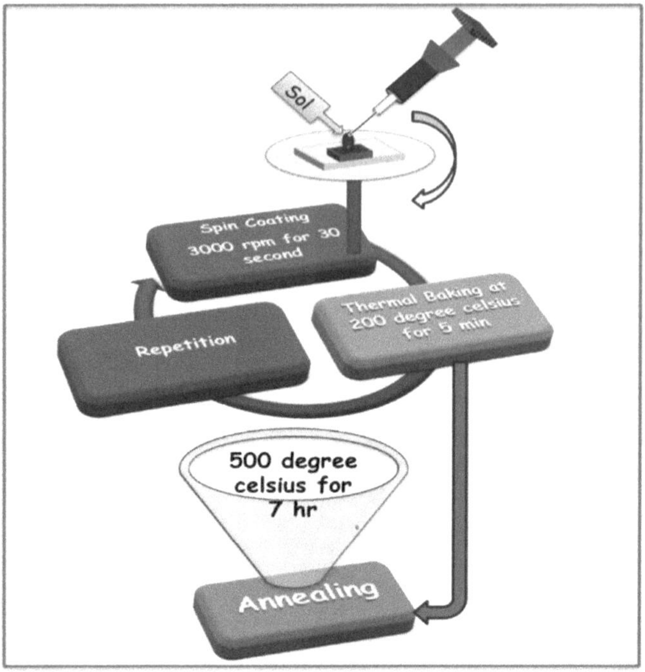


FIGURE 8.4 Spin coating technique.

**Spin coating:** The proper amount of 'Sol' is dropped on the surface of the substrate as shown in Figure 8.4. The substrate is rotated at a particular rotation speed and for a particular time. Then, the substrate is kept for baking at a particular temperature for the required time. This completes one cycle of coating. A suitable number of coats are repeated to achieve the desired thickness of the film. The fabricated thin film of the required composition is then kept in the furnace for annealing which enhances the crystallinity of the material.

**Pulsed laser deposition:** The PLD technique has advantages over other thin film deposition techniques due to the provision of selection for laser wavelength and scope for controlling synthesis parameters like; dimensions and crystallinity [46]. The schematic diagram of the PLD technique is shown in Figure 8.5. The experimental setup for synthesizing Thin Films of $CoFe_2O_4$ using PLD is shown in Figure 8.5. It consists of a Laser, vacuum chamber, and Gas Inlet pipe. The vacuum chamber consists of a target holder, target, substrate holder, and substrate with a heating arrangement. The distance between the target and substrate can be varied for the uniform thin film of $CoFe_2O_4$. The gas inlet pipe is generally employed for

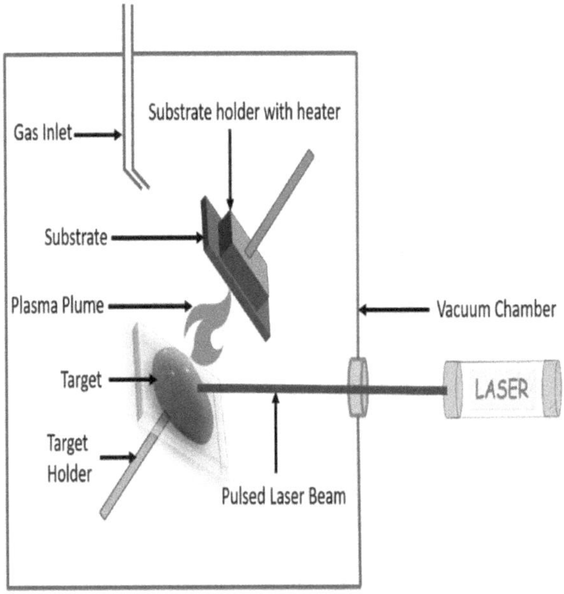


FIGURE 8.5 Schematic diagram of pulsed laser deposition technique [47].

maintaining the pressure inside the vacuum chamber. The laser beam of a suitable wavelength is allowed to strike on the target. The target generally comprises constituent elements for $CoFe_2O_4$. When the target is irradiated with the laser beam, it emits a plum of plasma consisting of atoms and molecules required to form a $CoFe_2O_4$ thin film—the plum of plasma forms the thin film on the substrate. The temperature of the substrate can be adjusted using a heating system.

## 8.8 PULSED LASER DEPOSITED THIN FILMS OF $CoFe_2O_4$

The synthesis technique parameters for $CoFe_2O_4$ thin films using PLD are explored in the following Table 8.5.

Depending upon the application, one can choose the laser wavelength, laser pulse duration, substrate temperature, and background gas pressure in the chamber. Various laser sources like; Nd: YAG, ArF-excimer, Ti-Sa laser, and KrF-excimer can be employed for irridating the target. The striking of the laser beam on the target ignites the plum of plasma which can be settled down at the substrate and leads to the generation of a thin film of $CoFe_2O_4$ on the substrate. Thus formed $CoFe_2O_4$ thin film finds applications in modern science and technology.

TABLE 8.5 Pulsed Laser Deposited $CoFe_2O_4$ Thin Films and Experimental Parameters

| Sr. No. | Target | Laser | Laser Wave Length | Laser Pulse Duration | Substrate | Substrate to Target Distance | Substrate Temp. | Thin Film Thickness | Application | Ref. |
|---|---|---|---|---|---|---|---|---|---|---|
| 1. | Sintered $CoFe_2O_4$ comprises $\alpha$-$Fe_2O_3$ and $Co_3O_4$ | Nd: YAG | 355 nm | 6 nS | Fused quartz | 45 mm | 450 °C | 200 nm | Supercapacitor Electrode | [48] |
| 2. | $Co_{1-x}Ru_xFe_2O_4$ | ArF-excimer | 193 nm | – | $\alpha$-$Al_2O_3$ (001) | – | 500 °C | Between140 nm and 190 nm | Spintronic devices | [49] |
| 3. | $CoFe_2O_4$ | Nd:YAG | 1064 nm | 10 nS | Si(100) | 33 mm | Non-heated | – | Magnetic Recording, Permanent Magnets etc., | [50] |
| 4. | $CoFe_2O_4$ | Excimer | 248 nm | – | MgO(100) and STO(100) | 60 mm | 450 °C | 200 nm | Magnetic Switching | [51] |
| 5. | $CoFe_2O_4$ | Ti: Sa laser | 800 nm | 40 fS | Si(100) | 16 mm | 300 °C | 200 nm | Recording media | [52] |
| 6. | $CoFe_2O_4$ | Nd: YAG | 532 nm | 10 nS | Si(100) | 50 mm | – | – | Sustainable hydrogen production | [53] |
| 7. | $CoFe_2O_4$ | KrF-excimer | 248 nm | – | MgO(001) | 65 mm | 500 °C | 130 nm | spintronics | [54] |
| 8. | $CoFe_2O_4$ | KrF-excimer | 248 nm | – | NSTO(002) | – | 625 °C | 150 nm | Resistive Switching | [55] |

## 8.9 CONCLUSIONS

Magnetic spinel coatings popularly known as spinel ferrite thin/ thick films can be fabricated using deposition techniques namely pulsed laser deposition(PLD), reactive co-sputtering, hydrothermal synthesis, spin coating method, dip coating, spray pyrolysis, and radio frequency magnetron sputtering. The uniform coatings of cobalt ferrite have been obtained using laser beams of Nd: YAG laser, ArF-excimer laser, KrF-excimer laser, and Ti: Sa lasers. Spin-coated magnetic spinels find novel applications in memristive devices for neuromorphic computing, high-performance electrodes in supercapacitors, and Li-ion batteries. Furthermore, spray-deposited magnetic spinels are useful for gas sensing and semiconductor technology. The extraordinary properties of magnetic spinels make them eligible for diverse applications of emerging technology.

## REFERENCES


1. J. Haetge, C. Suchomski, and T. Brezesinski, Inorg Chem **49**, 11619 (2010).
2. S.P. Dalawai, S. Kumar, M.A.S. Aly, M.Z.H. Khan, R. Xing, P.N. Vasambekar, and S. Liu, J Mater Sci: Mater Electron **30**, 7752 (2019).
3. S.B. Narang and K. Pubby, J Magn Magn Mater **519**, 167163 (2021).
4. W. Hu, N. Qin, G. Wu, Y. Lin, S. Li, and D. Bao, J Am Chem Soc **134**, 14658 (2012).
5. K. Gayakvad, K. Somdatta, V. Mathe, T. Dongale, M. W, and K. Patankar, Emergent Mater **7**, 103 (2024).
6. R. Abari, T. Chen, O. Malik, J. Anderson, T.G. Croda, S. Nahavandi, S. Basu, S. Farshchi, M.S. Newman, A. Chatterjee, B.M. Hammerli, W. Reeve, K. Moore, S. Welch, and J. Audino, Introduction to Magnetic Materials (n.d.).
7. H. Jiao, G. Jiao, and J. Wang, Synthesis and Reactivity in Inorganic, Met-Org Nano-Met Chem **43**, 131 (2013).
8. V.A. Jundale and A.A. Yadav, J Mater Sci: Mater Electron **33**, 19612 (2022).
9. H. Bayrakdar, O. Yalçın, S. Özüm, and U. Cengiz, J Alloys Compd **869**, 159334 (2021).
10. S. Robbennolt, E. Menéndez, A. Quintana, A. Gómez, S. Auffret, V. Baltz, E. Pellicer, and J. Sort, Sci Rep **9**, 10804 (2019).
11. L.A. Dubraja, C. Reitz, L. Velasco, R. Witte, R. Kruk, H. Hahn, and T. Brezesinski, ACS Appl Nano Mater **1**, 65 (2018).
12. A. Alazmi, V. Singaravelu, N.M. Batra, J. Smajic, M. Alyami, N.M. Khashab, and P.M.F.J. Costa, RSC Adv **9**, 6299 (2019).
13. Q. Wang, Y. Zhu, X. Liu, M. Zhao, M. Wei, F. Zhang, Y. Zhang, B. Sun, and M. Li, Appl Phys Lett **107**, 063502 (2015).
14. M. Kamran and M. Anis-ur-Rehman, Ceram Int **48**, 16912 (2022).
15. F. Tudorache, P.D. Popa, M. Dobromir, and F. Iacomi, Mater Sci Eng: B **178**, 1334 (2013).

16. S. Munjal and N. Khare, Sci Rep **7**, 12427 (2017).
17. S. Munjal and N. Khare, Nanotechnology **32**, 185204 (2021).
18. M. Mustaqima, P. Yoo, W. Huang, B.W. Lee, and C. Liu, Nanoscale Res Lett **10**, 168 (2015).
19. W. Hu, L. Zou, R. Chen, W. Xie, X. Chen, N. Qin, S. Li, G. Yang, and D. Bao, Appl Phys Lett **104**, 143502 (2014).
20. S. Munjal and N. Khare, Appl Phys Lett **113**, 243501 (2018).
21. Z. Xiahou, D.H. Kim, H. Xu, Y. Li, B.W. Lee, and C. Liu, J Mater Sci: Mater Electron **27**, 2255 (2016).
22. T.D. Dongale, A.A. Bagade, S.V. Mohite, A.D. Rananavare, M.K. Orlowski, R.K. Kamat, and K.Y. Rajpure, J Mater Sci: Mater Electron **29**, 3231 (2018).
23. J. Li, C. Yao, W. Huang, N. Qin, and D. Bao, J Alloys Compd **890**, 161814 (2022).
24. S. Munjal and N. Khare, in *3Rd International Conference on Condensed Matter and Applied Physics (Icc-2019)* (2020), p. 020171.
25. A. Hao, D. Jia, M. Ismail, R. Chen, and D. Bao, J Alloys Compd **790**, 70 (2019).
26. L. Wu, C. Dong, X. Wang, J. Li, and M. Li, J Alloys Compd **779**, 794 (2019).
27. A. Hao, D. Jia, M. Ismail, W. Huang, R. Chen, and D. Bao, Appl Phys Lett **114**, 203502 (2019).
28. L. Zhang, B. Xie, W. Chen, L. Fan, H. Zheng, Q. Wu, P. Zheng, L. Zheng, and Y. Zhang, Ceram Int **49**, 2991 (2023).
29. J. Li, C. Yao, Y. Ke, W. Huang, S.K. Thatikonda, N. Qin, and D. Bao, Appl Phys Lett **120**, 133501 (2022).
30. S. Wang, X. Ning, A. Hao, and R. Chen, J Alloys Compd **908**, 164569 (2022).
31. T.D. Dongale, S.S. Khot, A.A. Patil, S.V. Wagh, P.B. Patil, D.P. Dubal, and T.G. Kim, Mater Des **201**, 109493 (2021).
32. S.K. Tong, J.H. Chang, Y.H. Hao, M.R. Wu, D.H. Wei, and Y.L. Chueh, Appl Surf Sci **564,** 150091 (2021).
33. C. Jin, E.Y. Jiang, and H.L. Bai, Appl Surf Sci **257**, 8998 (2011).
34. W. Hu, X. Chen, G. Wu, Y. Lin, N. Qin, and D. Bao, Appl Phys Lett **101**, 063501 (2012).
35. G. Gong, C. Wu, P. Hu, Y. Li, N. Kwon, and C. Liu, Curr Appl Phys **16**, 980 (2016).
36. B. Sun, X. Zhang, G. Zhou, C. Zhang, P. Li, Y. Xia, and Y. Zhao, J Alloys Compd **694**, 464 (2017).
37. P. Ghosh, R.N. Bhowmik, M.R. Das, and P. Mitra, Physica E Low Dimens Syst Nanostruct **88**, 218 (2017).
38. A. Hao, M. Ismail, S. He, N. Qin, W. Huang, J. Wu, and D. Bao, RSC Adv **7**, 46665 (2017).
39. Q. Wang, Y. Zhu, X. Liu, M. Zhao, M. Wei, F. Zhang, Y. Zhang, M. Li, and M. Li, J Alloys Compd **693**, 945 (2017).
40. A. Hao, S. He, N. Qin, R. Chen, and D. Bao, Ceram Int **43**, S481 (2017).
41. A. Hao, M. Ismail, S. He, N. Qin, R. Chen, A.M. Rana, and D. Bao, Mater Sci Eng B Solid State Mater Adv Technol **229**, 86 (2018).

42. A. Hao, M. Ismail, S. He, N. Qin, W. Huang, J. Wu, and D. Bao, J Alloys Compd **732**, 573 (2018).
43. A. Hao, M. Ismail, S. He, W. Huang, N. Qin, and D. Bao, J Appl Phys **123,** 085108 (2018).
44. M. Ismail, A. Hao, W. Huang, J. Lee, S. Kim, and D. Bao, Appl Phys Lett **113,** 152103 (2018).
45. M. Ismail, A. Hao, S. He, W. Huang, N. Qin, and D. Bao, J Alloys Compd **753**, 100 (2018).
46. M. Sánchez-Arenillas, M. Oujja, F. Moutinho, J. De La Figuera, M. V. Cañamares, A. Quesada, M. Castillejo, and J.F. Marco, *Bulk and Surface Characterisation of Micrometer-Thick Cobalt Ferrite Films Grown by IR PLD* (n.d.).
47. J. Schou, Appl Surf Sci **255**, 5191 (2009).
48. S.M. Nikam, A. Sharma, M. Rahaman, A.M. Teli, S.H. Mujawar, D.R.T. Zahn, P.S. Patil, S.C. Sahoo, G. Salvan, and P.B. Patil, RSC Adv **10**, 19353 (2020).
49. F. Iwamoto, M. Seki, and H. Tabata, J Appl Phys **112,** 103901 (2012).
50. F. Fabris, Y.T. Xing, D.F. Franceschini, D.R. Sanchez, M. Alzamora, and W.C. Nunes, J Appl Phys **122,** 063901 (2017).
51. T. Dhakal, D. Mukherjee, R. Hyde, P. Mukherjee, M.H. Phan, H. Srikanth, and S. Witanachchi, J Appl Phys **107,** 053914 (2010).
52. G. Dascalu, G. Pompilian, B. Chazallon, O.F. Caltun, S. Gurlui, and C. Focsa, in Appl Surf Sci (Elsevier B.V.), 38–42 (2013).
53. G. Bulai, V. Trandafir, S.A. Irimiciuc, L. Ursu, C. Focsa, and S. Gurlui, Ceram Int **45**, 20165 (2019).
54. F. Eskandari, P. Kameli, H. Salamati, and A.S. Esmaeily, J Magn Magn Mater **484**, 188 (2019).
55. Q. Wang, Y. Zhu, X. Liu, M. Zhao, M. Wei, F. Zhang, Y. Zhang, B. Sun, and M. Li, Appl Phys Lett **107,** 063502 (2015).

CHAPTER 9

# Advanced Nanomaterials and Characterization Techniques for Photovoltaic Applications

Aman Kumar

## 9.1 INTRODUCTION

The depletion of fossil fuels and the consequent harm that their burning causes to the environment force people to actively seek out green, renewable energy as a practical substitute for fossil fuels. Because solar energy is infinitely available and ecologically benign, it is a very promising choice [1]. A period of 1 hour's worth of solar radiation reaching Earth contains enough energy to meet every person's needs for a whole year [2–4]. Because of this, solar energy has a lot of potential to replace fossil fuels—that is, if its utilization efficiency is high enough. Photovoltaics (PVs) and photocatalysis are two examples of photon-enabled technologies that may transform solar energy into chemical and electrical energy. The basic ideas behind solar energy conversion suggest that several important factors affect its efficiency [4]. These include sunlight absorption, exciton production and dissociation, and charge carrier transit. Designers deliberately use cutting-edge functional materials to accentuate these attributes. Recent years have seen a considerable increase in interest in plasmonic nanomaterials due to their localized surface plasmon resonance (LSPR) [5–7]. LSPR occurs in metallic nanoparticles (NPs) when scattered electrons collectively oscillate at the metal/dielectric interface. This makes light easier to travel through or absorb. In addition, this action produces hot charge carriers


DOI: 10.1201/9781003499732-9

and strengthens the NP's surface electric field [8–10]. LSPR absorbs light, which greatly raises the photocurrent of PV cells. You can tune plasmonic NPs to exhibit a broad spectrum of visible to near-infrared (NIR) LSPR wavelengths. In PVs and photocatalysis, this property makes it easier for absorbers to take in light. It also lets photocatalysts use a wider range of wavelengths [11].

When it comes to photocatalysis, typical semiconductor-based photocatalysts mostly absorb ultraviolet (UV) light. This restricts the use of visible and NIR light, which together account for approximately 45% and 44% of the solar spectrum [12]. New materials called plasmonic NPs and semiconductors work together to absorb light, move hot charge carriers through LSPR, and interact with each other as dipoles. The goal of this combination is to produce a lot of photocatalysis [13]. In semiconductor materials, creating heterojunctions and/or homojunctions is a popular and efficient technique to physically divide electron-hole pairs and prevent charge carriers from recombining. As a result, PV systems and photocatalysis operate better [4, 14]. Characterization methodologies are required to understand the relationship between materials and performance. This knowledge is critical for developing novel functional materials for photon-enabled applications. Researchers have made significant progress in improving photoconversion efficiency, particularly in the field of solar cells. High-spatial-resolution imaging, in situ characterization, and ultrafast spectroscopies have made these breakthroughs possible by showing how things work at the molecular level. In advanced characterizations, theoretical computing and algorithmic data analysis are often crucial because they provide measurable facts and more profound understandings of physics. Our group is committed to improving solar energy use through logical material design and property optimization using contemporary characterization methods. This report primarily highlights our ongoing work on the synthesis of advanced functional materials, particularly heterojunctions and plasmonic NPs. We also investigate their uses in photon-based technology. Next, we employ various advanced techniques to conduct a comprehensive qualitative and quantitative analysis of advanced and composite nanosystem.

These methods seek to fully comprehend their working processes to provide insightful information for the future creation of materials with enhanced photoconversion capabilities. Finally, we offer a concise overview of potential future paths in the field of advanced functional materials and characterization.

## 9.2 ADVANCED FUNCTIONAL MATERIALS FOR SOLAR ENERGY UTILIZATION

PV and photocatalysis materials often face similar difficulties, such as inadequate optical absorption and unfavourable charge carrier dynamics. As a result, their performance in the process of converting photons is restricted. We have dedicated significant resources to the research and production of sophisticated nanomaterials with enhanced functionality to tackle the challenges. It presents a summary of the three studied materials and their respective contributions to the performance. These materials include design concepts such as incorporating plasmonic NPs as well as constructing heterojunctions and homojunctions.

### 9.2.1 Plasmonic Metallic Nanomaterials

Plasmonic NPs frequently contain metals or materials like metals with many free electrons and a negative real percentage of permittivity. Specifically, plasmonic metallic NPs such as gold (Au) and silver (Ag) have been widely employed to improve the efficiency of PV and photocatalytic processes. The LSPR phenomenon exhibited by plasmonic NPs is primarily influenced by factors such as the size, shape, composition, and dielectric surroundings of the nanostructure. Plasmonic nanomaterials that are well-designed act as efficient light antennas. They may transport the absorbed energy to nearby semiconductors through both radiative and nonradiative processes. These processes are crucial for the functioning of PV systems and photocatalysis [7, 23]. Specifically, gold NPs (Au NPs) with plasmonic properties have garnered significant interest in the fields of solar cells and photocatalysis. Their exceptional stability and potent LSPR effect in the visible-light spectrum are the primary causes [24]. We investigated and made several plasmonic Au-NP-based composites to help plasmon-enhanced solar cells do better at photocatalysis and solar energy harvesting. In this section, we highlight some important works on the subject.

#### *9.2.1.1 Plasmon Enhanced Solar Cells*

The short exciton lifetime and diffusion length in bulk heterojunction solar cells limit the thickness of the active layer, which means that photons can't be absorbed well, and the cells don't work as well as they could. Plasmonic NPs can improve the absorption of photons.

There are three main ways that plasmon-enhanced solar cells work: (i) far-field light scattering, which makes the optical path longer and boosts exciton generation in semiconductors; (ii) near-field surface plasmon resonance (SPR) effect, which boosts the electric field and boosts exciton generation rate; and (iii) surface plasmon polariton (SPP) effect, which makes the semiconductor absorb more light by making it easier for light to travel through the metal/dielectric interface (Figure 9.1a, left part) [25–26]. These boosting methods involve overlapping the spectrum between the SPR absorption of plasmonic NPs and the bandgap excitation of semiconductors. To increase photon, capture in solar cells without modifying the photoactive layer's thickness, we shaped Au NPs into various forms. We can use plasmonic Au NPs, to enhance semiconductor absorption. To better fit the absorption spectra of organic semiconductors, Au NPs can change their shape. This can also change the wavelength of their SPR from visible to NIR. As an illustration, we deliberately engineered (GNRs) that are covered with an extremely thin layer of polystyrene sulfonate (PSS) through electrostatic interactions.

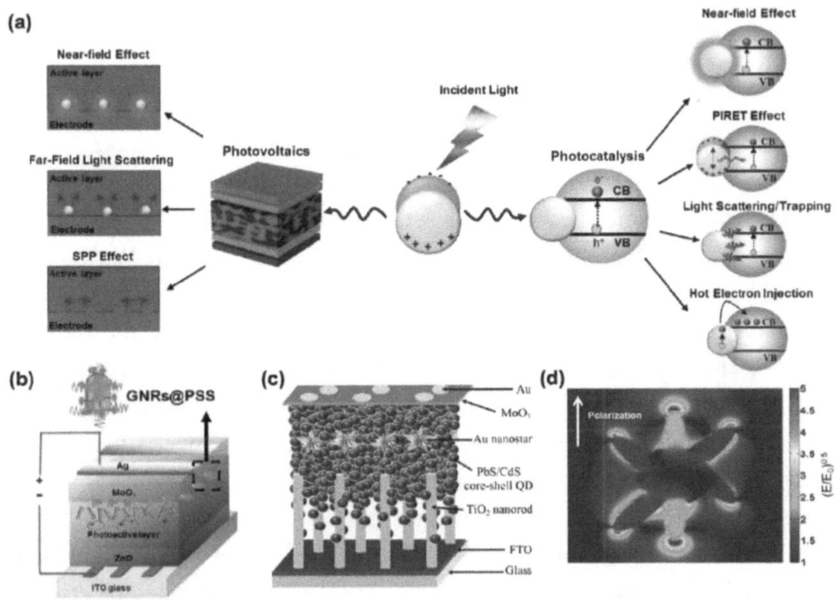


FIGURE 9.1 (a) Mechanisms of plasmon-enhanced solar cells and photocatalysis under solar irradiation. Plasmon-enhanced solar cells in our work: (b) organic solar cells with plasmon-enhanced GNR@PSS. (Reproduced with permission from ref. [7]. Copyright 2022 American Chemical Society.) (c) Quantum dot solar cells enhanced by plasmonic Au nanostars and (d) calculated electromagnetic field distribution surrounding an Au nanostar in (c) under an excitation wavelength of 980 nm by FDTD simulation. (Reproduced with permission from ref. [15]. Copyright 2016 Royal Society of Chemistry.)

We took this action to prevent charges from recombining on the uncoated GNR surface. PSS and (GNRs) are added to the non-fullerene photoactive layer of organic solar cells (OSCs) (Figure 9.1b) to change it.

This change increases power conversion efficiency (PCE) by more than 20% by making it easier for the sun to shine on the material and create electricity through a strong SPR effect and backscattering [7]. To improve the near-field effect in quantum dot (QD)-based solar cells, our lab created five-tipped Au nanostars. We then placed these on a spin-coated PbS/CdS core-shell QD film (Figure 9.1c). To increase the PCE, we precisely controlled the concentration and placement of plasmonic Au nanostars. We investigated the impact of plasmonic Au nanostars on the device's operation. Plasmon resonance's spectrum response is consistent with its higher external quantum efficiency. Simulations using finite-difference time domain (FDTD) (Figure 9.1d) [15] show that the SPR effect of the Au nanostars led to increases in the near field, which in turn led to higher PCE.

#### *9.2.1.2 Plasmon Enhanced Photocatalysis*

It works a lot like plasmon-enhanced solar cells and might make photocatalytic activity higher (Figure 9.1a, right part). It does this by moving energy from plasmonic metals to semiconductors through electromagnetic and scattering effects in the near field [27]. Moreover, Wu's group recently proposed [28, 29] a plasmon-induced resonance energy transfer (PIRET) mechanism. Wu's group has suggested the PIRET mechanism [28, 29]. Through dipole-dipole interactions, this process shows how plasmonic energy can be transferred to an absorber with a slight blue shift without radiation. However, there still needs to be some spectrums overlap for this transfer to occur. To make new photocatalysts, Xu and Zhang et al. mixed plasmonic Au NPs and the up-conversion material $NaYF_4$:$Yb^{3+}$, $Er^{3+}$, and $Tm^{3+}$ (NYF) with porous $TiO_2$ spheres and two-dimensional (2D) graphitic carbon nitride (g-$C_3N_4$) nanosheets (NSs). The objective of this synthesis was to gather UV, visible, and NIR photons for photocatalysis, as shown in Figure 9.2a–d. References [16, 17, 30] provide a full description of the research. It was the SPR-induced hot electron injection mechanism that did most of the work in both cases to break down methyl orange (MO) under visible light. Subsequently, semiconductors and Au NPs absorbed the UV and visible light that the NIR light had induced in the NYF. Therefore, we successfully collected NIR photons for photocatalysis. Furthermore, we discovered that when exposed to UV light, the Au

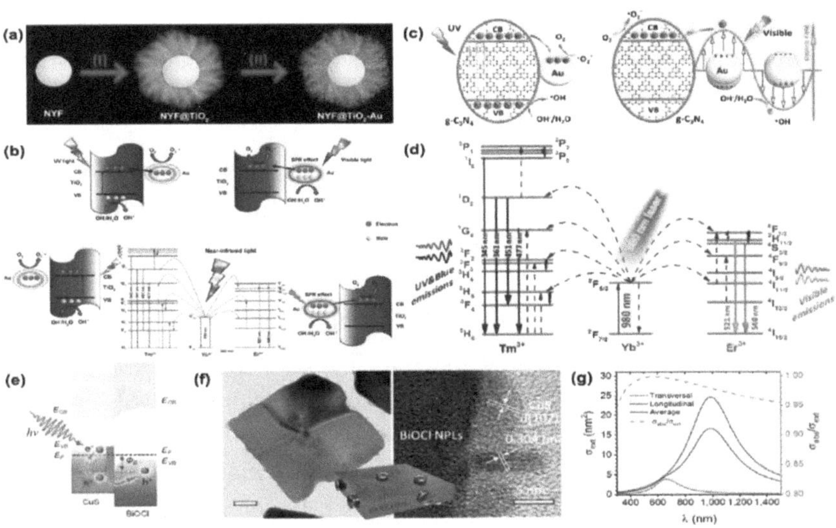


FIGURE 9.2 Plasmon-enhanced photocatalysis in our work: (a) schematic illustration of the formation process of the NYF@$TiO_2$-Au core@shell microspheres and (b) photocatalytic mechanisms of the NYF@$TiO_2$-Au photocatalyst under UV, visible, and NIR irradiation. (Reproduced with permission from ref. [16]. Copyright 2015 Wiley.) (c, d) Photocatalytic mechanisms of Au-NYF/g-$C_3N_4$ photocatalyst under UV, visible ($\lambda > 475$ nm), and NIR irradiation. (Reproduced with permission from ref. [17]. Copyright 2017 American Chemical Society.) (e) Schematic diagram depicting the injection of hot holes (glowing orange sphere) from CuS into BiOCl and (f) TEM images of CuS/BiOCl NPLs. The inset in (f) is the schematic diagram showing the facets of BiOCl and the positional configurations between CuS and BiOCl nanoplatelets) optical cross sections of the CuS nanoplatelet, showing its extinction cross section when exciting its two main plasmonic modes across and along the diameter of the particle, respectively. The theoretical calculations confirmed that the CuS nanoplatelets showed two plasmonic modes, that is, transversal and longitudinal (g), and the former was considered dominating in hot hole generation and contributing more than the latter to photocatalysis. (Reproduced with permission from ref. [9]. Copyright 2022 Elsevier.)

NPs functioned as electron receptors, absorbing energetic electrons from the conduction band (CB) of the semiconductors. This action effectively hindered the recombination of charge carriers, resulting in enhanced photocatalytic efficiency. In this arrangement, effective charge transfer between the semiconductor and plasmonic NPs necessitates close contact. We greatly seek plasmonic NPs with surfaces devoid of any additional substances, as the presence of organic ligands on the surface can hinder the transmission of charges. With this in mind, we have generated Au NPs without any ligands using a "pulsed laser ablation in liquid" technique [31]. These NPs can make close connections with various semiconductors, enhancing the passage of electrons between the semiconductors and Au NPs. In a similar manner, we also created plasmonic Au/Pt alloy NPs without ligands using a modified laser technique. The NPs that were made had good plasmonic properties and the metal alloying had good effects on photocatalysis [32].

A different system, consisting of plasmonic Ag NPs produced in situ and a metal-organic matrix, further validated the significance of the close interaction between components. This system demonstrated the very effective production of hydrogen using visible-light photocatalysis [19]. To improve the efficiency of light absorption [18], we made a porous, hollow $TiO_2$ microsphere that could be used as a base for plasmonic Au NPs. Theoretical research suggests that the unique hollow structure of the $TiO_2$ spheres facilitates the "nanonet effect," enhancing the excitation of the Au NPs on the shadow side of the spheres and significantly increasing the absorption of UV-vis light. The natural catalytic activity of Au NPs and SPR made the reaction between 4-nitrophenol and 4-aminophenol more effective when visible light was present.

### 9.2.2 Plasmonic Semiconductor

Plasmonic semiconductors have a high charge carrier density, which gives rise to a LSPR. Among the materials are metal nitrides, metal oxides, and metal chalcogenides. Methods such as defect doping can adjust its LSPR across a broad range. We observed a prominent LSPR in covellite CuS nanocrystals at NIR wavelengths, on the extraction of heated electrons from plasmonic metallic NPs. However, it is more challenging to extract hot holes from plasmonic semiconductors for photocatalysis because of their very short lifetime—roughly 100 picoseconds.

This is due to the collective coherent oscillations of free holes induced by photons [33]. However, the current paradigm of plasmon-boosted photocatalysis still centres on another semiconductor to create heterojunctions and extract hot holes from CuS; it has been confirmed by other researchers and our lab that this process can lead to a drop in the density of free holes, which can reduce or eliminate the LSPR of CuS when exposed to light [33, 34]. For photocatalysis, we specifically chose p-type BiOCl nanostructures with lower valence band (VB) locations to interface with CuS.

This allowed us to maintain the strong LSPR of CuS nanoplatelets. This prevents the loss of CuS holes during coupling before irradiation (Figure 9.2e) [9] yet allows the injection of hot holes from the LSPR-excited CuS into the connected BiOCl during irradiation. Because they are higher in the VB, the {1 1 0} facets are better at putting hot holes into the CuS that are made by LSPR than the facets in tetragonal BiOCl. The scientists chose tetragonal BiOCl nanoplates (NPLs) and NSs with different proportions of exposed {1 1 0} and {0 0 1} facets. The scientists then combined them with CuS nanoplatelets, utilizing a brand-new in situ ligand exchange/coupling technique (Figure 9.2f). The CuS/BiOCl NPLs broke down MO and sulfamethoxazole faster than the CuS/BiOCl NSs when light in the NIR range was used. This was because the NPLs had more exposed {1 1 0} to {0 0 1} facets. This suggests that hot hole extraction was successful and emphasizes the significance of facet engineering. Theoretical studies reveal that CuS nanoplatelets have two plasmonic modes: one transverse and one longitudinal (Figure 9.2g). When compared to the longitudinal mode, it is believed that the transversal mode contributes more to photocatalysis and is the dominant mode for generating hot holes.

### 9.2.3 Heterojunction and Homojunction

Achieving high photocatalytic performance under sunlight requires the fulfilment of several conditions. These include the ability to absorb a wide range of light, efficiently generate and separate electron-hole pairs, have an abundance of reaction sites, and possess a high redox capability [4, 14]. It is easier for photogenerated electrons and holes to separate spatially in heterojunctions than in single component photocatalysts. This makes it less likely that they will combine again. Furthermore, we can build the heterojunction in a way that expands the range of light absorption from UV to visible and even NIR areas. Carefully selecting two semiconductors with similar bandgaps achieves this. As seen in Figure 9.3a, conventional heterojunctions fall into four types: Z-scheme, type I (straddling bandgap type), type II (staggered bandgap type), and type III (broken bandgap type). Variations in band alignments and charge transfer routes distinguish these types. Z-scheme and type II heterojunction arrangements can effectively achieve charge carrier division in space. On the other hand, since type III heterojunctions have nonoverlapping energy bands made up of two semiconductors, they are unable to separate the pairs of electrons and holes [35]. Certain photocatalytic scenarios use the type I heterojunction, which lacks charge separation, to enhance the

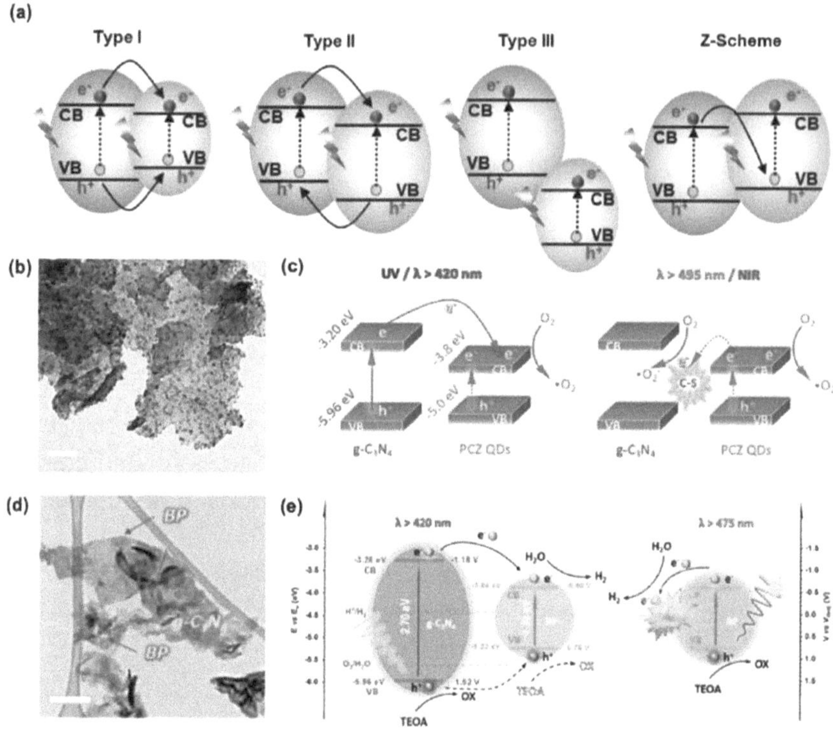


FIGURE 9.3 (a) Schematic mechanisms of different types of conventional heterojunction photocatalysts (type I, type II, type III, and Z-scheme). Heterojunction photocatalysts in our work: (b) TEM image of 0D/2D PCZ QDs/g-$C_3N_4$ and (c) proposed mechanism for the photocatalysis over PCZ QDs/g-$C_3N_4$. (Reproduced with permission from ref. [20]. Copyright 2020 Elsevier.) (d) TEM image of BP/g-$C_3N_4$ and (e) proposed mechanisms of BP/g-$C_3N_4$ photocatalyst for the photocatalytic $H_2$ evolution. (Reproduced with permission from ref. [21]. Copyright 2019 Wiley.)

stability of the primary semiconductor photocatalyst. While the Scheme and type II heterojunctions share a staggered band orientation, they exhibit distinct charge transfer pathways. What's more, charge transfer along the Z-scheme direction makes the heterojunctions stronger than type II heterojunctions by increasing their redox capacities [36]. Although g-$C_3N_4$ has been thoroughly investigated as a photocatalyst that is active under visible light [12], its drawbacks include the rapid recombination of charge carriers and its small light-response range ($\lambda < 460$ nm). Creating heterojunctions between g-$C_3N_4$ and narrow bandgap semiconductors with compatible band structures is another way to get around the problems that come with g-$C_3N_4$. In our study, Zhang et al. created a novel 0D/2D heterojunction that is responsive to UV, visible, and NIR light. The heterojunction consists of 0D PbS@CdS@ZnS core@shell@shell quantum dots (QDs) (PCZ QDs) and 2D g-$C_3N_4$ NSs [20]. We used a basic microwave-assisted method to create the water-dispersible PCZ QDs [37]. The QDs have a well-defined shape. The 2D g-$C_3N_4$ NSs have a lot of surface area that reactants can stick to, and they make it easier to load and spread out PCZ QDs (Figure 9.3b). Due to the wide optical absorption of the uniformly dispersed ultrasmall PCZ QDs, the strong interactions between PCZ QDs and g-$C_3N_4$ made possible by the formation of C-S bonds, and the effective charge transfer at the interface, the heterojunction is an excellent photocatalytic system for degrading MO in the UV to NIR ranges. When put next to other systems of the same type, the sample with 7% PCZ QDs/g-$C_3N_4$ showed amazing NIR photocatalytic activity. The experimental results have shown a potential mechanism of photocatalysis under various light irradiations, as seen in Figure 9.3c. Adding PCZ QDs to g-$C_3N_4$ makes it easier for electrons and holes to separate in space, which lets different charge transport routes work under different irradiation conditions. The combination achieved strong and broadband photocatalytic activity as a result. The use of heavy-metal-containing materials for environmental cleanup still raises valid issues, even if the ultrathin ZnS shell can effectively restrict the release of $Pb^{2+}$ and $Cd^{2+}$ ions into the aqueous solution. So, we made a 2D-on-2D heterostructure without any metals by stacking black phosphorus (BP) on top of g-$C_3N_4$ NSs (Figure 9.3d) [21]. One major barrier to using BP is its lack of stability in conditions with moisture and oxygen.

Additionally, it is difficult to exfoliate bulk BP to create a few-layer BP NS, and ensuring their long-term stability is also a challenge. We were able to successfully use a new technique that uses "organic ice" to help with the exfoliation process. Consequently, we can generate wide, flawless, few-layer BP NSs with about six layers rapidly. This method significantly reduces the possibility of blood pressure worsening during preparation by significantly reducing the time and effort required for liquid exfoliation. Combining BP with g-$C_3N_4$ enhanced its stability.

The BP/g-$C_3N_4$ photocatalyst, which is entirely free of metal, exhibited a significant rate of $H_2$ production (384.17 $\mu$-mol $g_1$ $h_1$) when exposed to light with a wavelength greater than 420 nm.

This rate is comparable to the previously disclosed Pt-containing system. Despite the contact being a type I heterojunction, BP and g-$C_3N_4$

successfully transferred charge. Both experimental and theoretical methods verified the enhanced light absorption resulting from the inclusion of BP. This contributed to the efficient and wide-ranging synthesis of $H_2$ by photocatalysis. We created an energy diagram of the BP/g-$C_3N_4$ photocatalyst using experimental analysis. The figure illustrates the potential process of $H_2$ evolution under various light irradiations (Figure 9.3e). When g-$C_3N_4$ is exposed to light with a wavelength greater than 420 nm, the electrons in the CB become excited. This makes it easier to make $H_2$ through photocatalysis. Since their energy levels are compatible, we can then transport these excited electrons to the CB of BP. Furthermore, exposure to light with a wavelength greater than 475 nm enhances the process of generating hydrogen ($H_2$) through photocatalysis. This enhancement is due to the presence of interfacial trap sites between BP and g-$C_3N_4$, which are formed by the N-P bond. These trap sites make it easier for excited electrons to move from BP's CB to the trapping sites. This makes it much less likely for charge carriers to mix up again, which speeds up the photocatalytic production of $H_2$.

Unlike heterojunctions, homojunctions consist of semiconductors with nearly identical chemical compositions, yet distinct energy levels and crystal forms. These differences make the homojunctions' tight and chemically similar surfaces more conducive to interfacial charge transfer. Recently, our group has made significant strides in synthesizing homojunction photocatalysts for $CO_2$ photocatalytic reduction. We have discovered a new kind of hierarchical metal-organic framework (MOF) photocatalyst with a homojunction structure. This photocatalyst has shown a much better ability to use visible light to promote $CO_2$ reduction than MOFs without a homojunction. Hollow Ag/Au NPs, acting as catalysts in a one-pot process, formed coordination bonds between cobalt and COOH in the organic linker of tetracids (4-carboxyphenyl) porphyrin (TCPP) to create cobalt-MOF (Co-MOF). We form the homojunction by stacking a larger MOF nanoplate, known as MOF(l), on top of a smaller MOF nanoplate, known as MOF(s). We used transmission electron microscopy (TEM) and scanning electron microscopy (SEM) images to confirm the structural integrity of Co-MOF nanostacks, specifically the sample known as Co-MOF-3. The pictures showed concentrically aligned, very consistently oriented nanoplates (Figure 9.4a, b). This alignment resulted in the creation of significant interfacial contact regions in the vertical direction. Moreover, the removal of the tiny nanoplate lowers the height of the MOF(l) in the Co-MOF-3 by around 27 nm, according to atomic force microscopy (AFM) pictures

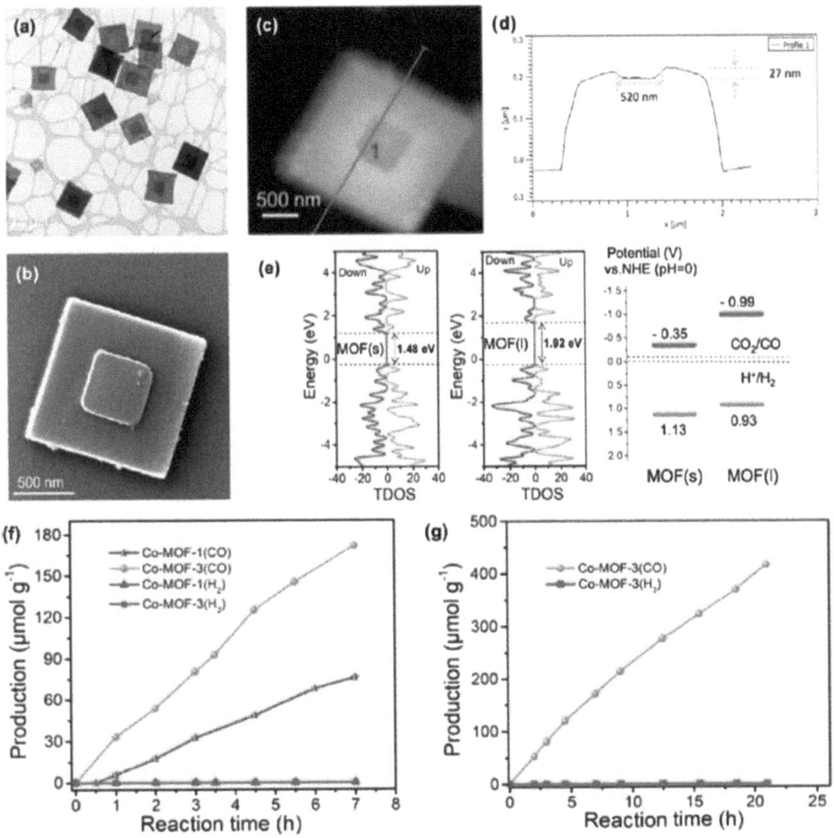


FIGURE 9.4 Homojunction photocatalyst in our work: (a) TEM image and (b) SEM image of the Co-MOF homojunction, (c) AFM image of Co-MOF, showing a pit in the centre and the corresponding height profile (d), (e) TDOS plots of monolayer MOF(s) and MOF(l) in nanostacks and calculated energy levels of monolayer MOF(s) and MOF(l) in nanostacks [potential vs NHE (pH = 0)], (f) evolution of CO and $H_2$ as a function of reaction time during the $CO_2$ reduction in aqueous solution under visible-light illumination for the Co-MOF photocatalysts, and (g) stability of Co-MOF during the continuous $CO_2$ reduction reaction. (Reproduced with permission from ref. [22]. Copyright 2021 Springer Nature.)

(Figure 9.4c, d). This implies that the MOF(s) are submerged within the MOF(l). To study the bandgaps, VB edges, and CB edges of two stacked MOFs in Co-MOF, MOF(l), and MOF(s), that was the main goal of the study. For this study, we used the total density of states (TDOS). The results demonstrated the creation of a type II homojunction, as shown in Figure 9.4e, which permits the efficient separation and transfer of photogenerated charges. When exposed to visible light, the Co-MOF-3 homojunction in water reacted with $CO_2$ far more quickly than it did with CO. Figure 9.4f illustrates this rate, which was around 2.5 times greater than that of the single MOF nanoplate. The experiment demonstrated exceptional stability under continuous photocatalytic $CO_2$ reduction over 22 hours (Figure 9.4g). We should not disregard the noteworthy selectivity of around 100% for CO generation that the Co-MOF homojunction demonstrated. Our calculations using density functional theory indicate that this could be due to the distinct composition and structure of Co-MOF, as well as the lowered energy barrier for CO synthesis.

## 9.3 ADVANCED CHARACTERIZATION IN PHOTON-ENABLED APPLICATIONS

### 9.3.1 Structure Characterization

The development of more complex photoconversion devices and processes requires a precise understanding of the local chemical environments and nanoscale structures of the nanomaterials. Conventional techniques such as X-ray diffraction and classical electron imaging are useful, but they cannot fully explain complicated structures and the physics behind them. This chapter highlights the importance of advanced characterizations in deepening our knowledge of materials and technology. These techniques perform better than standard procedures and provide valuable knowledge that ultimately enhances electronics and materials.

#### *9.3.1.1 Synchrotron Radiation-Based X-Ray Absorption Spectroscopy (XAS)*

X-ray absorption spectroscopy (XAS) directly studies how electrons are excited and move from one state to another by measuring how many X-ray photons a material absorbs as a function of the energy of the X-rays [38, 39]. Sometimes, the XAS spectrum shows three main parts: the pre-edge; the absorption edge, which is also called the X-ray absorption near edge structure (XANES) or near edge X-ray absorption fine structure (NEXAFS); and the post-edge, which is also called the extended X-ray absorption fine structure (EXAFS). We examined a specific type of homojunction nanostructure of Co-MOFs using XAS. Figure 9.5a shows that the nanostructure in Figure 9.4a consists of nanoplates of MOF(s) and MOF(l) with identical composition components. We found that $Co^{2+}$ ions were present in the Co-MOF homojunction by comparing the XANES

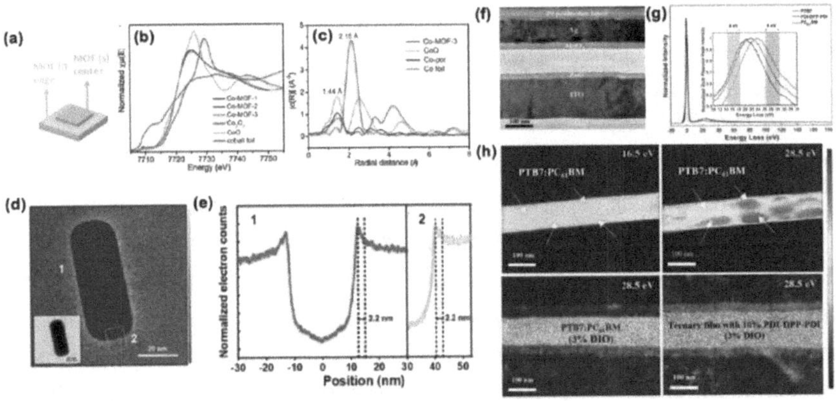


FIGURE 9.5 (a) Schematic illustration of Co-MOF nanostacks, (b) determination of Co oxidation states in Co-MOF by normalized XANES spectra, and (c) characterization of Co coordination number in Co-MOF by Fourier-transformed k2-weighted EXAFS. (Reproduced with permission from ref. [22]. Copyright 2021 Springer Nature.) (d) EF-TEM image of a single GNR@PSS NP (inset grey EF-TEM image of the same NP), and (e) related normalized electron counts profile, indicating that the uniform thickness of the ultrathin PSS shell is ~2.2 (±0.3) nm. (Reproduced with permission from ref. [7]. Copyright 2022 American Chemical Society.) (f) Bright-field TEM image of a cross-sectional photovoltaic device, (g) reference bulk plasmon peak of PTB7, PDI-DPP-PDI, and PC61BM at 21, 23, and 24 eV, respectively, and (h) EF-TEM images of composition and morphology characterization in the cross-sectional binary (PC61BM/PTB7) and ternary devices (PC61BM/PTB7/10% PDI-DPP-PDI) with/without 1,8-diiodooctane (DIO) additive. (Reproduced with permission from ref. [40]. Copyright 2022 Wiley.)

spectra (Figure 9.5b) of the Co pre-edge and the absorption K-edge with those of other materials, like Co foil, CoO, and $Co_3O_4$. This suggests that Co had the same valence in each of the nanoplates that make up the homojunction. The Fourier-transformed k2-weighted EXAFS analysis (Figure 9.5c) unequivocally demonstrates that Co-N/O coordination constitutes the first shell of Co-MOFs. This is supported by the observation that the Co coordination centre is located at 1.44 Å, which is comparable to the distances discovered in $CoN_4$ and CoO. The Co-MOF has a coordination number of 5 for the Co atom. Therefore, XAS is a highly effective method for investigating the oxidation state, coordination number, and chemical bonding. Significantly, the discrimination of light elements using XAS (namely O and N in this case) continues to provide a challenge in this study. Identifying the aforesaid information is valuable as it forms the foundation for creating structural models and comprehending the associated photocatalytic action.

*9.3.1.2 Energy-Filtered Transmission Electron Microscopy (EF-TEM)*

To capture images, energy-filtered transmission electron microscopy (EF-TEM) uses the distinctive characteristics seen in a material's electron energy loss spectroscopy (EELS) spectra. By creating an energy slit, the device can gather electrons at specific energy levels. Because carbon-based materials have low atomic numbers among their component atoms and a low image contrast, they are difficult to identify with traditional TEM imaging. Despite its potential for characterization, the signal from the carbon K-edge (284 eV) alone presents a challenge in distinguishing different carbon components in organic systems. You can see contrast pictures of carbon-based materials with EF-TEM by changing the bulk plasmon resonance of the materials in the low-loss region ($10^{-40}$ eV) of their EELS spectra. Our study's characterization of a very thin PSS shell coated with plasmonic GNRs [7] provides an example. We used EF-TEM to create a carbon contrast image based on PSS's bulk plasmon peak (around 25 eV) and a 10-eV slit width ($25 \pm 5$ eV). We cantered the energy loss at 25 eV. We used the electron intensity profiles to verify the homogeneity of a thin PSS layer on the surface of a single GNR (Figure 9.5e). The results clearly showed that the PSS coating was deposited evenly around the GNRs, with a thickness of about $2.2 \pm 0.3$ nm. It is challenging to get structural information regarding the complex structure of GNRs@PSS using conventional methods; EF-TEM technology offers a more direct means of obtaining this data. By measuring the PSS layer's thickness and homogeneity, we were able to rule out charge recombination on the GNRs. This provided more insight into the intimate connection between the active layer and GNRs@PSS in OSCs. This has a significant impact on solar cell design. We might use EF-TEM to investigate the distribution of different organic domains in the active layer of OSCs. This technique could help optimize the nanoscale form of the materials. We used EF-TEM to look at how the guest non-fullerene acceptor PDI-DPP-PDI changes the shape of the photoactive layer (PTB7/PC61BM) in OSCs, focusing on the complex ternary blend film.

The number is 40. We looked at the cross-sectional OSCs with a bright-field TEM and saw that the active layer did not have a clear phase shape (Figure 9.5f). On the other hand, EF-TEM produced clear and distinct images of the internal structure of binary and ternary OSCs. Figure 9.5g shows the EELS spectra of the host donor PTB7 and the host acceptor PC61BM. To do this, we looked at the big difference in electron strengths at lower and higher energy levels of the reference bulk plasmon peaks. We were able to see the structure (Figure 9.5h) thanks to EF-TEM images, which clearly showed that the 10% PDI-DPP-PDI acceptor can stop fullerene molecules from clustering and make a more uniform structure with more molecular mixing. This helps to prevent the loss of charges in the photoactive layer.

### 9.3.2 Optical Properties Characterization

An ideal photoconversion material should possess a high level of photon absorption across a wide range of wavelengths and a low exciton binding energy. The wavelength-dependent optical properties alter at the plasmonic metal-dielectric interface, producing a noticeably amplified localized electric field. The absorption of photons may increase due to this larger electric field, which may increase photon absorption. The diffraction limit establishes the wavelength of the light source, which limits the resolution of ellipsometry and other optical techniques [41]. We can bypass the diffraction limit by combining imaging techniques with spectroscopic techniques, such as scattering-type scanning and near-field optical microscopy [42].

#### *9.3.2.1 Photoinduced Force Microscopy (PiFM)*

Photoinduced force microscopy (PiFM) is a wavelength-tenable laser and AFM-combined imaging-spectroscopic technique. If you carefully move and measure the dipole-dipole forces between the AFM tip and the material, you can see the topography, phase, chemical properties, and nanophotonics. This is achieved by adjusting the laser wavelength, resulting in highly accurate measurements with exceptional spatial resolution and surface sensitivity [43]. Specifically, PiFM offers a straightforward and effective way to observe the near-field impact of plasmonic nanomaterials with a strong signal-to-noise ratio (SNR). Visualizing certain phenomena in TEM can be quite challenging, especially when studying plasmonic nanomaterials on a relatively thick organic layer.

This is because the organic layer can mask the detected signal. In our lab, Yu and Du et al. used PiFM measurements to look at how GNRs@PSS changed the photoactive layer in OSCs through plasmons. We conducted analysis on the photoactive layer (Figure 9.6a). It was around 100 nm thick

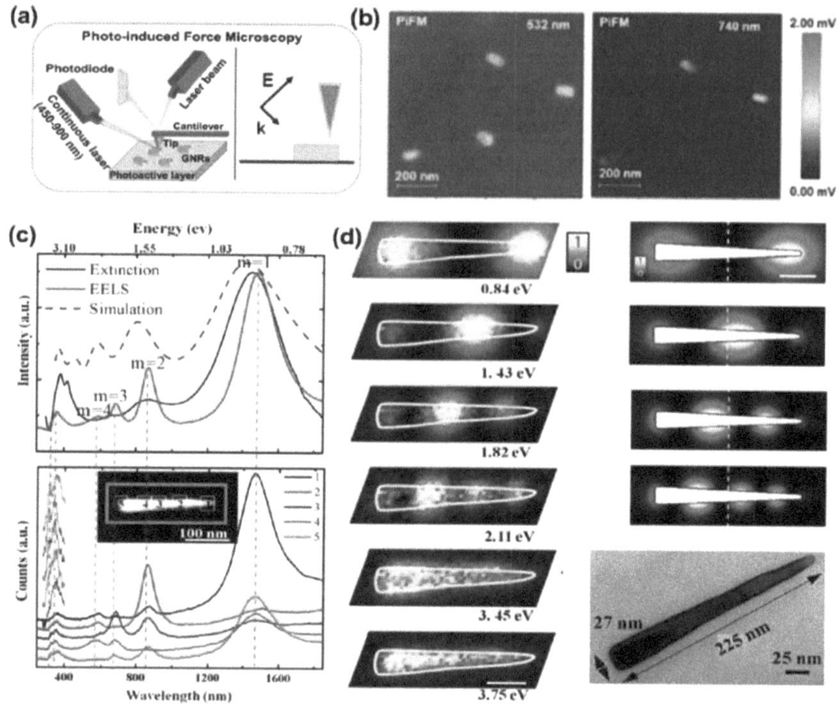


FIGURE 9.6 (a) Schematic illustration of the PiFM setup and (b) PiFM images of the localized electric field distribution of a single GNR@PSS on the PBDB-T/ITIC-2F photoactive layer excited by 532 and 740 nm lasers. (Reproduced with permission from ref. [7]. Copyright 2022 American Chemical Society.) (c) Comparison of the EELS spectrum with the optical extinction spectrum of a nano carrot ensemble and EELS signals acquired at five different locations around the nano carrot by STEM/EELS and (d) experimental and simulated maps of multiple plasmon resonances excited in the nano carrot extracted at energies belonging to the peaks in the EELS signal. (Reproduced with permission from ref. [45]. Copyright 2013 American Chemical Society.)

and made of PBDB-T/ITIC-2F. In Figure 9.6b, there are seven PiFM images that demonstrate the close-range concentration of GNRs@PSS when combined with the photoactive layer. We used two distinct wavelengths (520 and 740 nm) to excite the GNR@PSS, leading to the transverse (515 nm) and longitudinal (724 nm) electron oscillation modes. Understanding the SPR effect of plasmonic GNRs@PSS on the photoactive layer helps to clarify the reasons behind the improved performance. This is due, at least in part, to the increased absorption of sunlight and the generation of charges caused by the plasmonic effect of GNRs@PSS.

#### *9.3.2.2 Scanning Transmission Electron Microscopy or Electron Energy Loss Spectroscopy (STEM/EELS)*

Plasmonic NPs can have their resonance frequency altered by changing their size, shape, and chemical composition. UV-visible-NIR absorption spectroscopy easily reveals this alteration, but it only provides information on a larger scale. For investigating plasmonic behaviour on a nanoscale, scanning transmission electron microscopy or electron energy loss spectroscopy (STEM/EELS) proves to be highly valuable. This technique allows for the excitation and imaging of plasmonic oscillations using electrons. This technique can effectively analyse and display the specific resonance characteristics of plasmonic nanomaterials across a wide range of energies. It offers high resolution in both energy and space, allowing for detailed examination of these materials, even when their properties differ from those of bulk samples [44]. Understanding the SPR spectral broadening behaviour is facilitated by such observations. When considering PiFM, STEM/EELS offers a distinct advantage with its unique frequency profile and superior spatial resolution for SPR imaging. Figure 9.6c and d showcases an illustration of SPR imaging using STEM/EELS for a 200-nm Ag nano carrot NP [45]. The EELS spectra obtained from five distinct locations on a single Ag nano carrot displayed multiple resonances at various energy levels. These resonances were observed at 0.84, 1.43, 1.82, 2.11, 3.45, and 3.75 eV, respectively. Interestingly, the experimental extinction spectrum of the bulk Ag nano carrot sample and these resonances exhibit similar characteristics. Additionally, the EELS mapping demonstrates an uneven dispersion of the localized surface plasmon along the Ag nano carrot, as confirmed by the boundary element approach calculation. Therefore, STEM/EELS demonstrates that the tunability of SPR is influenced by the specific structural characteristics of individual NPs. The nano carrot is attractive for use in biological and sensing applications because, in these circumstances, it demonstrated a tenable longitudinal SPR in the NIR band.

### 9.3.3 Charge Carrier Transfer Dynamics

Functional photoactive materials produce free charge carriers through the absorption of photons, which are then involved in subsequent dynamic processes. It is crucial to comprehend the basic characteristics of photogenerated charge carrier dynamics. We studied how charges move at the interface of 2D-BP and g-$C_3N_{4.21}$ by using photoluminescence spectroscopy and transient photocurrent response experiments. We specifically examine in situ approaches in this section to uncover the dynamics of charge carriers.

#### *9.3.3.1 In Situ Irradiated X-Ray Photoelectron Spectroscopy (XPS)*

Using the photoelectric effect, X-ray photoelectron spectroscopy (XPS) is a fundamental technique that ascertains the elemental bonding and composition of a material. Moreover, by combining XPS with various light irradiation circumstances (e.g., on/off, and varied wavelengths), one can investigate peak shifts arising from differences in the ground and excited states' electron densities of various atoms. This method can determine the direction of the charge transfer carriers in a complex photocatalytic system [46]. Recently, we used the in situ irradiation XPS technique to investigate charge separation and transfer in a composite photocatalyst. In one of our investigations, we examined the charge transfer of a $ZnIn_2S_4/TiO_2$ heterojunction. We used XPS in this study both with and without in situ laser irradiation, as Figure 9.7a [10] illustrates.

A 380-nm light source subjected the $ZnIn_2S_4/TiO_2$ heterojunction, causing the Ti $2p^{3/2}$ (458.6 eV) and Ti $2p^{1/2}$ (464.2 eV) peaks to shift concurrently by −0.2 eV, while the In $3d^{5/2}$ (445.0 eV) and In $3d^{3/2}$ (452.6 eV) peaks shifted simultaneously by +0.2 eV. The former observation indicated a higher concentration of electrons in $TiO_2$, while the latter observation showed a lower concentration of electrons in $ZnIn_2S_4$. It is indicated that the energized electrons in the CB of $ZnIn_2S_4$ were transported to the CB of $TiO_2$. The findings conclusively demonstrate the movement of charge carriers from $ZnIn_2S_4$ to $TiO_2$ at their interface and establish a type II heterojunction

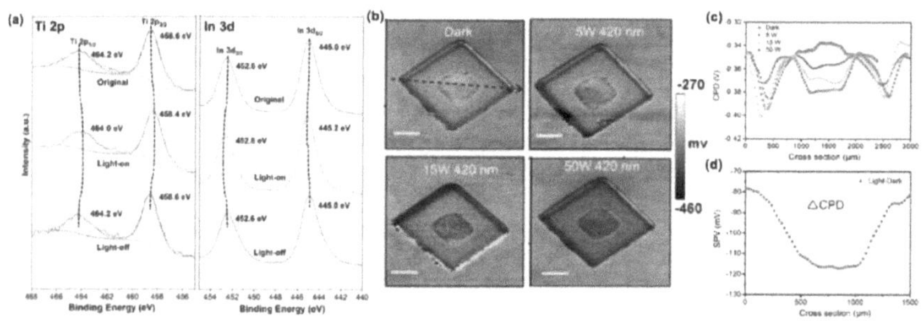


FIGURE 9.7 (a) XPS spectra of Ti 2p and In 3d in $ZnIn_2S_4/TiO_2$ nano tree powders (NtPs) acquired in the dark and under light. (Reproduced with permission from ref. [10]. Copyright 2022 American Chemical Society.) (b) SPVM analysis of the Co-MOF homojunction, (c) the corresponding normalized CPD profiles across the centre of Co-MOF [see the arrow line in (b)] under the dark and irradiation conditions, and (d) the SPV profile. (Reproduced with permission from ref. [22]. Copyright 2021 Springer Nature.)

between the two components. Nevertheless, it is crucial to consider other significant elements that can cause errors, including the stability of the specimen during irradiation, while conducting in situ irradiated XPS. This is necessary to eliminate the possibility of additional photoinduced reactions, such as photoreduction and photooxidation. These processes, while not directly related to the charge transfer process, have the potential to influence the no reversible binding energy of electrons.

#### *9.3.3.2 SPVM*

With a mix of lighting systems (for example, different wavelengths and intensities) and Kelvin probe force microscopy (KPFM), the SPVM method checks how light changes the surface potential [47].

The scanning probe voltage microscopy (SPVM) technique offers a precise resolution at the nanoscale and a high sensitivity for detecting voltage changes at the sub-millivolt level. This enables the investigation of local charge separation and transfer at the nanocomposites' surface or interface, providing crucial new insights into how they function. We investigated the distance between charge carriers made by light at the interface of the homojunction in the Co-MOF nanostack photocatalyst we talked about earlier (Figure 9.5a). Scanning probe voltage microscopy (SPVM) (Figure 9.7b–d) [22] achieved this for the first time. When the light is off, MOF(l) has a lower contact potential difference (CPD) value than MOF(s) (Figures 9.7b and 9.5a). This suggests that the work function of MOF(l) is larger than that of MOF(s). On the other hand, MOF(l) CPD value was greater than MOF(s) when the light was turned on, indicating that MOF(l) had a lower work function than MOF(s). At the same time, as the light intensity increased, the CPD value decreased. The outcome was negative SPV values (Figure 9.7c, d). The CPD readings during the dark might be subtracted to get these ($\Delta$CPD = CPD light on − CPD light off). Furthermore, the MOF(s) had a much larger negative value. The SPVM revealed that the two-MOF nanostack has an internal electric field that pushes photogenerated electrons from MOF(l) to MOF(s) across the homojunction interface. This matches the results of our calculations.

### 9.3.4 Towards Algorithm and Data-Driven Analytical Characterization Techniques

Understanding pertinent chemical and physical characteristics requires information beyond the raw data. Machine learning algorithms have proven effective in analysing the gathered data sets. We can employ convolutional neural networks for picture denoising and principal component analysis for identifying dependent features [48]. In our Co-MOF homojunction study (Figure 9.5a), it is hard to tell the difference between the small chemical differences between MOF (edge) and MOF (centre) because their compositions are so similar. We used a method called non-negative matrix factorization (NMF) to separate the signals from two stacked nanoplates that are cantered in Co-MOF. We were able to do this with the low-loss EELS data set, as Figure 9.8a shows. Through experimentation, we were able to ascertain the difference in the bulk plasmon

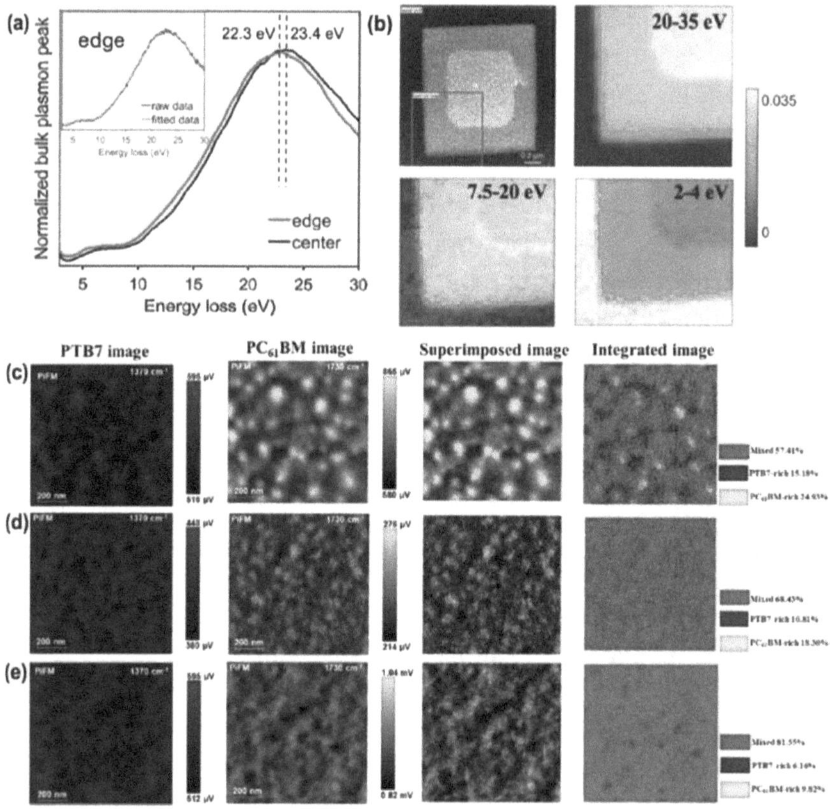


FIGURE 9.8 (a) Bulk plasmon peak of MOF (centre) and MOF (edge) at 23.4 and 22.3 eV, respectively; (b) component analysis by NMF in the low-loss EELS data set of Co-MOFs. (Reproduced with permission from ref. [22]. Copyright 2021 Springer Nature.) Quantitative analysis of domain size distribution and composition on (c) the binary film of PC61BM/PTB7, (d) ternary films of PC61BM/PTB7/10% PDI-DPP-PDI, and (e) PC61BM/PTB7/10% PDI-DPP-PDI (with 3% DIO), respectively. (Reproduced with permission from ref. [40]. Copyright 2022 Wiley.)

peak between two stacked nanoplates. The MOF (edge) measured 22.3 eV, whereas the MOF (centre) measured 23.4 eV. We found important patterns in the relative strength of two parts across a few energy ranges (20–35, 7.5–20, and 2–4 eV) by putting all this data together using non-NMF.

These findings corroborated the presence of a differential in valence electron density between the centre and edge regions, as shown in Figure 9.8b. Our lab also examines the distribution and composition of the PTB7 and PC61BM components in binary and ternary blend films using AFM and infrared (IR) spectroscopy in PiFM, with a focus on the distribution of domain sizes. Their unique excitation wavenumbers of 1370 $cm^{-1}$ and 1730 $cm^{-1}$ were first seen in PiFM pictures (Figure 9.8c). These wavenumbers can be used to subjectively deduce the domain size and material aggregation information. We determined the diameters of the acceptor-rich and donor-rich zones in a bulk heterojunction film by analysing 40 PiFM pictures using a segmentation analysis tool. We then measured the compatibility of these domains using OpenCV-Python image analysis and a "size calculator" tool (Figure 9.8d). The results indicate a decrease in the size of regions with a high concentration of donors and regions with a high concentration of acceptors, and an increase in the proportion of regions where donors and acceptors mix. This study proved that adding the guest non-fullerene acceptor PDI-DPP-PDI makes it easier for donors and acceptors to mix, which leads to better mixing.

## 9.4 CONCLUSIONS AND PERSPECTIVE

This account provides a concise overview of the latest research conducted by our group on improved functional materials used in PVs and photocatalysis. These materials show great potential for harnessing solar energy. The discussion has focused on three techniques for material design: (i) the utilization of plasmonic nanomaterials, (ii) the creation of 2D semiconductor-based heterojunctions, and (iii) the construction of MOF-based homojunctions. The focus of this discussion has been on how these strategies contribute to improving the performance of photon conversion. Noteworthy among our works is the thorough examination of intricate structures, made possible by using advanced characterization tools, and the analysis of the physical properties of materials and their connection to photon conversion performance. This research serves as the experimental foundation for optimizing materials and enhancing performance. We illustrate, for instance, how spectroscopic and imaging methods based on electrons and X-rays may help address a range of enquiries about elemental compositions, material structures, and physical properties. Additionally, we study the dynamics of charge carriers produced by light in solar cells and photocatalysis using in situ techniques. Finally, we provide a summary of the data-driven analytical characterization achieved by using mathematical algorithms on the raw data collected throughout our investigation. With this technique, we can extract more comprehensive insights from the raw data than we could with only basic data processing. Research on PV and photocatalytic processes has significantly advanced thanks to data processing, experimental exploration, and rational material design. Even though several techniques have been used to increase photon conversion efficiency, there are still significant obstacles and underlying problems that need more investigation. We propose a variety of strategies to enable the next generation of photon-enabled applications through the development of new functional materials. Initially, it is necessary to develop innovative designs for highly sophisticated functional materials to broaden the range of wavelengths they can absorb, enhance the separation of charges, increase their mobility, and reduce the likelihood of charge carriers recombining. Despite their capacity to gather a diverse range of solar photons, narrow-bandgap semiconductors typically limit their light-response range to wavelengths under 600 nm. Furthermore, the narrowing of the bandgap might decrease the redox levels, which is detrimental to certain photocatalytic activities. Plasmonic nanomaterials have the potential for efficient solar photon harvesting across a wide range of wavelengths without the requirement for decreasing the bandgap. This is because they can tune the wavelengths of LSPR and generate higher energy charge carriers. Nevertheless, the plasmonic metallic elements that are frequently utilized, such as Au, Ag, and Cu, are afflicted by issues of exorbitant expense, toxicity, and/or instability. Non-noble metal nitrides, which are common in the earth and do not contain noble metals, are particularly appealing as alternative options for nanophotonic and photocatalytic applications in the future. Nevertheless, at present, these materials are primarily produced using high-temperature or pulsed laser synthesis techniques. However, these approaches not only consume a significant amount of energy but also result in the formation of clusters with a low yield. Therefore, it is necessary to establish a fabrication method that is both cost-effective and efficient. Furthermore, to be useful in real-world scenarios, it is necessary to carry out synthesis on a large scale. However, the lack of consistency and uniformity in the size, shape, and quality of NPs across different batches hinders the expansion of production for most plasmonic nanomaterials. This is primarily due to the intricate nature of the synthesis process and the non-uniform distribution of temperature and concentration in large-scale systems. Strong partnerships between academics and industry are greatly sought after to address this challenging objective. Furthermore, the process of colloidal plasmonic NPs spontaneously organizing themselves into a larger, more structured form presents the possibility of exploring new avenues for controlling their properties. This involves manipulating the organization of the NPs in complex structures to modulate their plasmonic properties. We still need to thoroughly investigate much of this world. One of the final obstacles to using plasmonic materials for photon-enabled applications is the limited duration and distance that hot charge carriers can travel, which hinders their effective extraction and utilization. Constructing the metal-semiconductor interface in a way that maximizes the extraction of hot carriers is still a significant difficulty. Furthermore, numerous existing photoactive materials have inherent flaws, such as instability, that impose restrictions on the practical utilization of PVs and photocatalysis. Therefore, in order to ensure their suitability for practical use in PV and photocatalytic applications, it is crucial to conduct a comprehensive investigation of their stability under different conditions. This involves closely monitoring any changes in their structure and any decline in their performance. Accurate characterizations can unveil the precise structural details of nanomaterials and nano systems, offering a more profound understanding of the links between structures and properties. By combining the physical model and algorithm-driven data analysis, innovative characterization approaches that provide space- and time-resolved information can offer unparalleled observations and insights into the fundamental physics of functional materials. This account is anticipated to offer novel and helpful insights and guidance for research areas that utilize photons.

## REFERENCES

1. Kabir, E.; Kumar, P.; Kumar, S.; Adelodun, A. A.; Kim, K.-H. Solar Energy: Potential and Future Prospects. Renewable Sustainable Energy Rev. 2018, 82, 894–900.
2. Kannan, N.; Vakeesan, D. Solar Energy for Future World: A Review. Renewable Sustainable Energy Rev. 2016, 62, 1092–1105.
3. Zhang, Q.; Yang, F.; Xu, Z.; Chaker, M.; Ma, D. Are Lanthanide-Doped Upconversion Materials Good Candidates for Photocatalysis? Nanoscale Horiz. 2019, 4 (3), 579–591.

4. Liu, X.; Zhang, Q.; Ma, D. Advances in 2D/2D Z-Scheme Heterojunctions for Photocatalytic Applications. Solar RRL 2021, 5 (2), 2000397.
5. Wang, Y.; Zhang, Q.; Wang, Y.; Besteiro, L. V.; Liu, Y.; Tan, H.; Wang, Z. M.; Govorov, A. O.; Zhang, J. Z.; Cooper, J. K. Ultrastable Plasmonic Cu-Based Core-Shell Nanoparticles. Chem. Mater. 2021, 33 (2), 695–705.
6. Xu, Z.; Liu, Y.; Ren, F.; Yang, F.; Ma, D. Development of Functional Nanostructures and Their Applications in Catalysis and Solar Cells. Coord. Chem. Rev. 2016, 320, 153–180.
7. Du, Z.; Yu, T.; He, W.; Yurtsever, A.; Izquierdo, R.; Jafari, M.; Siaj, M.; Ma, D. Enhancing Efficiency of Nonfullerene Organic Solar Cells via Using Polyelectrolyte-Coated Plasmonic Gold Nanorods as Rear Interfacial Modifiers. ACS Appl. Mater. Interfaces 2022, 14 (14), 16185–16196.
8. Zada, A.; Muhammad, P.; Ahmad, W.; Hussain, Z.; Ali, S.; Khan, M.; Khan, Q.; Maqbool, M. Surface Plasmonic Assisted Photocatalysis and Optoelectronic Devices with Noble Metal Nanocrystals: Design, Synthesis, and Applications. Adv. Funct. Mater. 2020, 30 (7), 1906744.
9. Zhang, Q.; Mirzaei, A.; Wang, Y.; Song, G.; Wang, C.; Besteiro, L. V.; Govorov, A. O.; Chaker, M.; Ma, D. Extracting Hot Holes from Plasmonic Semiconductors for Photocatalysis. Appl. Catal., B 2022, 317, 121792.
10. Song, G.; Zhang, Q.; Liu, F.; Wang, C.; Yang, R.; Chen, Z.; Ma, D. Mixed-Phase $ZnIn_2S_4$ Nanosheets Grown on $TiO_2$ Nanotrees for the Visible-Light Photocatalytic Degradation of Organic Dyes. ACS Appl. Nano Mater. 2022, 5 (1), 380–390.
11. Thrithamarassery Gangadharan, D.; Xu, Z.; Liu, Y.; Izquierdo, R.; Ma, D. Recent Advancements in Plasmon-Enhanced Promising Third-Generation Solar Cells. Nanophotonics 2017, 6 (1), 153–175.
12. Zhang, Q.; Liu, X.; Chaker, M.; Ma, D. Advancing Graphitic Carbon Nitride-Based Photocatalysts toward Broadband Solar Energy Harvesting. ACS Mater. Lett. 2021, 3 (6), 663–697.
13. Wu, N. Plasmonic Metal-Semiconductor Photocatalysts and Photo-electrochemical Cells: A Review. Nanoscale 2018, 10 (6), 2679–2696.
14. Low, J.; Yu, J.; Jaroniec, M.; Wageh, S.; Al-Ghamdi, A. A. Heterojunction Photocatalysts. Adv. Mater. 2017, 29 (20), 1601694.
15. Gonfa, B. A.; Kim, M. R.; Zheng, P.; Cushing, S.; Qiao, Q.; Wu, N.; El Khakani, M. A.; Ma, D. Investigation of the Plasmonic Effect in Air-Processed PbS/CdS Core-Shell Quantum Dot Based Solar Cells. J. Mater. Chem. A 2016, 4 (34), 13071–13080.
16. Xu, Z.; Quintanilla, M.; Vetrone, F.; Govorov, A. O.; Chaker, M.; Ma, D. Harvesting Lost Photons: Plasmon and Upconversion Enhanced Broadband Photocatalytic Activity in Core@Shell Microspheres Based on Lanthanide-Doped $NaYF_4$, $TiO_2$, and Au. Adv. Funct. Mater. 2015, 25 (20), 2950–2960.
17. Zhang, Q.; Deng, J.; Xu, Z.; Chaker, M.; Ma, D. High-Efficiency Broadband $C_3N_4$ Photocatalysts: Synergistic Effects from Upconversion and Plasmons. ACS Catal. 2017, 7 (9), 6225–6234.

18. Zhang, Q.; Jin, X.; Xu, Z.; Zhang, J.; Rendón, U. F.; Razzari, L.; Chaker, M.; Ma, D. Plasmonic Au-Loaded Hierarchical Hollow Porous $TiO_2$ Spheres: Synergistic Catalysts for Nitroaromatic Reduction. J. Phys. Chem. Lett. 2018, 9 (18), 5317–5326.
19. Liu, Y.; Liu, C.-H.; Debnath, T.; Wang, Y.; Pohl, D.; Besteiro, L. V.; Meira, D. M.; Huang, S.; Yang, F.; Rellinghaus, B.; Chaker, M.; Perepichka, D. F.; Ma, D. Silver Nanoparticle Enhanced Metalorganic Matrix with Interface-Engineering for Efficient Photocatalytic Hydrogen Evolution. Nat. Commun. 2023, 14, 541.
20. Zhang, Q.; Yang, F.; Zhou, S.; Bao, N.; Xu, Z.; Chaker, M.; Ma, D. Broadband Photocatalysts Enabled by 0D/2D Heterojunctions of Near-Infrared Quantum Dots/Graphitic Carbon Nitride Nanosheets. Appl. Catal., B 2020, 270, 118879.
21. Zhang, Q.; Huang, S.; Deng, J.; Gangadharan, D. T.; Yang, F.; Xu, Z.; Giorgi, G.; Palummo, M.; Chaker, M.; Ma, D. Ice-Assisted Synthesis of Black Phosphorus Nanosheets as a Metal-Free Photocatalyst: 2D/2D Heterostructure for Broadband $H_2$ Evolution. Adv. Funct. Mater. 2019, 29 (28), 1902486.
22. Liu, Y.; Chen, C.; Valdez, J.; Motta Meira, D.; He, W.; Wang, Y.; Harnagea, C.; Lu, Q.; Guner, T.; Wang, H. Phase-Enabled Metal-Organic Framework Homojunction for Highly Selective $CO_2$ Photoreduction. Nat. Commun. 2021, 12, 1231.
23. Zhang, Q.; Gangadharan, T. D.; Liu, Y.; Xu, Z.; Chaker, M.; Ma, D. Recent Advancements in Plasmon-Enhanced Visible Light-Driven Water Splitting. J. Materiomics 2017, 3 (1), 33–50.
24. Kim, S.; Kim, J. M.; Park, J. E.; Nam, J. M. Nonnoble-Metal-Based Plasmonic Nanomaterials: Recent Advances and Future Perspectives. Adv. Mater. 2018, 30 (42), 1704528.
25. Atwater, H. A.; Polman, A. Plasmonics for Improved Photovoltaic Devices. Nat. Mater. 2010, 9 (3), 205–213.
26. Feng, L.; Niu, M.; Wen, Z.; Hao, X. Recent Advances of Plasmonic Organic Solar Cells: Photophysical Investigations. Polymers 2018, 10 (2), 123.
27. Linic, S.; Christopher, P.; Ingram, D. B. Plasmonic-Metal Nanostructures for Efficient Conversion of Solar to Chemical Energy. Nat. Mater. 2011, 10 (12), 911–921.
28. Cushing, S. K.; Li, J.; Meng, F.; Senty, T. R.; Suri, S.; Zhi, M.; Li, M.; Bristow, A. D.; Wu, N. Photocatalytic Activity Enhanced by Plasmonic Resonant Energy Transfer from Metal to Semiconductor. J. Am. Chem. Soc. 2012, 134 (36), 15033–15041.
29. Cushing, S. K.; Bristow, A. D.; Wu, N. Theoretical Maximum Efficiency of Solar Energy Conversion in Plasmonic Metal-Semiconductor Heterojunctions. Phys. Chem. Chem. Phys. 2015, 17 (44), 30013–30022.
30. Zhang, Q.; Liu, Y.; Xu, Z.; Zhao, Y.; Chaker, M.; Ma, D. Optimized Design and Mechanistic Understanding of Plasmon and Upconversion Enhanced Broadband Photocatalysts. Catal. Today 2020, 350, 25–32.

31. Zhang, J.; Chen, G.; Chaker, M.; Rosei, F.; Ma, D. Gold Nanoparticle Decorated Ceria Nanotubes with Significantly High Catalytic Activity for the Reduction of Nitrophenol and Mechanism Study. Appl. Catal., B 2013, 132, 107–115.
32. Xu, Z.; Kibria, M. G.; AlOtaibi, B.; Duchesne, P. N.; Besteiro, L. V.; Gao, Y.; Zhang, Q.; Mi, Z.; Zhang, P.; Govorov, A. O. Towards Enhancing Photocatalytic Hydrogen Generation: Which Is More Important, Alloy Synergistic Effect or Plasmonic Effect? Appl. Catal., B 2018, 221, 77–85.
33. Ludwig, J.; An, L.; Pattengale, B.; Kong, Q.; Zhang, X.; Xi, P.; Huang, J. Ultrafast Hole Trapping and Relaxation Dynamics in p-Type CuS Nanodisks. J. Phys. Chem. Lett. 2015, 6 (14), 2671–2675.
34. Kim, M. R.; Hafez, H. A.; Chai, X.; Besteiro, L. V.; Tan, L.; Ozaki, T.; Govorov, A. O.; Izquierdo, R.; Ma, D. Covellite CuS Nanocrystals: Realizing Rapid Microwave-Assisted Synthesis in Air and Unravelling the Disappearance of Their Plasmon Resonance after Coupling with Carbon Nanotubes. Nanoscale 2016, 8 (26), 12946–12957.
35. Su, Q.; Li, Y.; Hu, R.; Song, F.; Liu, S.; Guo, C.; Zhu, S.; Liu, W.; Pan, J. Heterojunction Photocatalysts Based on 2D Materials: The Role of Configuration. Adv. Sustainable Syst. 2020, 4 (9),2000130.
36. Low, J.; Jiang, C.; Cheng, B.; Wageh, S.; Al-Ghamdi, A. A.; Yu, J. A Review of Direct Z-Scheme Photocatalysts. Small Methods 2017, 1 (5), 1700080.
37. Benayas, A.; Ren, F.; Carrasco, E.; Marzal, V.; Del Rosal, B.; Gonfa, B. A.; Juarranz, Á.; Sanz-Rodriguez, F.; Jaque, D.; Garcia-Sole, J.; Ma, D.; Vetrone, F. PbS/CdS/ZnS Quantum Dots: A Multifunctional Platform for in Vivo Near-Infrared Low-Dose Fluorescence Imaging. Adv. Funct. Mater. 2015, 25, 6650–6659.
38. Cotte, M.; Susini, J.; Dik, J.; Janssens, K. Synchrotron-Based X-ray Absorption Spectroscopy for Art Conservation: Looking Back and Looking Forward. Acc. Chem. Res. 2010, 43 (6), 705–714.
39. Deng, J.; Zhang, Q.; Lv, X.; Zhang, D.; Xu, H.; Ma, D.; Zhong, J. Understanding Photoelectrochemical Water Oxidation with X-ray Absorption Spectroscopy. ACS Energy Lett. 2020, 5 (3), 975–993.
40. Yu, T.; He, W.; Jafari, M.; Guner, T.; Li, P.; Siaj, M.; Izquierdo, R.; Sun, B.; Welch, G. C.; Yurtsever, A.; Ma, D. 3D Nanoscale Morphology Characterization of Ternary Organic Solar Cells. Small Methods 2022, 6 (1), 2100916.
41. Zheludev, N. I. What Diffraction Limit? Nat. Mater. 2008, 7 (6), 420–422.
42. Li, P.; Fang, J.; Wang, Y.; Manzhos, S.; Cai, L.; Song, Z.; Li, Y.; Song, T.; Wang, X.; Guo, X.; Zhang, M.; Ma, D.; Sun, B. Synergistic Effect of Dielectric Property and Energy Transfer on Charge Separation in Non-Fullerene-Based Solar Cells. Angew. Chem., Int. Ed. 2021, 60 (27), 15054–15062.
43. Nowak, D.; Morrison, W.; Wickramasinghe, H. K.; Jahng, J.; Potma, E.; Wan, L.; Ruiz, R.; Albrecht, T. R.; Schmidt, K.; Frommer, J. Nanoscale Chemical Imaging by Photoinduced Force Microscopy. Sci. Adv. 2016, 2 (3), e1501571.

44. Rossouw, D.; Couillard, M.; Vickery, J.; Kumacheva, E.; Botton, G. Multipolar plasmonic resonances in silver nanowire antennas imaged with a subnanometer electron probe. Nano Lett. 2011, 11 (4), 1499–1504.
45. Liang, H.; Rossouw, D.; Zhao, H.; Cushing, S. K.; Shi, H.; Korinek, A.; Xu, H.; Rosei, F.; Wang, W.; Wu, N.; Botton, G. A.; Ma, D. Asymmetric Silver "Nanocarrot" Structures: Solution Synthesis and Their Asymmetric Plasmonic Resonances. J. Am. Chem. Soc. 2013, 135 (26), 9616–9619.
46. Liu, X.; Dong, G.; Li, S.; Lu, G.; Bi, Y. Direct observation of charge separation on anatase $TiO_2$ crystals with selectively etched {001} facets. J. Am. Chem. Soc. 2016, 138 (9), 2917–2920.
47. Chen, R.; Fan, F.; Dittrich, T.; Li, C. Imaging photogenerated charge carriers on surfaces and interfaces of photocatalysts with surface photovoltage microscopy. Chem. Soc. Rev. 2018, 47 (22), 8238–8262.
48. Kalinin, S. V.; Ophus, C.; Voyles, P. M.; Erni, R.; Kepaptsoglou, D.; Grillo, V.; Lupini, A. R.; Oxley, M. P.; Schwenker, E.; Chan, M. K. Machine learning in scanning transmission electron microscopy. Nat. Rev. Methods Primers 2022, 2 (1), 11.

CHAPTER 10

# Exploring the Potential of $TiO_2$ Thin Films for Photovoltaic Use

## *Mesoporous and Compact Layers Coating*

Sana Handor, Mohamed Sahlaoui, Laura Hrostea, Abderrahman Abbassi, Liviu Leontie, Mohammed Sajieddine, and Abdelati Razouk

### 10.1 INTRODUCTION

To enhance the perovskite solar cells (PSCs) efficiently, scientists have been focused on improving their quality and physical stability. This may involve exploring different compositions, crystal structures and manufacturing techniques to achieve higher performance [1] and to improve charge transport in the device by optimizing the charge transport layers, i.e., the electron and hole transport layers. These layers facilitate the efficient extraction and collection of charges generated by the perovskite layer [2, 3]. A mesoporous oxide layer, such as $Al_2O_3$, $ZrO_2$ and $TiO_2$, presents a unique structural advantage, with interconnected nanopores in the film. These nanopores offer high surface area and provide a scaffold to support infiltration of the perovskite absorber and selectively extract electrons from it, to enhance the conversion efficiency of the PSCs, improving light absorption and charge separation [4, 5]. In addition, the use of mesoporous

 DOI: 10.1201/9781003499732-10

$TiO_2$ in PSCs offers several advantages [6]. First, it demonstrates favorable band alignment with the perovskite layer, meaning that the energy levels of the two materials are well-matched. This alignment facilitates efficient electron transfer from the perovskite layer to the $TiO_2$ layer. Additionally, mesoporous $TiO_2$ (mp-$TiO_2$) helps to prolong the lifetime of electrons in the solar cell [7]. This extended electron lifetime is beneficial as it reduces the occurrence of hysteresis, which is a phenomenon where the device's performance varies depending on the direction of the voltage sweep during testing [8]. By minimizing hysteresis, mp-$TiO_2$ contributes to more stable and reliable operation of PSCs [9, 10]. The mp-mesoporous PSCs (mp-PSCs) are generally integrated with a compact electron transport layer, which effectively transports the electrons generated from the perovskite layer to the external circuit and also acts as a barrier/blocking layer preventing the photogenerated holes from reaching the opposite electrode, i.e., the cathode [11, 12]. The compact $TiO_2$ is also capable of reducing hysteresis effects in the device by reducing the defect concentration at the interface between the perovskite layer and the indium tin oxide (ITO) substrate [13–15]. Thin film deposition techniques, such as spin coating, have become immensely popular in the manufacturing of $TiO_2$ films due to their simplicity, scalability and their compatibility with large-scale production [16–19]. The controlled deposition of $TiO_2$ layers on substrates allows to accurately regulate film properties such as thickness, porosity and morphology, which directly influence the performance of photovoltaic devices [20–22].

In this work, we aim for a comprehensive exploration of spin-coated mesoporous (mp-$TiO_2$) and compact (c$TiO_2$) titanium dioxide thin films, with particular emphasis on their potential applications in solar energy. Our study begins with an experimental section in which we describe the chemicals, materials and substrates used in this process, followed by the spin coating operation. Once the thin films have been obtained, they will be characterized by X-ray diffraction (XRD) and atomic force microscopy (AFM) for a solid assessment of their intrinsic properties, as well as by a DektakXT profilometer for a transmittance test. The results obtained will be presented and discussed in the following section. Theoretical calculations of the electronic and optical properties will be carried out using the full potential integrated into density functional theory (DFT) to verify the behavior of this material and compare it with the experimental study. All this is followed by a conclusion summarizing the results obtained.

## 10.2 EXPERIMENTAL SECTION

The objective of the experimental study was to elaborate thin films of titanium dioxide ($TiO_2$) using the spin coating technique, a well-established method for depositing thin films onto a substrate while maintaining control over film thickness and ensuring uniformity. This work was carried out by adjusting and experimenting with various process parameters.

The substrates utilized in this study consisted of glass and ITO glass, both measuring 2.54 cm × 2.54 cm size. The commercial ITO glass had a surface resistivity of 20 Ω/sq and a thickness of 100 nm. To etch the ITO glass substrates, they were placed in a Petri dish with a small amount of HCl (37%) for 5 minutes. Subsequently, the samples were soaked in a beaker containing 99.9% absolute ethyl alcohol while being cleaned in an ultrasonic bath filled with distilled water, which was operated for 10 minutes. It is crucial to emphasize the significance of thorough substrate cleaning processes, as even trace amounts of contaminants on the glass and ITO samples can adversely impact the quality of the fabricated film.

### 10.2.1 Materials Preparation

Absolute ethanol was placed in two 5 ml glass containers, followed by 0.63 ml of titanium isopropoxide (TTIP) in one container and 0.07 ml of HCl (2M) in the other. After that these two containers were placed in an ice bath magnetic stirrer for 30 min. The two solutions were then combined, and the ice bath magnetic stirrer was running for another 30 minutes before the final solution was placed in the refrigerator. The compact $TiO_2$ solution has a colorless appearance.

In order to prepare the mesoporous $TiO_2$, 1 g of $TiO_2$ powder was dissolved in 5 ml of absolute ethanol, the resulted solution was then stirred for 20 hours at room temperature and the final solution appeared to be dense with milky in appearance with a concentration labeled D = 0.2 g/ml with the same process of stirring a lighter solution has been prepared with a lower concentration labeled L = 0.1 g/ml.

### 10.2.2 Spin-Coated Films

The spin coating process was employed to deposit the compact $TiO_2$ thin films. The deposition was carried out at a spinning speed of 4000 rpm for a timing of 30 and 60 seconds, depending on specific samples. Following the spin coating step, the samples underwent an annealing process at 450°C for an additional 1 hour, followed by natural cooling. Similarly, mp-$TiO_2$

TABLE 10.1 Spin-Coated Samples

| Sample Name | Material | Substrate | Concentration (g/ml) | Spinning Time (seconds) |
|---|---|---|---|---|
| Dgl 30 | mp-$TiO_2$ | Glass | 0.2 | 30 |
| Dgl 60 | mp-$TiO_2$ | Glass | 0.2 | 60 |
| Lgl 30 | mp-$TiO_2$ | Glass | 0.1 | 30 |
| Lgl 60 | mp-$TiO_2$ | Glass | 0.1 | 60 |
| c-$TiO_2$ | c-$TiO_2$ | Glass | – | 30 |
| Dc 30 | mp-$TiO_2$ | c-$TiO_2$ + glass | – | 30 |
| Dc 60 | mp-$TiO_2$ | c-$TiO_2$ + glass | – | 60 |
| (c+m)Dgl 30 | c$TiO_2$ + mp-$TiO_2$ | Glass | 0.2 | 30 |
| (c+m)$D_{ITO}$30 | c$TiO_2$ + mp-$TiO_2$ | ITO | 0.2 | 30 |

nanoparticles dispersed in ethanol (1:3 wt.%) were spin-coated onto the glass, ITO and compact $TiO_2$ (c-$TiO_2$) with the same deposition conditions of compact $TiO_2$ films. Table 10.1 shows all the produced samples.

## 10.3 RESULTS AND DISCUSSION

### 10.3.1 Phase Structure of Samples Spin Coated on Glass

Figure 10.1 shows the XRD pattern of five spin-coated $TiO_2$ films on glass substrate; the results indicated that the deposited samples annealed at $T_r$ = 450°C for 1 hour start to crystallize with the presence of a specific directional alignment along the (1 0 1) plane. Same result was found by Lin et al. [23] for $TiO_2$ films under a thermal analysis at various temperatures. In the case of mesoporous $TiO_2$ films (Figure 10.1b–e), the porous structure may contribute to the absence of sharp peaks. The presence of pores and voids can perturb the long-range order of the crystal lattice, leading to a diffraction pattern devoid of distinct peaks [24]. However, for a temperature value of 450°C, a distinct peak related to the crystalline phase of anatase at 25.2° (1 0 1) starts to become visible, signifying the initiation of crystallization in spin-coated films on glass for compact $TiO_2$ (c$TiO_2$) and mesoporous $TiO_2$ coated with a lighter solution on glass for 30 and 60 seconds (Lgl 30, Lgl 60), with a denser solution coated on glass for 30 seconds (Dgl 30), and also denser solution $TiO_2$ on c$TiO_2$ film for 30 seconds (Dc), for a spin time of 30 and 60 seconds. This result suggests that the thermal energy, which is equivalent to the annealing temperature at 450°C, is not sufficient to transform the amorphous phase into the anatase phase [25–27].

Therefore, amorphous or nanocrystalline materials, low crystallinity, mesoporous structure and the capabilities of the XRD instrument used

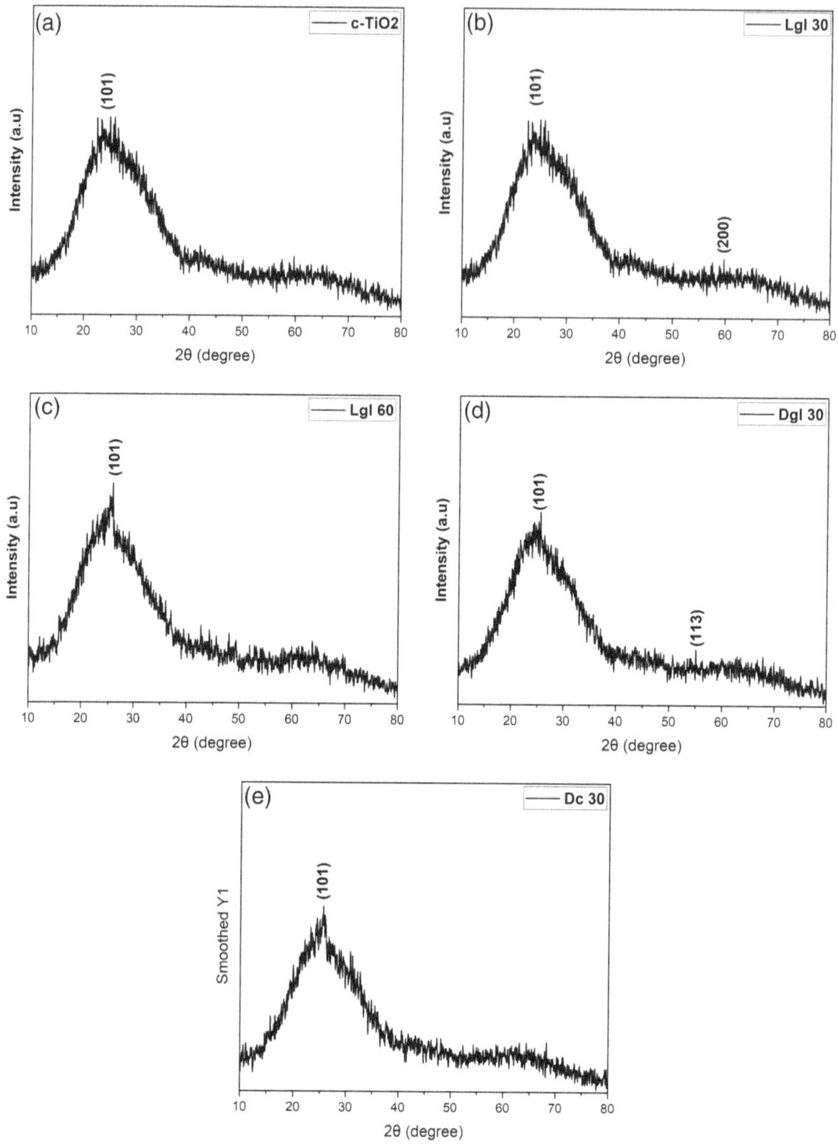


FIGURE 10.1 (a) XRD of c$TiO_2$ sample on glass, (b) XRD spin-coated film of light solution on glass for 30 seconds (Lgl30), (c) XRD of spin-coated film of light solution on glass for 60 seconds (Lgl60), (d) XRD of $TiO_2$ thin film of m$TiO_2$ dense solution spin coated on glass for 30 seconds (Dgl 30), (e) XRD of spin-coated film of mp-$TiO_2$ on c$TiO_2$ (Dc30).

can all be responsible for the similarity of the XRD patterns of the compact and mesoporous $TiO_2$ films we fabricated, which are characterized by the absence of clear and distinct peaks. The significance of complementary characterization techniques in obtaining a thorough comprehension of the structural characteristics of these materials is highlighted by these results [28].

### 10.3.2 Surface Morphology

To evaluate the surface roughness and morphology of the films according to solution concentration, spinning time and coating substrate, the AFM measurement was performed. Comparing to the mp-$TiO_2$, the c$TiO_2$ film surface is found to be more homogeneous, without any void and crack; also the surface roughness of all mp-$TiO_2$ samples is higher than that of the c$TiO_2$. In the grown sample particles have middle shape formation of different sizes. The average surface roughness of c$TiO_2$ (Figure 10.2a) is small and found to be 20.4 nm. For Figure 10.2b and c mp-$TiO_2$ is coated on glass using low concentration solution with a value of Cm=0.05 g/mol, showing big size grains and an average surface roughness of 117.1 and 90.7 nm, respectively, for a spinning time of 30 and 60 seconds. Figure 10.2d and e showed that different pore sizes were observed on the surface of mp-$TiO_2$ and c/mp-$TiO_2$ with an average surface roughness of 116.1 and 139.3 nm, respectively, indicating that the device is influenced by the deposition substrate, where the c/mp-$TiO_2$ film exhibits larger pore sizes and more significant height variations than the mp-$TiO_2$ film. Consequently, the slightly rougher surface of the c/mp-$TiO_2$ ETL (electron transport layer) could be beneficial in terms of electron transport [28]. Although the c/mp-$TiO_2$ layer is known to play an important role in reducing carrier recombination at the ITO/perovskite interface, its transmittance value was not very promising due to the high thickness that still need to be reduced.

TABLE 10.2 Thickness and Roughness of $TiO_2$ Samples

| Sample | Thickness (nm) | R (nm) |
|---|---|---|
| c$TiO_2$ 30 | 116 | 20.4 |
| Lgl 30 | 262 | 117.1 |
| Lgl 60 | 160 | 90.7 |
| Dgl 30 | 473 | 116.1 |
| Dgl 60 | 287 | – |
| Dc 30 | 997 | 139.3 |

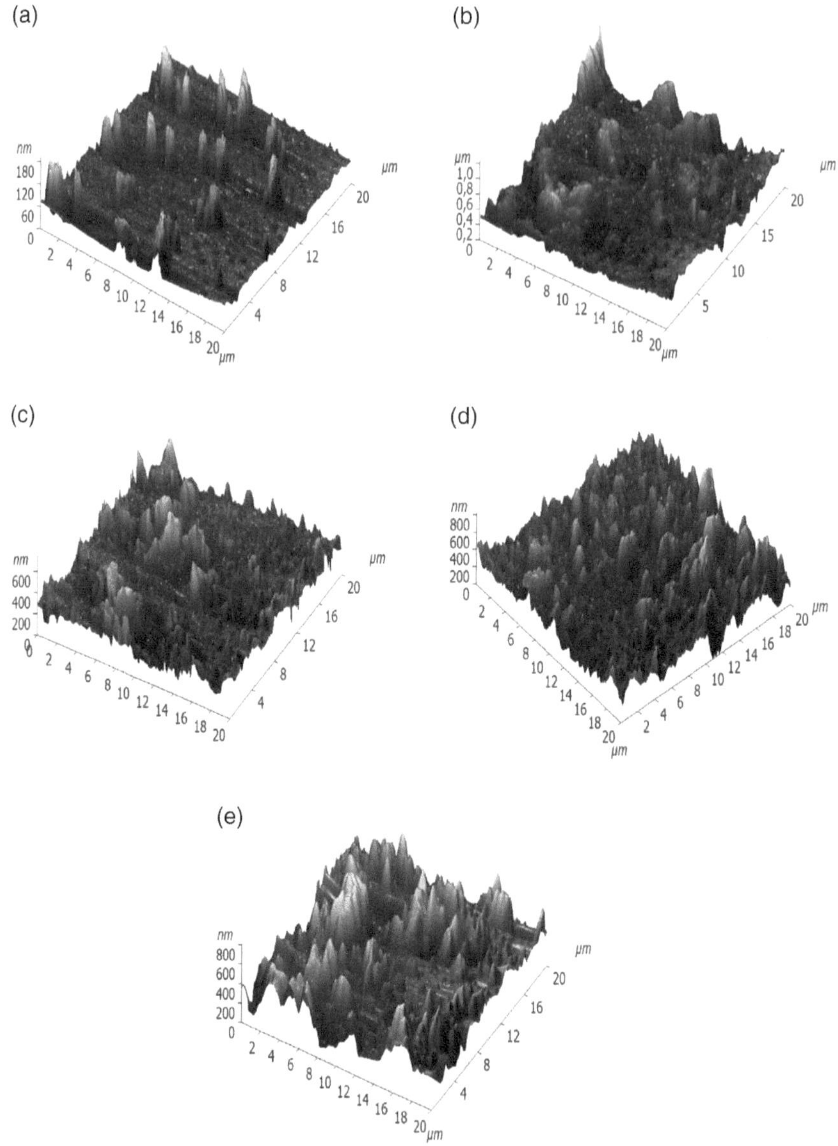


FIGURE 10.2 (a) AFM of $cTiO_2$ sample on glass, (b) AFM spin-coated film of light solution on glass for 30 seconds (Lgl30), (c) AFM of spin-coated film of light solution on glass for 60 seconds (Lgl60), (d) AFM of $TiO_2$ thin film of $mTiO_2$ dense solution spin coated on glass for 30 seconds (Dgl30), (e) AFM of spin-coated film of mp-$TiO_2$ on $cTiO_2$ (Dc30).

### 10.3.3 Thermogravimetric (TGA)

Thermogravimetric (TGA) measurements of $TiO_2$ powder used in these experiments were done by exposing the powder sample to controlled heating while simultaneously recording its weight variation as a function of temperature. This technique provides valuable information on the thermal stability, decomposition behavior and weight loss/gain of $TiO_2$ powder. As shown in Figure 10.3, three main stages in the TGA curve of the $TiO_2$ sample were observed as a function of the thermal profile. The temperature increases from 20 to 180°C as the first step, which is attributed to the removal of remaining ethanol and water, and a weight loss of around 4% occurred. In the temperature range from 180 to 400°C, a weight loss of 2% is attributed, which may correspond to the thermal decomposition of residual organic groups in as-synthesized $TiO_2$ [29]. Indeed, at temperatures up to 400°C, $TiO_2$ powder does not undergo decomposition or significant phase change. Also, we notice that the sensitivity of the $TiO_2$ powder to incident light decreased from 2 to 1 (μV/mW) while the temperature increased from 20 to 800°C, which means that factors such as changes in band gap energy, carrier mobility and surface properties are influencing the material's response to light. From these TGA data, we can conclude that the sample tested is of high purity without any impurities that might decompose or volatilize at lower temperatures, and it is highly stable thermally.

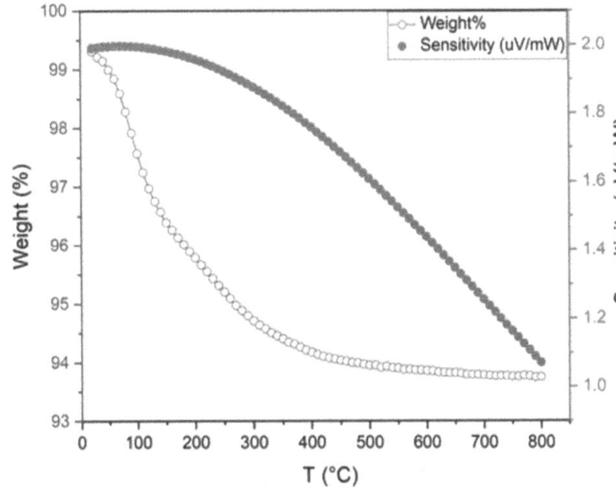


FIGURE 10.3 Thermogravimetric measurements of $TiO_2$ powder.

### 10.3.4 Thickness and Transmittance Evaluation

The thickness of the samples determined by profilometry using a DektakXT Stylus profilometer showed that there is an important influence of the spinning time on the thickness of mp-$TiO_2$; it decreases for Lgl thin films from 262 to 160 nm and for Dgl from 473 to 287 nm for 30 and 60 seconds, respectively. We can conclude that longer spin coating time allows the solution to spread over the substrate; it also provides more time for solvent evaporation and material redistribution, resulting in a thinner film [30]. Also, by increasing the concentration of material solution, the film thickness increases due to the higher viscosity of the coated solution.

The samples were subjected to an optical transmittance analysis using a TEC5 spectrophotometer, covering wavelengths ranging from 250 to 1100 nm. The results were then categorized to highlight variations in the fabrication parameters of the $TiO_2$ thin film. The analysis of transmittance levels among c$TiO_2$, mp-$TiO_2$ and mp/c$TiO_2$ thin films deposited on glass and ITO, Figure 10.4 shows that the transmittance of the samples is lower in the 300–400 nm UV region and higher in the visible spectral region of 400–800 nm, indicating a high UV absorption due to the wide band gap of titanium dioxide [31]. The large difference between compact and mesoporous $TiO_2$ concerning transmittance values with almost

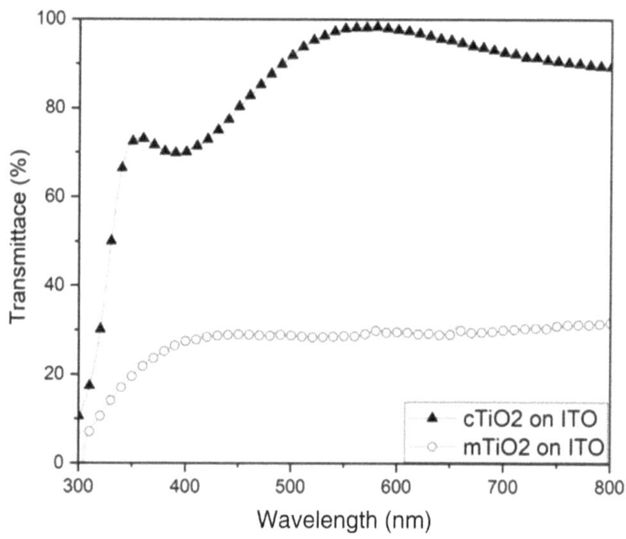


FIGURE 10.4 Transmittance of c$TiO_2$ and mp-$TiO_2$.

90 and 30%, respectively, reveals that there is a reciprocal relation between surface roughness and material transmittance, whereby the higher the roughness, the lower the transmittance. The surface roughness of c$TiO_2$ is about 20.4 nm, and 116.1 nm for mp-$TiO_2$ (*Dgl 30*).

As a result, the transmittance of mp-$TiO_2$/ITO is low due to the light confinement via multiple diffractions by its rough surface. The transmittance of the c$TiO_2$/ITO increased because of the reduced surface roughness. However, mp-$TiO_2$ was fabricated using solutions of different concentrations and varying spin coating durations, clearly demonstrating that samples made from concentrated solutions exhibit significantly lower transmittance values as shown in Figure 10.5. These values fall below 50%, regardless of whether the spin coating time was 30 or 60 seconds. Besides that, the deposition of a less concentrated solution of $TiO_2$ on glass indicates results in $TiO_2$ thin films with a good transmittance value; over 60% for a number of spinning times of 30 seconds. The increase in the number of spins during the preparation of a thin $TiO_2$ film leads to the

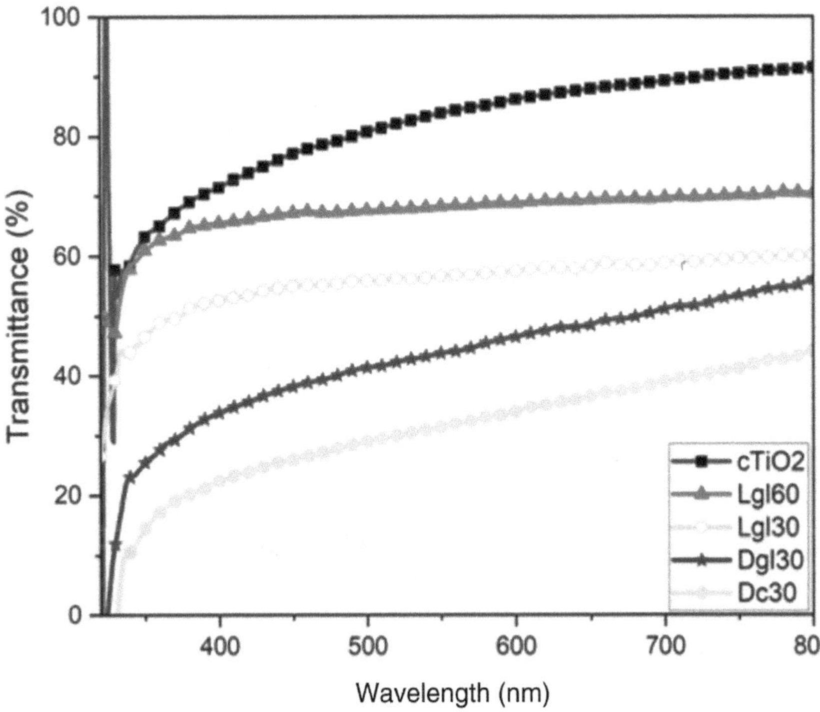


FIGURE 10.5 Transmittance of $TiO_2$ thin films produced using solutions of varying concentration and different spin coating durations coated on glass.

emergence of small pores in the film. Despite a reduction in film thickness associated with these pores, an expected reduction in absorption levels is observed, resulting in higher transmittance, which indicates that more light can pass through the thin film, clearly seen in samples Lgl 30 and Lgl60. However, the transmittance of c/mp-$TiO_2$ (Dc30) showed a trading relationship with thickness, although the thickness of c/mp-$TiO_2$ was higher (Figure 10.6), around 1µm; due to the growth of mp-$TiO_2$ on c$TiO_2$, the transmittance was remarkably low ranging between 20 and 40%. Similar results have also been reported previously. In comparison with the theoretical [32] results obtained by DFT calculation and reported subsequently, it's clear that compact $TiO_2$ films are closer in terms of transmittance, with very similar values ranging from 80 to 98%, while mesoporous $TiO_2$ showed a lower transmittance of 50–60%, but still valuable for photovoltaic applications. The combination of the two was not very promising regarding the transmittance. The c/mp-$TiO_2$ layer was thicker, with a large grain size, resulting in a dense film. The difference in grain size can be explained by a nucleation and growth mechanism [33].

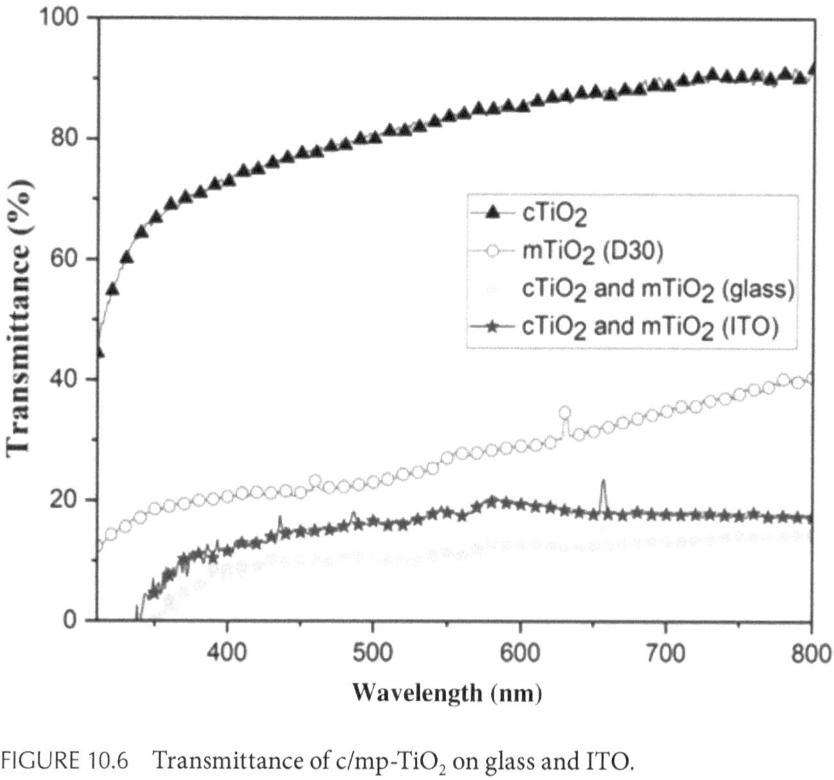


FIGURE 10.6 Transmittance of c/mp-$TiO_2$ on glass and ITO.

## 10.4 THEORETICAL ANALYSIS OF ANATASE $TiO_2$

The structure of $TiO_2$ is studied using DFT with the modified Becke-Johnson correction, which is considered as an approach to improve the underestimation of the band gap and the accuracy of electronic structure calculations on WIEN2k software. The aim is to study and compare the optoelectronic properties with experimental results.

The band structure shown in Figure 10.7 defines the main characteristics of pure $TiO_2$. Specifically, at the Γ point, pure $TiO_2$ exhibits a direct band gap of 2.14 eV. Although the calculated band gap is in line with other theoretical results, it is slightly lower than the experimental value of 3.2 eV. We note that the Fermi level is closer to the valence band. Consequently, a predominance of holes in the valence band indicates P-type conduction. In addition, an electron dispersion analysis reveals increased electron dispersion between the Γ point and the X point, while dispersion between the X and M points is characterized by relative stability. The density of states is plotted in Figure 10.8. The 3s and 3p states of Ti are not shown in this figure as they are located at much lower energies and do not contribute directly to the bonding between Ti and O. The peak centered at

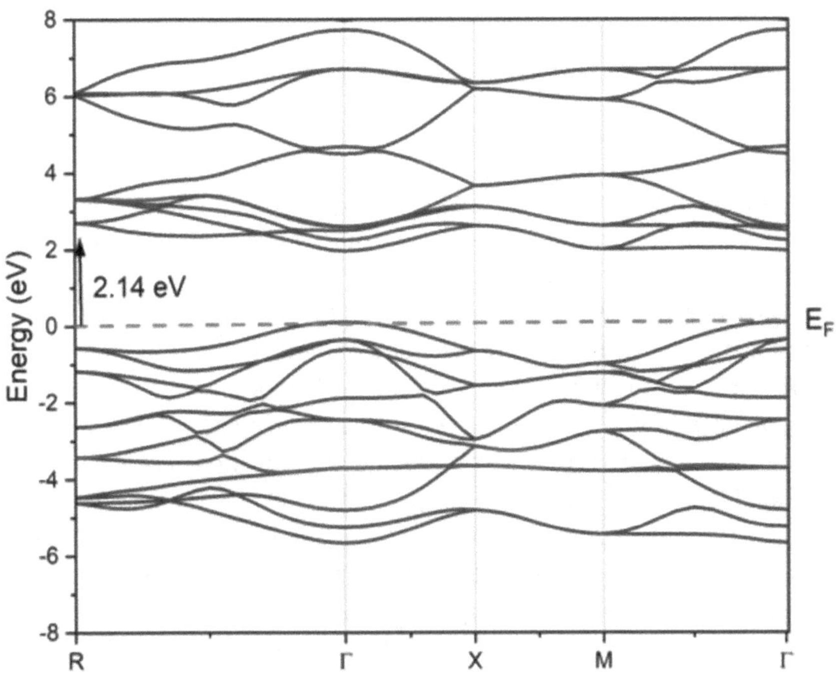


FIGURE 10.7 Band structure of pure $TiO_2$.

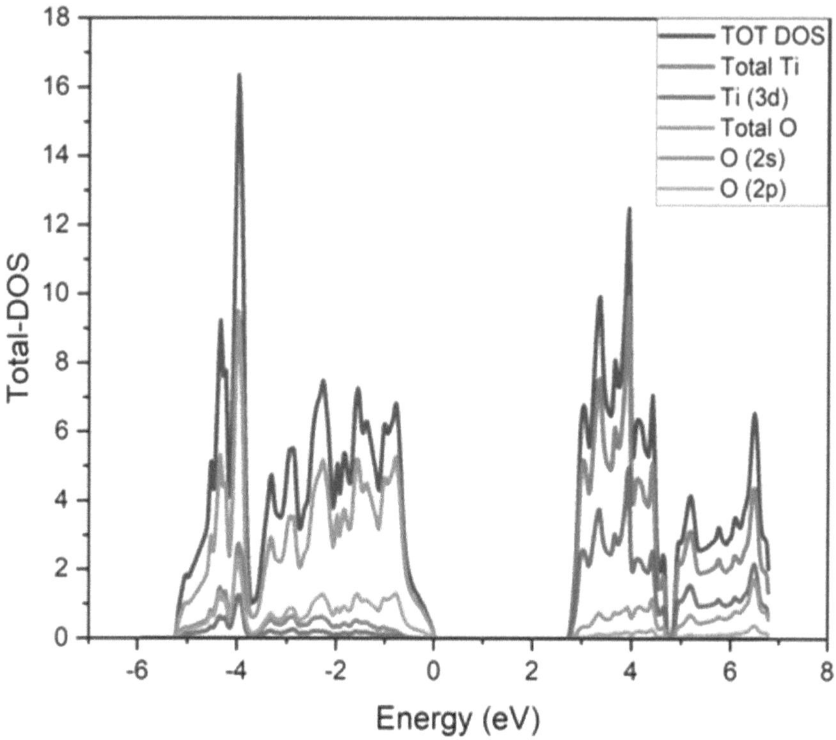


FIGURE 10.8 PDOS and TDOS of $TiO_2$ structure.

16.35 eV belongs to total density of state. The valence band from 0 to −5.25 eV shows a strong hybridization between the O (2p) and the Ti (3d) orbitals. The bottom of the conduction band from 2.7 to 4.7 eV is dominated by Ti (3d) orbitals and suggests a low hybridization between the O (2p) and the Ti (3d) orbitals, while the top of the conduction band indicates strong hybridization between these two orbitals. We note that the calculated electronic structures of pure $TiO_2$ with the mBJ (modified Becke-Johnson) approach are in good agreement with other works cited in [34, 35] and Figure 10.9 shows that the energy-dependent absorption coefficients (eV) of pure $TiO_2$ exhibit an absorption peak at 8.5 eV for the $x$ and $z$ components along the *OX* and *OZ* axes of the crystal structure, respectively. This corresponds to a wavelength of 145 nm, and it is interesting to note that absorption in the z direction is higher than in the x and y directions, which are similar due to the shape of the anatase structure of $TiO_2$. The variation in absorption values for these components is attributed to the orientation and crystalline structure of the material. Figure 10.10 shows

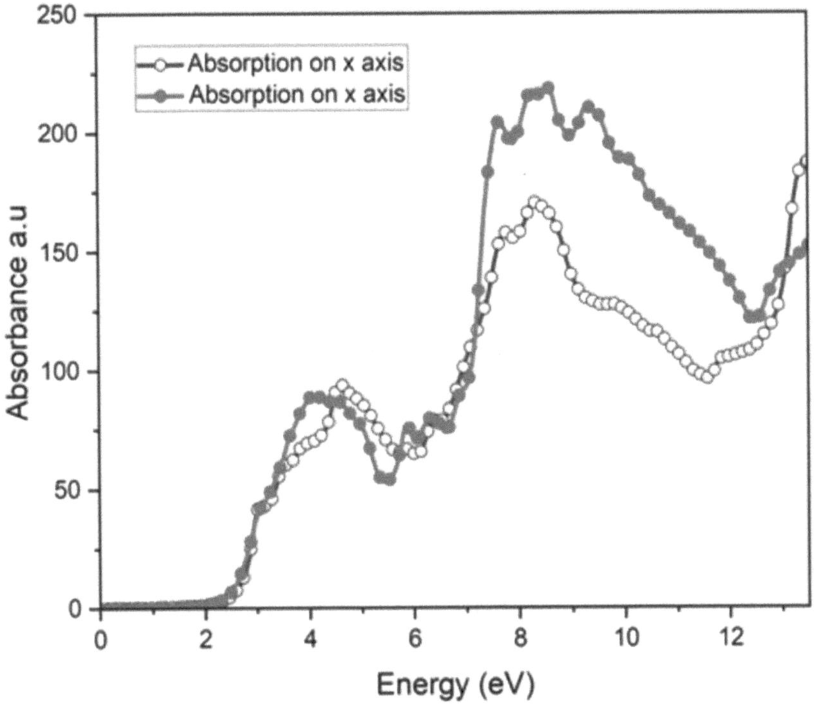


FIGURE 10.9 Absorption spectrum of $TiO_2$ as function of photon energy.

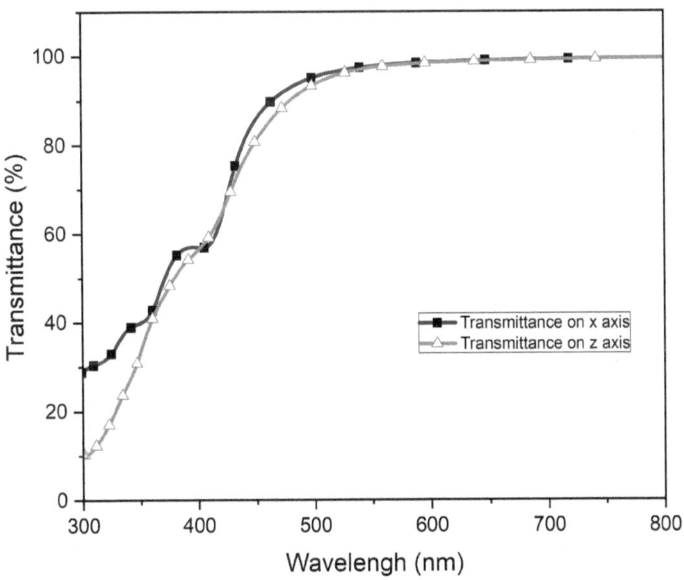


FIGURE 10.10 Optical transmittance spectra of pure anatase $TiO_2$.

the high transmittance value for the *x* and *z* components of the pure $TiO_2$ which indicates efficient light transmission in the visible region of the electromagnetic spectrum. These can be explained by the following: The wide band gap energy which mean that $TiO_2$ absorbs very little light in the visible spectrum. As a result, most visible light passes through the $TiO_2$ thin film, contributing to its high transmittance. These help minimize reflection at the interface of the thin film. When light encounters the $TiO_2$ thin film, instead of being reflected back, it tends to propagate through the film, increasing transmittance. We can also consider the thickness and uniformity as a crucial factor. Thin films with uniform thickness across the entire surface experience less light scattering and interference, leading to higher transmittance. The smooth surfaces on $TiO_2$ thin films reduce light scattering, allowing more light to pass through the film. Rough or uneven surfaces can cause light to scatter in various directions, decreasing transmittance. Finally, the optical interference can occur within the $TiO_2$ thin film, enhancing the transmission of certain wavelengths of light. This phenomenon can be optimized by controlling the thickness of the thin film to create constructive interference at specific wavelengths, further increasing transmittance.

Specific crystallographic orientations or directions can lead to different optical properties, due to factors such as the anisotropy of the material's electrical structure [36]. Additionally, the direction of the incident light can affect the way the material interacts with photons.

## 10.5 CONCLUSION

In summary, the surface morphology, structural and optical properties of mesoporous (mp-$TiO_2$) and compact (c$TiO_2$) films, produced by sol-gel coating, are influenced by the type of $TiO_2$ solution used and their concentration. The highest transmittance was achieved for compact $TiO_2$, reaching around 90%, consistent with theoretical calculations employing DFT. XRD analysis revealed that $TiO_2$ films did not crystallize in the anatase phase at an annealing temperature of 450°C. AFM studies demonstrated that the structure of $TiO_2$ films was non-crystalline, with roughness dependent on the type of thin film, ranging from 20.4 to 139.3 nm. This comprehensive analysis underscores the pivotal role of precursor characteristics in dictating $TiO_2$ thin film properties, offering valuable insights for tailored applications, particularly in photovoltaic devices where precise control of film properties is paramount for optimal performance.

## CONFLICT OF INTEREST

On behalf of all the authors, the corresponding author states that there are no conflicts of interest.

## REFERENCES


1. Y. Wu, X. Yang, W. Chen, Y. Yue, M. Cai, F. Xie, E. Bi, A. Islam, L. Han, Perovskite solar cells with 18.21% efficiency and area over 1 $cm^2$ fabricated by heterojunction engineering, Nat. Energy 1 (2016) 1–7. https://doi.org/10.1038/nenergy.2016.148
2. V.M. Le Corre, M. Stolterfoht, L. Perdigón Toro, M. Feuerstein, C. Wolff, L. Gil-Escrig, H.J. Bolink, D. Neher, L.J.A. Koster, Charge transport layers limiting the efficiency of perovskite solar cells: how to optimize conductivity, doping, and thickness, ACS Appl. Energy Mater. 2 (2019) 6280–6287. https://doi.org/10.1021/acsaem.9b00856
3. S.S. Shin, S.J. Lee, S. Il Seok, Metal oxide charge transport layers for efficient and stable perovskite solar cells, Adv. Funct. Mater. 29 (2019) 1–30. https://doi.org/10.1002/adfm.201900455
4. B. Electrochemical, A. Overview, An Overview, (n.d.).
5. M.M. Lee, J. Teuscher, T. Miyasaka, T.N. Murakami, H.J. Snaith, Efficient hybrid solar cells based on meso-superstructured organometal halide perovskites, Science (80-.). 338 (2012) 643–647. https://doi.org/10.1126/science.1228604
6. A. Hernández-Granados, A.N. Corpus-Mendoza, P.M. Moreno-Romero, C.A. Rodríguez-Castañeda, J.E. Pascoe-Sussoni, O.A. Castelo-González, E.C. Menchaca-Campos, J. Escorcia-García, H. Hu, Optically uniform thin films of mesoporous TiO2 for perovskite solar cell applications, Opt. Mater. (Amst). 88 (2019) 695–703. https://doi.org/10.1016/j.optmat.2018.12.044
7. G. Liu, B. Yang, H. Chen, Y. Zhao, H. Xie, Y. Yuan, Y. Gao, C. Zhou, In situ surface modification of TiO2 by CaTiO3 to improve the UV stability and power conversion efficiency of perovskite solar cells, Appl. Phys. Lett. 115 (2019). https://doi.org/10.1063/1.5131300
8. N.K. Elumalai, A. Uddin, Hysteresis in organic-inorganic hybrid perovskite solar cells, Sol. Energy Mater. Sol. Cells 157 (2016) 476–509. https://doi.org/10.1016/j.solmat.2016.06.025
9. N.J. Jeon, J.H. Noh, Y.C. Kim, W.S. Yang, S. Ryu, S. Il Seok, Solvent engineering for high-performance inorganic-organic hybrid perovskite solar cells, Nat. Mater. 13 (2014) 897–903. https://doi.org/10.1038/nmat4014
10. H. Park, Y. Park, W. Kim, W. Choi, Surface modification of TiO2 photocatalyst for environmental applications, J. Photochem. Photobiol. C Photochem. Rev. 15 (2013) 1–20. https://doi.org/10.1016/j.jphotochemrev.2012.10.001

11. M. Shahiduzzaman, S. Fukaya, E.Y. Muslih, L. Wang, M. Nakano, M. Akhtaruzzaman, M. Karakawa, K. Takahashi, J.M. Nunzi, T. Taima, Metal oxide compact electron transport layer modification for efficient and stable perovskite solar cells, Materials (Basel). 13 (2020). https://doi.org/10.3390/ma13092207
12. A. Lewis, J.R. Troughton, B. Smith, J. McGettrick, T. Dunlop, F. De Rossi, A. Pockett, M. Spence, M.J. Carnie, T.M. Watson, C. Charbonneau, In-depth analysis of defects in $TiO_2$ compact electron transport layers and impact on performance and hysteresis of planar perovskite devices at low light, Sol. Energy Mater. Sol. Cells 209 (2020). https://doi.org/10.1016/j.solmat.2020.110448
13. S. Van Reenen, M. Kemerink, H.J. Snaith, Modeling anomalous hysteresis in perovskite solar cells, J. Phys. Chem. Lett. 6 (2015) 3808–3814. https://doi.org/10.1021/acs.jpclett.5b01645
14. J. Shao, S. Yang, L. Lei, Q. Cao, Y. Yu, Y. Liu, Pore size dependent hysteresis elimination in perovskite solar cells based on highly porous TiO2 films with widely tunable pores of 15-34 nm, Chem. Mater. 28 (2016) 7134–7144. https://doi.org/10.1021/acs.chemmater.6b03445
15. C. Zhen, T. Wu, R. Chen, L. Wang, G. Liu, H.M. Cheng, Strategies for modifying $TiO_2$ based electron transport layers to boost perovskite solar cells, ACS Sustain. Chem. Eng. 7 (2019) 4586–4618. https://doi.org/10.1021/acssuschemeng.8b06580
16. R. Wu, B. Yang, J. Xiong, C. Cao, Y. Huang, F. Wu, J. Sun, C. Zhou, H. Huang, J. Yang, Dependence of device performance on the thickness of compact $TiO_2$ layer in perovskite/$TiO_2$ planar heterojunction solar cells, J. Renew. Sustain. Energy 7 (2015). https://doi.org/10.1063/1.4926578
17. M.D. Tyona, A comprehensive study of spin coating as a thin film deposition technique and spin coating equipment, Adv. Mater. Res. 2 (2013) 181–193. https://doi.org/10.12989/amr.2013.2.4.181
18. M.M. Ferdaus, M.M. Rashid, M.A. Rahman, Design and fabrication of a simple cost effective spin coater for deposition of thin film, Adv. Environ. Biol. 8 (2014) 729–733.
19. A. Hosseini, K. Içli, M. Özenbaş, Erçelebi, Fabrication and characterization of spin-coated $TiO_2$ films, Energy Procedia 60 (2014) 191–198. https://doi.org/10.1016/j.egypro.2014.12.332
20. J. Ma, J. Chang, Z. Lin, X. Guo, L. Zhou, Z. Liu, H. Xi, D. Chen, C. Zhang, Y. Hao, Elucidating the roles of TiCl4 and PCBM fullerene treatment on TiO2 electron transporting layer for highly efficient planar perovskite solar cells, J. Phys. Chem. C 122 (2018) 1044–1053. https://doi.org/10.1021/acs.jpcc.7b09537.
21. F. Scarpelli, T.F. Mastropietro, T. Poerio, N. Godbert, Mesoporous $TiO_2$ thin films: state of the art, Titan. Dioxide - Mater. a Sustain. Environ. (2018). https://doi.org/10.5772/intechopen.74244
22. C. Zhou, Z. Xi, D.J. Stacchiola, M. Liu, Application of ultrathin $TiO_2$ layers in solar energy conversion devices, Energy Sci. Eng. 10 (2022) 1614–1629. https://doi.org/10.1002/ese3.1142.

23. C.P. Lin, H. Chen, A. Nakaruk, P. Koshy, C.C. Sorrell, Effect of annealing temperature on the photocatalytic activity of $TiO_2$ thin films, Energy Procedia 34 (2013) 627–636. https://doi.org/10.1016/j.egypro.2013.06.794.
24. A. Orendorz, A. Brodyanski, J. Lösch, L.H. Bai, Z.H. Chen, Y.K. Le, C. Ziegler, H. Gnaser, Phase transformation and particle growth in nanocrystalline anatase $TiO_2$ films analyzed by X-ray diffraction and Raman spectroscopy, Surf. Sci. 601 (2007) 4390–4394. https://doi.org/10.1016/j.susc.2007.04.127.
25. A.S. Bakri, M.Z. Sahdan, F. Adriyanto, N.A. Raship, N.D.M. Said, S.A. Abdullah, M.S. Rahim, Effect of annealing temperature of titanium dioxide thin films on structural and electrical properties, AIP Conf. Proc. 1788 (2017). https://doi.org/10.1063/1.4968283
26. M.M. Hasan, A.S.M.A. Haseeb, R. Saidur, H.H. Masjuki, M. Hamdi, Influence of substrate and annealing temperatures on optical properties of RF-sputtered $TiO_2$ thin films, Opt. Mater. (Amst). 32 (2010) 690–695. https://doi.org/10.1016/j.optmat.2009.07.011
27. K. Yanagisawa, J. Ovenstone, Crystallization of anatase from amorphous titania using the hydrothermal technique: effects of starting material and temperature, J. Phys. Chem. B 103 (1999) 7781–7787. https://doi.org/10.1021/jp990521c
28. H. Elbushra, M. Ahmed, H. Wardi, N. Eassa, Synthesis and characterization of TiO2 using sol-gel method at different annealing temperatures, MRS Adv. 3 (2018) 2527–2535. https://doi.org/10.1557/adv.2018.230
29. E.C. Muniz, M.S. Góes, J.J. Silva, J.A. Varela, E. Joanni, R. Parra, P.R. Bueno, Synthesis and characterization of mesoporous $TiO_2$ nanostructured films prepared by a modified sol-gel method for application in dye solar cells, Ceram. Int. 37 (2011) 1017–1024. https://doi.org/10.1016/j.ceramint.2010.11.014
30. P. Kumar, A. Sharma, R.K. Karn, S.K. Pandiyan, Synthesis of $TiO_2$ nanoparticles by sol-gel method and their characterization synthesis of $TiO_2$ nanoparticles by sol-gel method and their characterization, J. Basic Appl. Eng. Res. 1 (2014) 1–5. http://www.krishisanskriti.org/jbaer.html
31. S.A. Ansari, M.H. Cho, Highly visible light responsive, narrow band gap $TiO_2$ nanoparticles modified by elemental red phosphorus for photocatalysis and photoelectrochemical applications, Sci. Rep. 6 (2016) 1–10. https://doi.org/10.1038/srep25405
32. C. Bi, Q. Wang, Y. Shao, Y. Yuan, Z. Xiao, J. Huang, Non-wetting surface-driven high-aspect-ratio crystalline grain growth for efficient hybrid perovskite solar cells, Nat. Commun. 6 (2015) 1–7. https://doi.org/10.1038/ncomms8747
33. M. Askari, M.S. Ghamsari, H. Mehranpour, Nucleation and growth of $TiO_2$ nanoparticles, Nanomaterials (2011) 2–26. https://doi.org/10.5772/25912
34. S.-Y. Wang, L.-Y. Chen, F. Pei, S. Wu, G. Wang, M. Xu, Theoretical Study on the Point Defects in N-Doped Anatase TiO2, October 53 (2008) 2292–2298. https://doi.org/10.3938/jkps.53.2292

35. Y.N. Wu, J.K. Wuenschell, R. Fryer, W.A. Saidi, P. Ohodnicki, B. Chorpening, Y. Duan, Theoretical and experimental study of temperature effect on electronic and optical properties of $TiO_2$: comparing rutile and anatase, J. Phys. Condens. Matter 32 (2020). https://doi.org/10.1088/1361-648X/ab9d4f
36. J.S. Lin, W.C. Chou, S.Y. Lu, G.J. Jang, B.R. Tseng, Y.T. Li, Density functional study of the interfacial electron transfer pathway for monolayer-adsorbed InN on the $TiO_2$ anatase (101) surface, J. Phys. Chem. B 110 (2006) 23460–23466. https://doi.org/10.1021/jp061975u

CHAPTER 11

# Theoretical Study of the Optical, Structural and Electronic Properties of Germanium Dioxide ($GeO_2$) Doped with Ti and Nb by the DFT Method with mBJ Correction

Younes Ziat, Hamza Belkhanchi, Zakaryaa Zarhri, and Charaf Laghlimi

## 11.1 INTRODUCTION

$GeO_2$, also known as germanium(II) oxide, is a promising material that is attracting growing interest in the field of materials science due to its unique properties and varied potential applications [1, 2]. $GeO_2$ is a chemical compound made up of germanium and oxygen, with a typical crystalline structure. It belongs to the class of semiconductors [3] and has several polymorphic phases, including the rutile phase [4], α-quartz [5] and the amorphous glass [6, 7], each with its own distinct characteristics and applications; of the various $GeO_2$ phases, the rutile phase is particularly interesting for its stability and favorable electrical and optical properties [8, 9]. $GeO_2$ rutile is widely used in various fields, including as a coating

DOI: 10.1201/9781003499732-11

material for optical glasses [10], in optoelectronics for the manufacture of photovoltaic devices [11], in catalysis and high surface area; it is used in physical and chemical applications [12–14].

To study the properties of $GeO_2$ rutile and other phases, scientists often turn to advanced techniques such as density functional theory (DFT). DFT is a powerful theoretical approach used to calculate the electronic and structural properties of materials at the atomic scale [15–17]. By combining DFT with generalized gradient approximation (GGA) and dispersion corrections such as the modified Becke-Johnson (mBJ) method, researchers can obtain more accurate and reliable results [18, 19], particularly for systems containing heavy elements such as germanium.

In addition, the doping effect is a commonly used strategy for enhancing $GeO_2$ properties, including electrical conductivity and light absorption. By doping impurities or foreign elements into the material's structure, researchers can modify its bandgap, conductivity and other key properties [20–22], paving the way for new applications such as high-performance electronic devices and efficient solar cells [23–25].

$GeO_2$, particularly in its rutile form, has great potential in a variety of technological fields, from optoelectronics to chemical catalysis. With advanced modeling tools such as DFT and innovative doping strategies, researchers are continually exploring the possibilities offered by this fascinating material to develop more advanced and sustainable technologies.

The study of titanium (Ti) and niobium (Nb) doping in $GeO_2$ offers a fascinating opportunity to explore new properties and applications for this material. By introducing these doping elements, we aim to understand their influence on the crystal structure, electronic properties and optical characteristics of $GeO_2$. In this research, we have used a combination of the GGA-mBJ computational methods, renowned for their ability to deliver accurate results in the study of doped materials. These approaches enable us to analyze in detail the changes induced by doping in terms of bandgap, electronic energy levels and optical transitions. By understanding these changes at the atomic scale, we aim to predict the potential applications of these doped materials in various technological fields, from electronics to optics.

## 11.2 CALCULATION METHOD

For our calculations on $GeO_2$, we explored both the pure state and configurations doped with titanium (Ti) and niobium (Nb) at a concentration of 6.25% in the germanium (Ge) position. These dopings are relevant because they can modify the electronic and magnetic properties of the material, opening the way to new applications in various fields. Wien2k package was used to process the calculations of materials. It is particularly appreciated for its ability to use the Full Potential Linearized Plane Wave (FP-LAPW) method, which is based on DFT [15–17]. We have used the GGA-mBJ calculation method, which combines GGA theory with mBJ dispersion correction [26, 27]. This hybrid approach is particularly well suited to studying the electronic and structural properties of materials containing heavy elements such as germanium.

For our simulations, we chose 2 × 2 × 1 $GeO_2$ rutile supercells. These extended supercells (Figure 11.1) are essential for realistically modeling the effects of doping on the material's crystal structure, and for capturing interactions between atoms on a significant scale. With regard to k-point

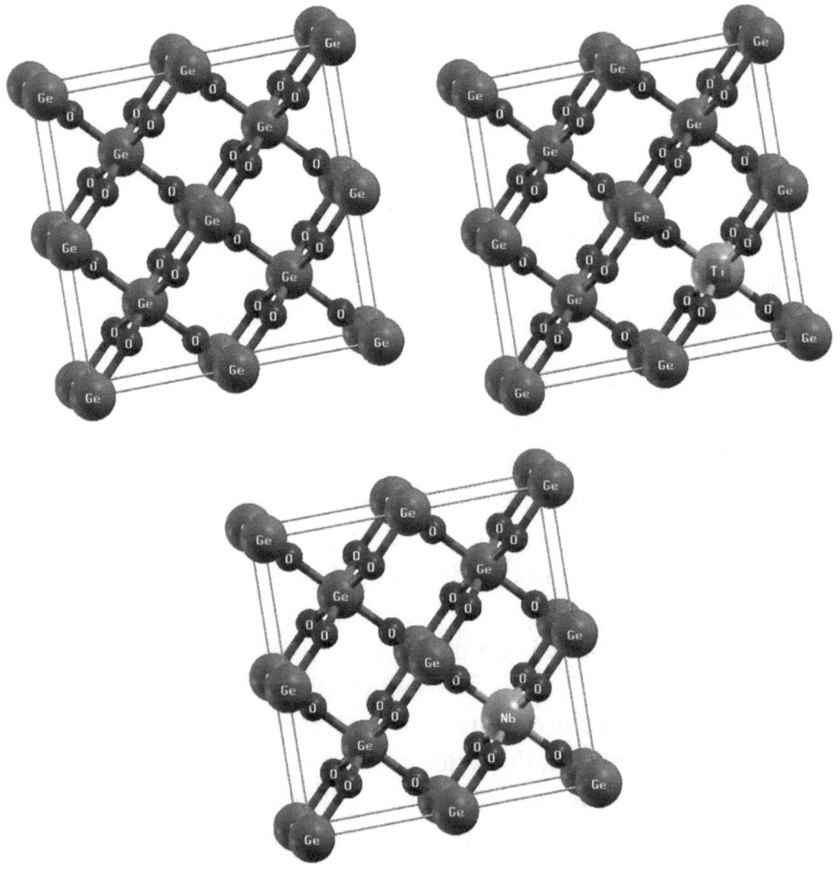


FIGURE 11.1 Structure of $Ge_{0.9375}X_{0.0625}\ O_2$ (X = Ti and Nb): (a) $GeO_2$ Pure, (b) $Ge_{0.9375}Ti_{0.0625}\ O_2$ and (c) $Ge_{0.9375}Nb_{0.0625}\ O_2$.

sampling in the first Brillouin zone, we have used a mesh of dimensions $10 \times 10 \times 10$ for both $GeO_2$ unit cell and supercell calculations. This fine mesh enables us to obtain an accurate description of electron behavior in the material, taking into account subtle variations in electron wave functions across the crystal structure. The total energy and charge convergence during self-coherent cycling were fixed at 0.0001 (Ry) and 0.001 (e).

The results of our calculations reveal calculated lattice constants for rutile $GeO_2$, with $a_0 = b_0 = 4.45$ Å and $c_0 = 2.92$ Å. These values are in excellent agreement with experimental results ($a_0 = b_0 = 4.395$ Å and $c_0 = 2.86$ Å) [5], testifying to the accuracy and reliability of our computational approach. This agreement reinforces our confidence in the data we have obtained and the conclusions we can draw regarding the properties of $GeO_2$, both in its pure and doped form.

In sum, the use of the GGA-mBJ method, combined with extended supercells and accurate k-point sampling, enables us to obtain reliable and accurate results on the properties of $GeO_2$, both in its pure and doped state, opening up new prospects for its use in a variety of technological applications.

## 11.3 ELECTRONIC PROPERTIES

The investigation of the impact of germanium vacancies on the electronic and optical properties of $GeO_2$ is crucial to understanding the behavior of this material in various applications. Introducing a germanium vacancy into a $2 \times 2 \times 1$ supercell, resulting in a germanium vacancy concentration of 6.25%, can significantly influence the material's properties.

Total density of states (TDOS) and partial density of states (PDOS) provide valuable information on the distribution of electronic states in pure $GeO_2$ as well as in samples doped with Ti and Nb, as shown in Figure 11.2. These data are essential for assessing the impact of dopants on the material's electronic structure, and for designing devices with optimized properties. Figure 11.2a, illustrating the pure structure of $GeO_2$, shows that the valence band (VB), located between −10 and 0 eV, is mainly made up of O-2p hybrid states, with small contributions from Ge-4p states. The predominance of O-2p derivative states in the VB suggests significant covalent bonding. In contrast, the states near the bottom of the conduction band (CB), covering the energy range from 0 to 7 eV, are predominantly made up of Ge-4d states, while the composition of states in the CB offers insights into the material's conductivity and its ability to carry charge.

Our results, obtained using the GGA-mBJ method, reveal a bandgap energy value (Eg) of 4.09 eV for $GeO_2$ in remarkable agreement with a

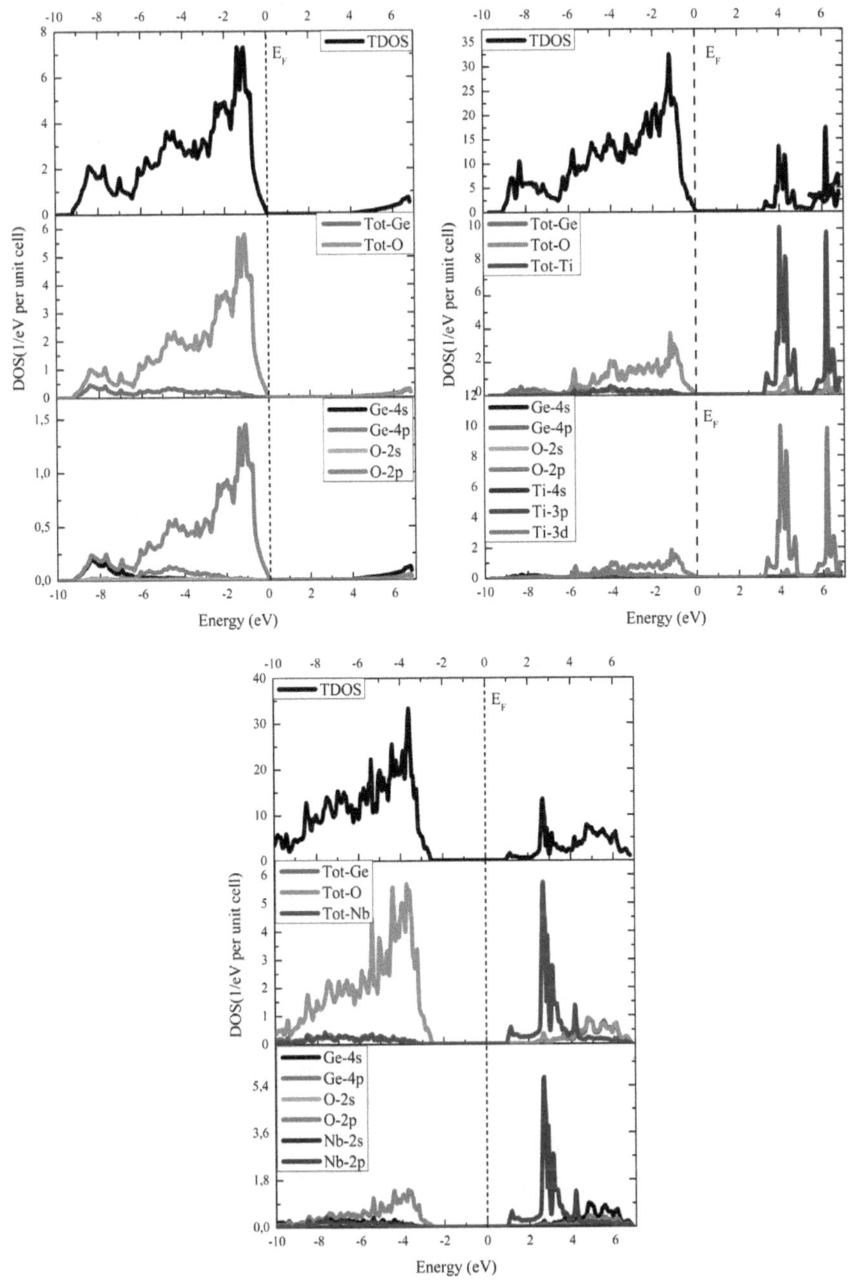


FIGURE 11.2 PDOS and TDOS of $Ge_{0.9375}X_{0.0625}$ $O_2$ (X = Ti and Nb): (a) $GeO_2$ Pure, (b) $Ge_{0.9375}Ti_{0.0625}$ $O_2$ and (c) $Ge_{0.9375}Nb_{0.0625}$ $O_2$.

previous study using the same method (4.10 eV [28]). In comparison with other studies, we observe some variation in the reported values for Eg. For example, calculations using the DFT-LDA method yielded lower values, ranging from 1.80 to 3.12 eV [4, 29, 30], while the DFT-GGA method produced slightly higher values, ranging from 1.25 to 1.91 eV [28, 31]. In contrast, more sophisticated methods such as DFT-PAW + HSE06 and DFT-LDA + $G_0W_0$ resulted in higher values for Eg, with 4.64 and 4.44 eV, respectively [32, 33]. This comparison highlights the sensitivity of the results to the calculation method used but also confirms the reliability of our approach in predicting the properties of doped $GeO_2$. However, these values are at odds with the experimental data, which indicate an Eg value of 4.68 eV [34]. This discrepancy between calculated results and experimental values can be attributed to several factors, including approximations in the theoretical model used, mesh size effects or complex interactions between crystal defects and impurity states. A better understanding of these phenomena requires further investigation, taking into account the subtleties of the electronic structure of dioxygen germanate and its interactions with crystal defects.

In the context of doped structures, for the $Ge_{0.9375}Ti_{0.0625}O_2$ (Figure 11.2b), we observe a particular construction of the VB and CB. The bottom of the VB is mainly made up of O-2p impurity states for $Ge_{0.9375}Ti_{0.0625}O_2$. These impurity states are partially occupied and can lead to further transitions between the intrinsic energy levels of Ti atom and allowed states below the Fermi level. As for the CB of $Ge_{0.9375}Ti_{0.0625}O_2$, it is mainly built up by Ti-3d impurity states. This particular electronic structure can have a significant impact on the electronic and optical properties of the doped material, influencing its conductivity and ability to absorb or emit light in certain energy ranges. These observations underline the importance of understanding the electronic structure of doped materials in detail, as this can lead to important discoveries for the development of high-performance electronic and optical devices. Furthermore, a thorough analysis of these properties will help optimize the performance of doped materials for specific applications, such as solar cells and semiconductor devices.

In our analysis of the $Ge_{0.9375}Nb_{0.0625}O_2$ doped structure (see Figure 11.2c), we find that the bottom of the VB is mainly formed by partially occupied O-2p impurity states. These states may catalyze further transitions from extrinsic Ti impurity energy levels to allowed states above the Fermi level. As for the CB, it is mainly made up of Nb-2p impurity state. The presence of these impurity levels alters the electronic structure of the material, causing the Fermi level to shift toward the CB as a result of the doping-induced increase in the number of carriers. This phenomenon is characteristic of n-type semiconductors, implying a close correlation between electrical conductivity and the width of the bandgap (Eg), determined by the concentration of Nb in the material. The observation of the Burstein-Moss effect [35, 36], typical of heavily doped semiconductors, can be used to explain the Eg broadening in the VB, which also occurs with Nb doping. By shifting the Fermi level into the CB, impurity carriers from the dopant atoms fill this band. As a result, the energy required for a photon to cause an electronic transition between the VB and an unoccupied state in the CB increases.

Incorporating dopants into $GeO_2$ can be an effective strategy for reducing its energy gap, resulting in a decrease in the minimum energy required for an electron to make a transition between electronic bands. The Eg, which represents this energy gap, decreases from 4.09 eV for pure $GeO_2$ to 3.11 eV for $Ge_{0.9375}Ti_{0.0625}O_2$, then to 3.49 eV for $Ge_{0.9375}Nb_{0.0625}O_2$. This reduction in the energy gap can potentially facilitate the passage of electrons from the VB to the CB, which could have a significant impact on the electrical conductivity of $GeO_2$ due to the induced changes in Fermi levels. Dopants have the ability to shift the Fermi level, which can influence the material's electrical conductivity as well as its ability to transport charges efficiently. By adjusting the electronic properties of $GeO_2$ through the incorporation of dopants, it is possible to modulate its electrical conductivity to meet specific application requirements. This approach offers promising possibilities for optimizing the performance of electronic devices and semiconductor materials using doped $GeO_2$ as a key component. These results highlight the importance of crystal defects in modulating the electronic and optical properties of materials, offering valuable leads for materials engineering for specific purposes, such as electronic devices and solar cells.

## 11.4 OPTICAL PROPERTIES

The electronic characteristics and crystalline structure of a material play a crucial role in its optical properties. Dielectric permittivity, conductivity, refractive index, absorption function, absorption and energy-loss spectrum are examples of interconnected optical properties. Dielectric permittivity can be expressed in complex form:

$$\varepsilon(\omega) = \varepsilon_1(\omega) + i\varepsilon_2(\omega) \tag{11.1}$$

This complex dielectric function, $\varepsilon(\omega)$, is a fundamental tool for evaluating the optical properties of materials. To obtain the imaginary part of this function, $\varepsilon_2(\omega)$, we resorted to FP-LAPW calculations [37, 38]. This method allows us to apply the appropriate formulas to extract the required data from our simulations. Thanks to this approach, we can gain a better understanding of the behavior of light as it interacts with the material under study, which is essential for the design and optimization of optoelectronic devices. Using the Kramers-Kronig relation, we can immediately extract the $\varepsilon_1(\omega)$ from the $\varepsilon_2(\omega)$ as follows [37, 38]:

$$\varepsilon_2(\omega)=\left(\frac{4\pi e^2}{\omega^2 m^2}\right)\sum_{i,j}\langle i|M|j\rangle^2 f_i\left(1-f_j\right)*\delta\left(E_f-E_i-\omega\right)d^3k \tag{11.2}$$

$$\varepsilon_1(\omega)=1+\frac{2}{\pi}\int_0^\infty\left(\frac{\omega'-\varepsilon_2(\omega)}{\omega'^2-\omega^2}\right) \tag{11.3}$$

where $f_i$ is the Fermi distribution for the $i$th state, $E_i$ is the energy of the electron in the state, $m$ is the mass, $M$ is the dipole matrix element, $e$ is the electron charge and $i$ and $j$ are the starting and final states, respectively.

Once both the $\varepsilon_1(\omega)$ and the $\varepsilon_2(\omega)$ are calculated, all other optical characteristics, including absorption $\alpha(\omega)$, reflectivity $R(\omega)$, the refractive index $n(\omega)$, the complex conductivity function $\sigma(\omega)$ and the energy-loss spectrum $L(\omega)$, are described by expressions (4, 5, 6 and 7) [37, 39].

$$\alpha(\omega)=\sqrt{2}\omega\left[\sqrt{\varepsilon_1^2(\omega)+\varepsilon_2^2(\omega)}-\varepsilon_1(\omega)\right]^{\frac{1}{2}} \tag{11.4}$$

$$n(\omega)=\frac{1}{\sqrt{2}}\left[\sqrt{\varepsilon_1^2(\omega)+\varepsilon_2^2(\omega)}+\varepsilon_1(\omega)\right]^{\frac{1}{2}} \tag{11.5}$$

$$\sigma(\omega)=\sigma_1(\omega)+i\sigma_2(\omega)=-i\frac{\omega}{4\pi}\left[\varepsilon(\omega)-1\right] \tag{11.6}$$

$$L(\omega)=Im\left(\frac{-1}{\varepsilon(\omega)}\right)=\frac{\varepsilon_2(\omega)}{\varepsilon_1^2(\omega)+\varepsilon_2^2(\omega)} \tag{11.7}$$

Figures 11.3–11.8 display the optical characteristics of both pure and $GeO_2$-doped materials. The $\varepsilon(\omega)$ obtained by utilizing xx directions (100) and zz directions (001) represents the system's linear reaction to an external electromagnetic field.

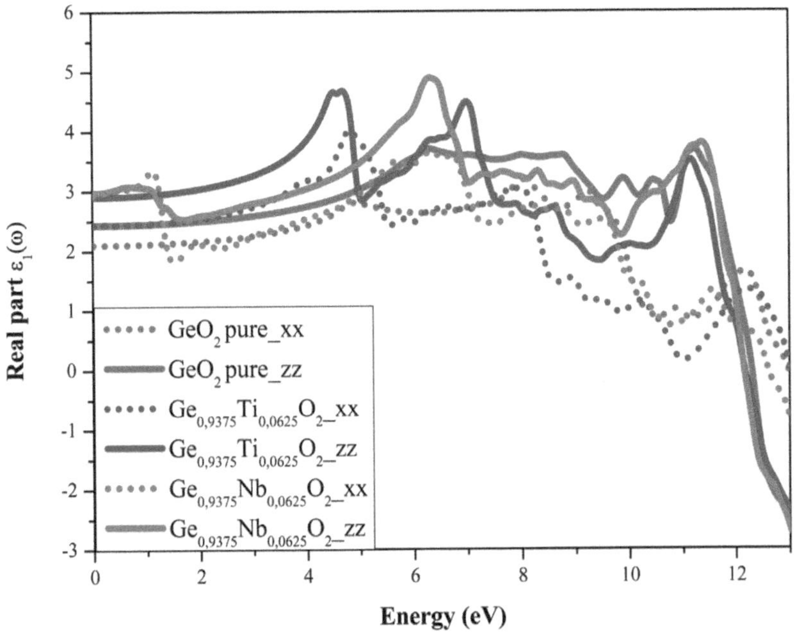


FIGURE 11.3 Real part of $Ge_{0.9375}X_{0.0625}$ $O_2$ (X = Ti and Nb).

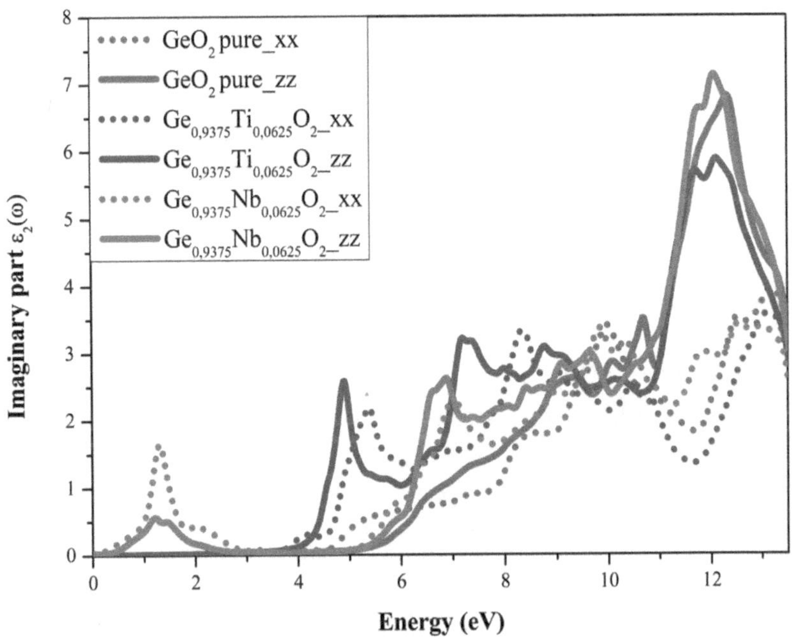


FIGURE 11.4 Imaginary part of $Ge_{0.9375}X_{0.0625}$ $O_2$ (X = Ti and Nb).

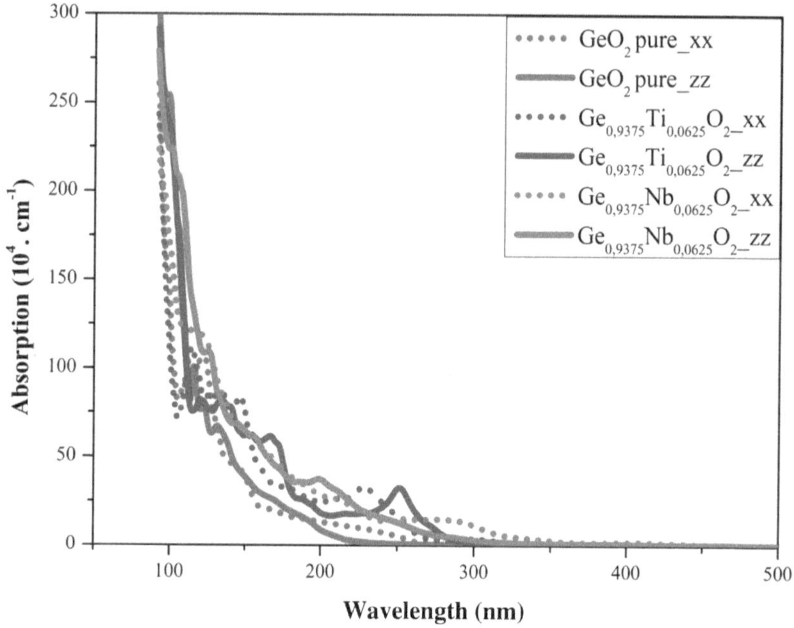


FIGURE 11.5 Absorption of $Ge_{0.9375}X_{0.0625}\ O_2$ (X = Ti and Nb).

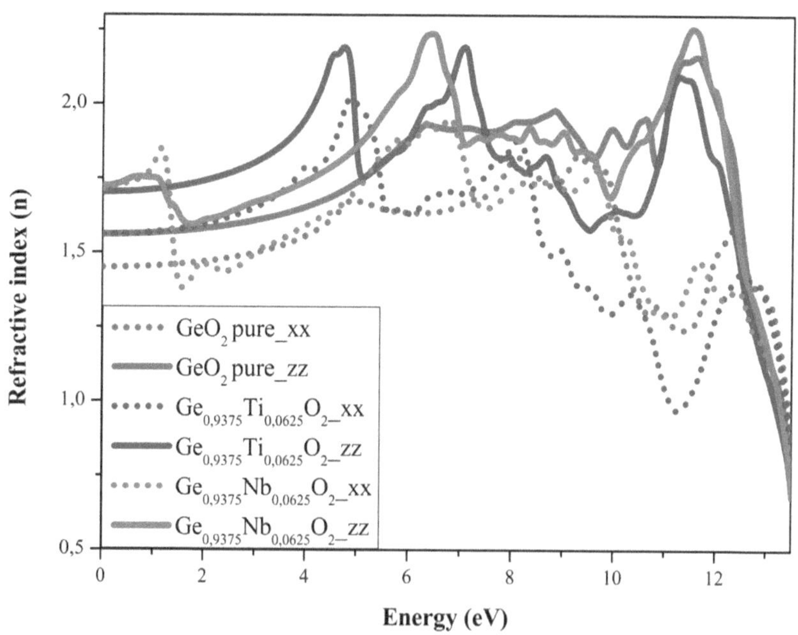


FIGURE 11.6 Refractive index of $Ge_{0.9375}X_{0.0625}\ O_2$ (X = Ti and Nb).

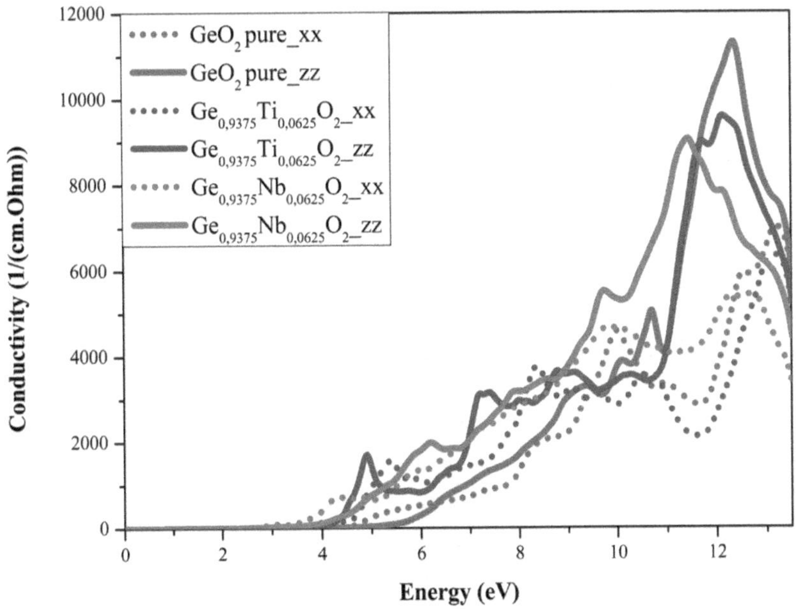


FIGURE 11.7 Conductivity of $Ge_{0.9375}X_{0.0625}$ $O_2$ (X = Ti and Nb).

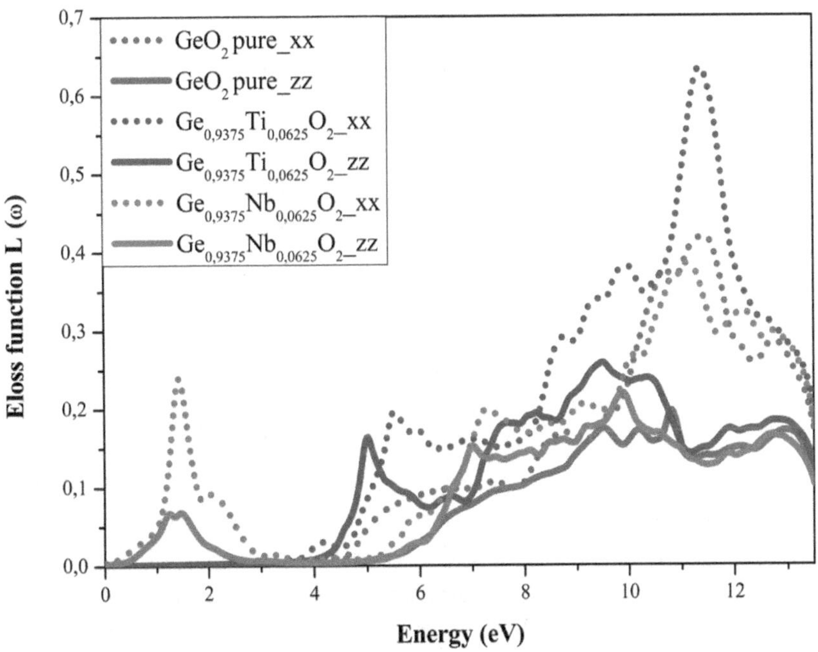


FIGURE 11.8 Eloss function of $Ge_{0.9375}X_{0.0625}$ $O_2$ (X = Ti and Nb).

The real $\varepsilon_1(\omega)$ and imaginary $\varepsilon_2(\omega)$ parts of the dielectric function calculated from (1 0 0) and (0 0 1) are shown in Figures 11.3 and 11.4, respectively. The values of $\varepsilon_2(0)$ are 0 eV for pure and doped $GeO_2$. After 0 eV, the $\varepsilon_2(\omega)$ spectra increase with increasing energy. The values of $\varepsilon_1(\omega)$ at 12 eV become negative for pure and $GeO_2$ doped, which confirm the metallic characteristics of the compound [40]. The $\varepsilon_2(\omega)$ has been calculated as a function of photon energy in Figure 11.4 and represents electron transitions between occupied VB states and unoccupied CB states. Analysis of the $\varepsilon_2(\omega)$ reveals the presence of various energy peaks, each with a particular significance in the context of our study. In structure (1 0 0), the most intense peak is observed, and it can be attributed to the interband transitions between the O-2p bands and the valence layer of our dopant. With the addition of the dopant, this peak shifts toward higher energies, reaching its maximum at an energy of 12 eV for a doping percentage of 6.25%. In structure (0 0 1), the intense peak is associated with the Nb dopant and is described by the interband transitions between the O-2p and Nb-3d bands. This observation is consistent with the Burstein-Moss effect [35, 36], a phenomenon observed in heavily doped semiconductors. This effect is caused by the filling of the conduction level with impurity carriers from the dopant atoms, pushing the Fermi level into the conduction level. This redistribution of charge carriers increases the energy required for a photon to generate an electronic transition between the VB and an unoccupied state in the CB, thus explaining the rise in energies observed in the dielectric function peaks. Indeed, electronic transitions from low energy levels can facilitate the movement of electric charges through the material, which is particularly advantageous for materials intended to be used as transparent conductors. Furthermore, Figure 11.5 shows that most of the absorption occurs in the ultraviolet region, while a smaller fraction is observed in the visible region. These results suggest that material Nb exhibits improved photocatalytic efficiency for visible light compared to pure $GeO_2$. Figure 11.6 shows the refractive index $n(\omega)$ has values of 1.45 and 1.55 from positions (1 0 0) and (0 0 1), respectively, for pure $GeO_2$. It then begins to increase with doping, reaching its maximum at 11.53 eV for the (0 0 1) direction with Nb doping. At this maximum, the refraction phenomenon disappears, as the $n(\omega)$ becomes almost equal to "1", and the material behaves like a free space.

The optical conductivity corresponds to the electron conduction induced when a material is exposed to a photon of a specific frequency. Figure 11.7 illustrates the optical conductivity, $\sigma(\omega)$, for $Ge_{0.9375}X_{0.0625}O_2$ (X = Ti and Nb) as a function of energy. In Figure 11.7, it is obvious that the compounds pure and doped $GeO_2$ exhibit high optical conductivity in ultraviolet range (3.61–13 eV). This observation suggests that pure and doped $GeO_2$ behaves as a semiconductor, and the optical conductivity phenomenon does not occur at a photon energy of 0 eV.

One excellent tool for examining the various characteristics of materials is the energy-loss function $L(\omega)$ or Eloss. Apart from photon absorption, there are numerous different techniques to excite electrons in a material. For instance, a fast-moving electron might excite other electrons and lose energy represented by the Eloss when it travels through a material. Electrons in the solid are excited by energy, which is represented by the Eloss. The Eloss spectrum is influenced by plasmonic excitations, intra- and inter-band transmissions and other potential excitations. Thus, by examining the Eloss, which is connected to the dielectric function as seen below, all excitations can be found by Equation 11.7. Eloss spectrum peaks can be attributed to interband transitions from the VB to the CB or to plasmons, which are collective oscillations of free electrons whose energy is dependent on the valence electron density. The intra- and interband excitations are associated with the peaks in the Eloss curves; the primary peak, shown in Figure 11.8, represents the energy of the plasmons of volume, which are polarized by electric fields in parallel (xx) and perpendicular (zz) directions. It is evident from both situations that the plasmonic frequency of $GeO_2$ compounds increases with thickness. Additionally, in the case of xx or zz, $Ge_{0.9375}Ti_{0.0625}O_2$ have a greater plasmonic frequency than pure $GeO_2$ or $Ge_{0.9375}Nb_{0.0625}O_2$.

## 11.5 CONCLUSION

In conclusion, the study of $GeO_2$ doped with Ti and Nb at 6.25% revealed substantial changes in its structural, electronic and optical properties, using the DFT method with the GGA-mBJ approximation. Density of states (DOS) peaks were assigned to Ti-3d orbitals for $Ge_{0.9375}Ti_{0.0625}O_2$ and Nb-2p for $Ge_{0.9375}Nb_{0.0625}O_2$, and the energy gap decreased from 4.09 eV for pure $GeO_2$ to 3.4 eV for $Ge_{0.9375}Nb_{0.0625}O_2$ and 3.11 eV for $Ge_{0.9375}Ti_{0.0625}O_2$. Optical properties showed an increase in refractive index with doping, strong absorption in the ultraviolet region and high optical conductivity in this range. The doped compounds also show metallic characteristics at 12 eV and increased plasmonic frequencies, particularly for $Ge_{0.9375}Ti_{0.0625}O_2$. These results suggest that doping $GeO_2$ with Ti and Nb improves its performance for applications in optoelectronic devices and as transparent conductors.

## REFERENCES


1. Sierra, J. H., Carvalho, D. O., Kassab, L. R., da Silva Bordon, C. D., Samad, R. E., Wetter, N. U., & Alayo, M. I. (2021). Pedestal waveguides based on $GeO_2$-$Bi_2O_3$, $GeO_2$-PbO, $Ta_2O_5$ and $SiO_xN_y$ cores as platforms for optical amplifiers and nonlinear, optics applications: Review of recent advances. *Journal of Luminescence, 236*, 118113. https://doi.org/10.1016/j.jlumin.2021.118113
2. Abdulghani, S. O., Salih, E. Y., & Mohammed, A. S. (2023). Fabrication and photo-responsive characteristics of $GeO_2$ doped $SnO_2$/porous Si film for ultraviolet photodetector application. *Materials Chemistry and Physics, 303*, 127859. https://doi.org/10.1016/j.matchemphys.2023.127859
3. Tsao, J. Y., Chowdhury, S., Hollis, M. A., Jena, D., Johnson, N. M., Jones, K. A., ... & Simmons, J. A. (2018). Ultrawide-bandgap semiconductors: Research opportunities and challenges. *Advanced Electronic Materials, 4*(1), 1600501.
4. Sahnoun, M., Daul, C., Khenata, R., & Baltache, H. (2005). Optical properties of germanium dioxide in the rutile structure. *The European Physical Journal B-Condensed Matter and Complex Systems, 45*, 455–458.
5. Grimsditch, M., Polian, A., Brazhkin, V., & Balitskii, D. (1998). Elastic constants of $\alpha$-$GeO_2$. *Journal of Applied Physics, 83*(6), 3018–3020.
6. Walker, B., Dharmawardhana, C. C., Dari, N., Rulis, P., & Ching, W. Y. (2015). Electronic structure and optical properties of amorphous $GeO_2$ in comparison to amorphous $SiO_2$. *Journal of Non-Crystalline Solids, 428*, 176–183.
7. Micoulaut, M., Cormier, L., & Henderson, G. S. (2006). The structure of amorphous, crystalline and liquid $GeO_2$. *Journal of Physics: Condensed Matter, 18*(45), R753.
8. Liu, Q. J., Liu, Z. T., Feng, L. P., & Tian, H. (2010). First-principles study of structural, elastic, electronic and optical properties of rutile $GeO_2$ and $\alpha$-quartz $GeO_2$. *Solid State Sciences, 12*(10), 1748–1755. https://doi.org/10.1016/j.solidstatesciences.2010.07.025
9. Trukhin, A., Kink, M., Maksimov, Y., Jansons, J., & Kink, R. (2006). Luminescence of $GeO_2$ glass, rutile-like and $\alpha$-quartz-like crystals. *Journal of non-Crystalline Solids, 352*(2), 160–166. https://doi.org/10.1016/j.jnoncrysol.2005.11.027
10. Sakaguchi, S., & Todoroki, S. I. (1997). Optical properties of $GeO_2$ glass and optical fibers. *Applied Optics, 36*(27), 6809–6814. https://doi.org/10.1364/AO.36.006809
11. Nejaty-Moghadam, L., Esmaeili-Bafghi-Karimabad, A., Gholamrezaei, S., Hamadanian, M., & Salavati-Niasari, M. (2015). Facile synthesis of $GeO_2$ nanostructures and measurement of photocatalytic, photovoltaic and photoluminescence properties. *Journal of Materials Science: Materials in Electronics, 26*, 6386–6394. https://doi.org/10.1007/s10854-015-3227-5

12. Peng, M., Meng, X., Qiu, J., Zhao, Q., & Zhu, C. (2005). $GeO_2$: Bi, M (M = Ga, B) glasses with super-wide infrared luminescence. *Chemical Physics Letters, 403*(4–6), 410–414.
13. Seal, M., Bose, N., & Mukherjee, S. (2018). Application of $GeO_2$ nanoparticle as electrically erasable memory and its photo catalytic behaviour. *Materials Research Express, 5*(6), 065007.
14. Kucheyev, S. O., Baumann, T. F., Wang, Y. M., van Buuren, T., Poco, J. F., Satcher, J. H., & Hamza, A. V. (2006). Monolithic, high surface area, three-dimensional $GeO_2$ nanostructures. *Applied Physics Letters, 88*(10) 103117. https://doi.org/10.1063/1.2182064.
15. Orio, M., Pantazis, D. A., & Neese, F. (2009). Density functional theory. *Photosynthesis Research, 102*, 443–453. https://doi.org/10.1007/s11120-009-9404-8
16. Argaman, N., & Makov, G. (2000). Density functional theory: An introduction. *American Journal of Physics, 68*(1), 69–79. https://doi.org/10.1119/1.19375
17. Blaha, P., Schwarz, K., Madsen, G. K. H., Kvasnicka, D., & Luitz, J. (2001). WIEN2k: An Augmented Plane Wave + Local Orbitals Program for Calculating Crystal Properties. Vienna University of Technology, Vienna, Austria.
18. Gilani, S. S., Tariq, S., Jamil, M. I., Tahir, B., & Faridi, M. A. (2018). Elucidating DFT study on structural, electronic, thermal and elastic properties of $SrTcO_3$ by using GGA and mBJ approach. *Chinese Journal of Physics, 56*(1), 308–314.
19. Kadim, G., Masrour, R., & Jabar, A. (2021). A comparative study between GGA, WC-GGA, TB-mBJ and GGA+ U approximations on magnetocaloric effect, electronic, optic and magnetic properties of $BaMnS_2$ compound: DFT calculations and Monte Carlo simulations. *Physica Scripta, 96*(4), 045804.
20. Shohany, B. G., & Zak, A. K. (2020). Doped ZnO nanostructures with selected elements-structural, morphology and optical properties: A review. *Ceramics International, 46*(5), 5507–5520.
21. Alsaad, A. M., Al-Bataineh, Q. M., Ahmad, A. A., Albataineh, Z., & Telfah, A. (2020). Optical band gap and refractive index dispersion parameters of boron-doped ZnO thin films: A novel derived mathematical model from the experimental transmission spectra. *Optik, 211*, 164641.
22. He, W., Wang, Z., Zheng, T., Wang, L., & Zheng, S. (2021). Origin of the band Gap reduction of In-doped $\beta$-$Ga_2O_3$. *Journal of Electronic Materials, 50*, 3856–3861.
23. Wang, J. T. W., Wang, Z., Pathak, S., Zhang, W., deQuilettes, D. W., Wisnivesky-Rocca-Rivarola, F., … & Snaith, H. J. (2016). Efficient perovskite solar cells by metal ion doping. *Energy & Environmental Science, 9*(9), 2892–2901.
24. Roose, B., Pathak, S., & Steiner, U. (2015). Doping of $TiO_2$ for sensitized solar cells. *Chemical Society Reviews, 44*(22), 8326–8349.

25. Al-Harbi, L. M., Alsulami, Q. A., Farea, M. O., & Rajeh, A. (2023). Tuning optical, dielectric, and electrical properties of polyethylene oxide/carboxymethyl cellulose doped with mixed metal oxide nanoparticles for flexible electronic devices. *Journal of Molecular Structure, 1272*, 134244.
26. Tran, F., & Blaha, P. (2009). Accurate band gaps of semiconductors and insulators with a semilocal exchange-correlation potential. *Physical Review Letters, 102*(22), 226401. https://doi.org/10.1103/PhysRevLett.102.226401
27. Nourozi, B., Aminian, A., Fili, N., Zangeneh, Y., Boochani, A., & Darabi, P. (2019). The electronic and optical properties of MgO mono-layer: Based on GGA-mBJ. *Results in Physics, 12*, 2038–2043.
28. Singh, H., Singh, M., Kumar, S., & Kashyap, M. K. (2011). Full potential calculation of electronic properties of rutile $RO_2$ (R=Si, Ge, Sn and Pb) compounds via modified Becke Johnson potential. *406*(20), 3825–3830. https://doi.org/10.1016/j.physb.2011.07.004
29. Christie, D. M., & Chelikowsky, J. R. (2000). Electronic and structural properties of germania polymorphs. *Physical Review B*, 62(22), 14703–14711. https://doi.org/10.1103/physrevb.62.14703
30. Sevik, C., & Bulutay, C. (2007). Theoretical study of the insulating oxides and nitrides: $SiO_2$, $GeO_2$, $Al_2\ O_3$, $Si_3NC$, and $Ge_3N_4$. *Journal of Materials Science, 42*, 6555–6565. https://doi.org/10.1007/s10853-007-1526-9
31. Liu, Q.-J., Liu, Z.-T., Feng, L.-P., & Tian, H. (2010). First-principles study of structural, elastic, electronic and optical properties of rutile $GeO_2$ and α-quartz $GeO_2$. *12*(10), 1748–1755. https://doi.org/10.1016/j.solidstatesciences.2010.07.025
32. Chae, S., Lee, J., Mengle, K. A., Heron, J. T., & Kioupakis, E. (2019). Rutile $GeO_2$: An ultrawide-band-gap semiconductor with ambipolar doping. *Applied Physics Letters, 114*(10), 102104. https://doi.org/10.1063/5.0056674
33. Chae, S., Mengle, K., Bushick, K., Lee, J., Sanders, N., Deng, Z., … & Kioupakis, E. (2021). Toward the predictive discovery of ambipolarly dopable ultra-wide-band-gap semiconductors: The case of rutile $GeO_2$. *Applied Physics Letters, 118*(26), 260501.
34. Stapelbroek, M., & Evans, B. D. (1978). Exciton structure in the UV-absorption edge of tetragonal $GeO_2$. *Solid State Communications, 25*(11), 959–962. https://doi.org/10.1016/0038-1098(78)90311-3
35. Feneberg, M., Osterburg, S., Lange, K., Lidig, C., Garke, B., Goldhahn, R., … & Krost, A. (2014). Band gap renormalization and Burstein-Moss effect in silicon-and germanium-doped wurtzite GaN up to $10^{20}$ $cm^{-3}$. *Physical Review B, 90*(7), 075203.
36. Kamat, P. V., Dimitrijevic, N. M., & Nozik, A. J. (1989). Dynamic Burstein-Moss shift in semiconductor colloids. *The Journal of Physical Chemistry, 93*(8), 2873–2875.
37. Ziat, Y., Abbassi, A., Slassi, A., Hammi, M., Raiss, A. A., Rhazouani, O. E., & Kenz, A. E. (2016). First-principles investigation of the electronic and optical properties of Al-doped $FeS_2$ pyrite for photovoltaic applications. *Optical and Quantum Electronics, 48*(1), 1–8. https://doi.org/10.1016/j.physb.2023.415216

38. Ziat, Y., Zarhri, Z., Belkhanchi, H., Ifguis, O., Cano, A. D., & Lazrak, C. (2022). Effect of Be and P doping on the electron density, electrical and optoelectronic conduct of half-Heusler LiMgN within ab initio scheme. *Physica Scripta*, *97*(10), 105802. https://doi.org/10.1088/1402-4896/ac8b40
39. Radjai, M., Bouhemadou, A., & Maouche, D. (2021). Structural, elastic, electronic and optical properties of the half-Heusler ScPtSb and YPtSb compounds under pressure. *Condensed Matter Physics*, *24*, 4. https://doi.org/10.48550/arXiv.2112.09940
40. Mubashir, S., Butt, M. K., Yaseen, M., Iqbal, J., Iqbal, M., Murtaza, A., & Laref, A. (2021). Pressure induced electronic, optical and thermoelectric properties of cubic BaZrO3: A first principle calculations. *Optik*, *239*, 166694. https://doi.org/10.1016/j.ijleo.2021.166694

CHAPTER 12

# Internet of Things for Photovoltaic Solar Energy Systems

Abdelmalek Mimouni, Aumeur El Amrani, Mohamed El Amraoui, and Lahcen Bejjit

## 12.1 INTRODUCTION

With the growing world's population, particularly in developing and emerging economies, energy demands have increased rapidly over the last few years. Therefore, it's important to use different energy sources, both renewable and non-renewable, in order to satisfy these demands (Destek & Sinha, 2020). However, non-renewable energy sources, including fossil fuels like oil, natural gas, coal, and nuclear energy, are diminishing day by day due to their finite nature and the environmental and human health impacts associated with their extraction and combustion, such as ecological degradation by releasing harmful pollutants into the environment, climate change due to the emission of greenhouse gases like $CO_2$, $CH_4$, and $N_2O$ (Nathaniel and Iheonu, 2019 2019), and adverse health impacts on individuals, including respiratory ailments, heart problems, and the development of cancer. Therefore, for reducing air pollution and greenhouse gas emissions, as well as combating climate change to minimize its most devastating impacts, renewable energy sources play a crucial role in addressing these challenges and meeting sustainable development goals (Güney, 2019).

Renewable energy sources such as solar, wind, biomass, hydroelectric, wave energy, and geothermal are sustainable choices for the environment because they are constantly replenished through natural processes and offer eco-friendly alternatives to fossil fuels. In recent years, many countries


DOI: 10.1201/9781003499732-12

around the world have become increasingly reliant on renewable energy sources, not only to reduce the impacts of non-renewable energy sources but also to decrease reliance on imported non-renewables and improve energy security, reducing a nation's susceptibility to geopolitical unrest and change in the price of energy around the world. As well as their economic benefits by generating employment in research and development, manufacture, installation, and maintenance (Østergaard et al., 2022).

Solar energy, or solar photovoltaic (PV), is one of the most important renewable energy sources, that converts a part of the solar radiation into electric energy using PV cells. These cells absorb photons from the sun, which then produce an electric current, through a phenomenon called photovoltaic effect. PV solar system demand is growing rapidly due to its easy installation, robustness, and low maintenance cost. Additionally, solar energy provides flexible, adaptable, and sustainable options to meet diverse energy needs and requirements across multiple industries and applications, including power generation, heating, water pumping, desalination, lighting, transportation, etc. (Kabir et al., 2018).

In addition, the Internet of Things (IoT) is one of the many groundbreaking results of technology development. It is a terminology to describe the interconnection of diverse and heterogeneous objects to both the internet and each other. Thus, IoT technology enables people and connected devices to establish connections with anyone as well as anything, using the available networks in any location and at any time. Moreover, the IoT has numerous benefits and advantages such as increased efficiency, productivity, and cost savings across various industries, real-time monitoring that includes the collection, processing, analyzing, visualizing, and recording of the data, and remote control in real-time. Consequently, the IoT has a wide range of applications across fields such as industry, renewable energy, agriculture, transportation, water management, healthcare, smart cities, and many more (Nižetić et al., 2020). However, it also involves many challenges and disadvantages including security and privacy risks, scalability challenges, compatibility and interoperability issues, and considering cost implications, that must be resolved in order to guarantee effective adoption and implementation (Srivastava & Pandey, 2022).

The performance of the PV system, including its power, current, and voltage output, is influenced by several environmental conditions, such as changes in irradiance levels, fluctuations in temperature, variations in humidity levels, and the accumulation of dust and soiling. The collection, real-time monitoring, supervision, and analysis of these environmental and electrical data are made possible by the integration of IoT technology with PV solar systems. Moreover, an IoT-based PV system allows real-time monitoring and data analytics, enabling users to monitor energy production and consumption as well as environmental data in real-time. This offers the ability to optimize energy usage, identify potential issues, and make well-informed decisions to improve system performance. Additionally, it offers remote tracking and control of the PV system, giving users the flexibility to adjust settings and enhance energy production without needing to be on-site. Furthermore, it helps in detecting and identifying multiple faults while also predicting the PV output power. This enables optimal performance, improved efficiency, ensured reliability, and decreased maintenance costs (Nath et al., 2023; Rao et al., 2024).

Despite the numerous benefits of IoT integration in PV solar energy systems, several challenges must be overcome for successful implementation. These include security and privacy risks, complexity in integrating IoT devices and protocols within existing PV infrastructure, compatibility issues, interoperability concerns, standardized communication protocols, as well as scalability and cost considerations, particularly in large-scale systems (Ansari et al., 2021).

In the literature, there are several review articles focusing on IoT system technology-based PV solar energy systems. Rahman et al. (2018) provided an overview of the architectures and features of various PV monitoring systems based on different methodologies. Ahmad et al. (2021) presented a comprehensive review of recent developments, including a comparative analysis of solar PV energy harvesters, and algorithms for self-powered, self-sustained IoT nodes, with a focus on maximum power point tracking (MPPT) techniques. Abir et al. (2021) presented an overview of existing studies on IoT applications to the smart grid system. Ansari et al. (2021) presented a survey on the monitoring system, classification, detailed description, and limitations of solar PV monitoring systems. Kirmani et al. (2022) presented an investigation on IoT-enabled smart grid systems, including features, architectures, prototypes, applications, advantages, and challenges associated with integrating IoT and smart grid technology. Nath et al. (2023) provided a state-of-the-art review on the application of IoT for effective solar energy utilization. This review discusses how IoT technologies are integrated into solar energy applications to enhance efficiency and management in the utilization of solar power. Qays et al. (2023) presented a comprehensive review that concern the communication technology, architectural design, cutting-edge applications, and protocols of

IoT-supported smart grids. Rao et al. (2024) provided an overview related to the techniques, methods, components, and approaches used in intelligent energy management for both independent and grid-connected hybrid renewable energy systems with a focus on IoT in PV power generation.

With the growth of IoT-based solar energy systems, new techniques, methods, components, and approaches are emerging every day. The main contributions of this chapter include examining the role of IoT in solar energy systems. Moreover, the particular architectures, potential software and hardware platforms, and communication network technologies for IoT-based PV systems.

The rest of the chapter is organized as follows: Section 12.2 presents the architectures of IoT-based PV systems; Section 12.3 provides the potential software and hardware IoT platforms, while Section 12.4 introduces the communication network technologies for IoT-based PV systems. The existing challenges and issues of IoT-based PV systems are discussed in Section 12.5. Finally, a conclusion is presented in Section 12.6.

## 12.2 ARCHITECTURES OF IoT-BASED PV SYSTEMS

### 12.2.1 Architectures of Internet of Things Systems

The design of an IoT system architecture depends mainly on the functionality of the system, as well as hardware and software requirements (Abir et al., 2021; Islam et al., 2023; Ud Din et al., 2019). The growth of IoT systems leads to an increase in the number of proposed architectures, which are layered, with different requirements for each layer. The fundamental and simple IoT architecture comprises three layers as shown in Figure 12.1 (Caraveo-Cacep et al., 2023): (1) the perception (or sensing) layer is the bottommost level of the architecture, where data is collected and acquired from the physical environment using various sensors and devices such as sensors, microcontrollers, actuators, and other IoT devices; (2) the network (or communication) layer handles the transmission and processing of the data acquired by the perception layer. It establishes connections between devices, other intelligent objects, servers, and network devices. Additionally, it is responsible for efficiently managing the transmission of all data across these interconnected elements; and (3) the application (or service) layer, which processes the data received from the network layer and presents them to the end-users through applications, dashboards, or user interfaces. However, recently several other more complex and multilayer architectures have been investigated. Figure 12.2 shows a four-layer

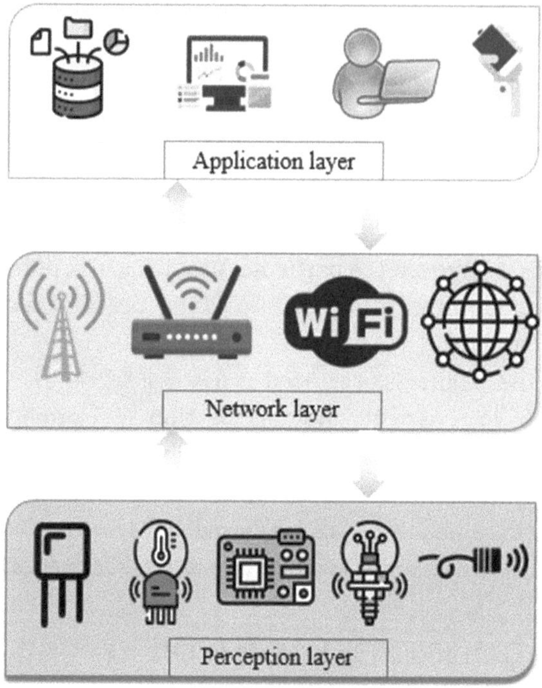


FIGURE 12.1 The three-layer architecture of an IoT system.

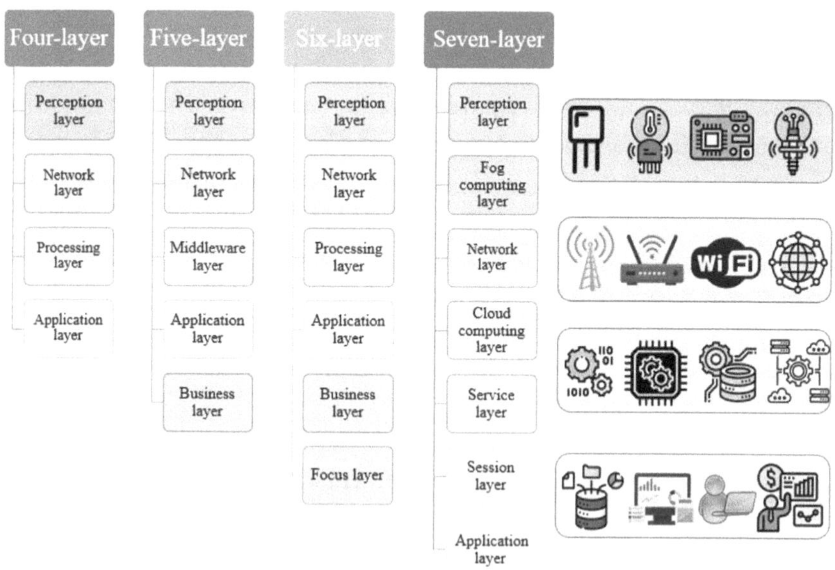


FIGURE 12.2 Multi-layer architectures for IoT systems.

architecture which consists of the following layers: perception, communication, processing, and application. This architecture is the most widely used for IoT devices today, enabling the processing of large amounts of sensor data and facilitating edge and cloud computing (Saleem et al., 2019). The five-layer architecture has been proposed to respond to the rapid growth of IoT systems. It comprises the middleware and business layers in addition to the three-layer architecture. The middleware layer is responsible for service management such as allowing the processing of the received data and making the decisions based on the results. While the business layer is accountable for managing the overall IoT system including the services and applications. It creates business models, charts, flow diagrams, etc., based on the data received from the previous layer (Al-Fuqaha et al., 2015; Khan et al., 2012). The six-layer IoT architecture is based on IoT framework and is divided into two groups. The first group, known as the internal layers group, consists of four layers: perception, communication, processing, and application layers, which are similar to the four-layer architecture mentioned above. The second group, known as the external layers group, consists of two layers: the focus layer, which helps to group the parameters and focus on the multidisciplinary aspects of the problem, and the business layer, which serves a similar function to the previous business layer in five-layer architectures (Kumar, Dash, et al., 2018). Finally, the seven-layer IoT architecture is based on the four-layer architecture, with the addition of IoT platform, management, and security layers (Domínguez-Bolaño et al., 2022). However, Mokhtari et al. (2019) proposed the addition of three layers: fog computing, service, and session layers, alongside the existing four-layer architecture.

### 12.2.2 PV Energy System Architecture

PV energy system architectures can be generally divided into three major classes (Madeti & Singh, 2017): stand-alone systems, which are designed to function independently without the need for a connection to a power grid. These systems comprise PV modules to generate electricity from solar radiation and a charge controller to regulate the voltage flow to and from storage batteries. These last are used during the lack sunlight. In addition, an inverter is used to convert the DC power into AC to supply the AC loads. Grid-connected systems involve connecting the PV panels or array to the utility grid. These systems often consist of PV modules, bidirectional converters, inverters, and other equipment for safety and metering (Argyrou et al., 2019). Hybrid PV systems integrate solar panels with other energy sources, such as wind turbines or batteries, for a more reliable power supply. Hybrid systems typically include PV modules, wind or other renewable energy sources, converters, and inverters, as well as interconnection equipment (Fathima & Palanisamy, 2015). These classifications describe the various configurations and modes of operation that PV systems can adopt to meet different energy needs, geographic locations, and environmental conditions. The selection of PV system architecture depends on various aspects, including the existence of alternative energy sources, the need for energy storage, the availability of grid infrastructure, and the particular requirements of the application.

### 12.2.3 Architectures of IoT-based PV System

The integration of the IoT into PV energy systems has introduced several features to the current solar field. These features include remote monitoring, control and optimization, data collection and analysis, and fault detection and maintenance. Many applications have been proposed to achieve these features, such as monitoring systems, remote control systems, MPPT, fault detection systems, power prediction systems, etc. These systems are based on diverse approaches using multiple methodologies and communication technologies (Nath et al., 2023).

As shown in Figure 12.3, the architecture of IoT-based PV monitoring systems can be categorized into three levels: collection and acquisition level, processing and analysis level, and storage and visualization level. At the collection and acquisition level, the environmental and electrical data

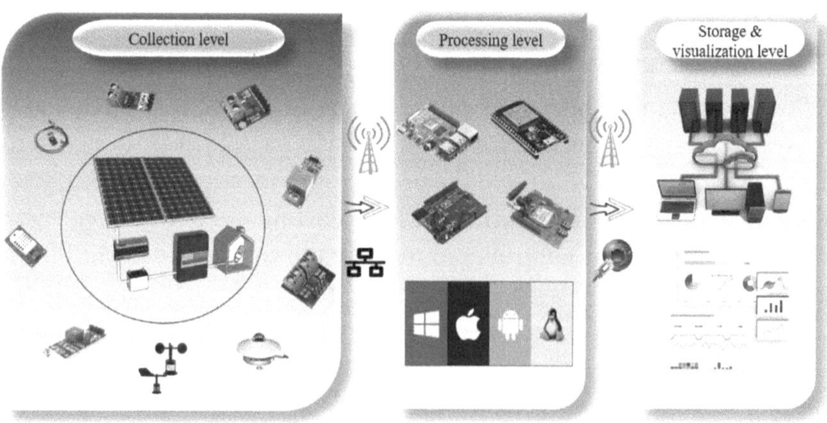


FIGURE 12.3 Basic architecture of an IoT-based PV system.

are collected from several sensors including power, voltage, current, temperature, humidity, solar irradiance, wind speed and direction, etc. These data are transmitted to the next level using wireless or wired protocols. At the processing and analysis level, the data received from the sensors are saved temporarily in process devices and sent to the last level using communication protocols. In the last level, the data are stored in databases and presented to end-users through applications, dashboards, or user interfaces (Rahman et al., 2018; Ansari et al., 2021). Over the last few years, several research works have been carried out related to IoT-based PV systems. An IoT-based remote monitoring system for solar power plants is suggested in (Adhya et al., 2016). The authors layered the proposed system into three basic layers, starting with the sensing layer, which comprises sensors and microcontrollers. Then, the network layer, where the data collected from sensors are carried out for real-time processing and storage in the databases. Finally, the application layer, where the processing and storage data are presented to end-users through the graphical user interfaces. Kumar, Atluri, et al. (2018) discussed a brief study on the use of IoT in PV systems. The authors proposed an IoT architecture for PV systems consisting of three different layers. The first layer contains the PV system design environment, where all components are interconnected according to the desired configurations. The second layer is referred as a gateway connection, and it connects the PV system's hardware designs to the web server via an internet-connected router. Finally, the remote monitoring and control layer, which supervises and stores the received data. De Arquer Fernández et al. (2021) proposed an IoT solution for PV plant monitoring by using open-source software. The suggested structure included the following layers: industrial devices, which are already equipped with sensors in the industrial environment; an IoT gateway, which provides the ability to collect, process, store, and export data from industrial devices; an IoT platform, that provides the management of IoT devices and centralization of communication and data; and finally, operation, which involves executing specific business related processing and human monitoring. Chinnathambi et al. (2022) developed an IoT-based energy management system for a grid-connected solar PV-powered DC residential building. This system is implemented for monitoring and controlling the applications and energy sources with five-layer architecture namely, perception layer, network layer, energy management and control layer, application layer, and IoT services layer. Panagopoulos and Argiriou (2022) designed a low-cost data acquisition system for solar thermal collector. In this work, the authors proposed an architecture comprising three layers: an ESP32 board and sensors as the acquisition layer, which upload measurements to the back-end layer. The back-end is responsible for processing and saving the received data. The last layer is the end-user, which offers data visualization. A design, deployment, implementation, and performance evaluation of an IoT-based smart energy management system proposed by M. U. Saleem et al. (2022), which the investigated architecture is divided into five layers, which are grouped into two groups. The first group namely, the consumer end, consists of the energy monitoring layer, energy control layer, and communication layer. The second group comprises the cloud analytics layer and application layer. Ul Mehmood et al. (2023) proposed a cloud-based IoT solution for remotely monitoring the amount of soiling on PV panels using an artificial neural network (ANN). The proposed system is based on a four-layer architecture: the perception layer, which contains sensors attached to the PV panels; the network layer, which establishes connections between devices, networks, and the cloud; the middleware layer, which acts as the brain of the IoT system, where the data received from the sensing layer are collected, processed, and stored; and finally, the application layer, which consists of user interfaces, including mobile applications, browser-based applications, or any other visualization tools.

## 12.3 POTENTIAL IoT PLATFORMS FOR PV SYSTEM

This section discusses potential hardware and software platforms suitable for IoT-based monitoring PV systems. The choice of these platforms depends on various aspects such as cost, scalability, integration capabilities, security, and the specific requirements of the PV system being monitored and controlled. Over the recent years, researchers have proposed different IoT-based PV systems, Table 12.1 presents a brief comparison between some IoT-based systems proposed in terms of data processing modules and data transmission modules.

### 12.3.1 Hardware

#### *12.3.1.1 Data Processing and Transmission Modules*

In IoT-based PV systems, a wide variety of data processing and transmission modules are available for consideration. However, the existing systems suggested in the literature are based on low-cost and low-power modules. This section describes the main data processing and transmission modules commonly used in developing IoT-based PV systems, focusing on their

TABLE 12.1 A Brief Comparison between Some IoT-Based Systems Proposed in Related Work

| References | Year | Data Processing and Transmission Modules | Communication Protocols | IoT Platforms | Achievements |
|---|---|---|---|---|---|
| Andreoni Lopez et al. (2012) | 2012 | DSPIC 30F3013 + Zigbee modules | Zigbee | LabView | Described an implementation of a wireless remote monitoring and control system of a solar photovoltaic distributed generator for microgrid applications. |
| Ngo et al. (2015) | 2015 | Beaglebone black + GSM + Arduino UNO | SMS | Web host application and Site web | Developed a real-time energy monitoring system for grid-tied photovoltaic installations. |
| Touati et al. (2016) | 2016 | Microcontroller Atmega32 + long-range XBee Pro wireless modules | XBee | LabView | Provided a customized and cost-effective solution for monitoring and predicting PV system performance. |
| Adhya et al. (2016) | 2016 | PIC18F46K22 + GSM/GPRS | GSM/GPRS | Web page | Implementation of a new cost-effective methodology based on IoT to remotely monitor a solar photovoltaic plant. |
| Han et al. (2017) | 2017 | PLC module + PLC gateway + Data logger | SUN module and internet | Web page | Proposed a low-cost power line communication module to monitor four PV panels simultaneously. |
| Pereira et al. (2018) | 2018 | PIC18Fxx5X + Raspberry Pi | HTTP | Online Web Monitor | Implemented and tested a renewable energy monitoring system, with a new concept on data acquisition and transmission systems applied to real-time cloud monitoring of a decentralized photovoltaic plant. |
| Samara and Natsheh (2019) | 2019 | Atmega2560 microcontroller + ESP8266 WiFi module | Not mentioned | Not mentioned | Presented a novel real-time monitoring system utilizing a small but efficient artificial neural network (ANN) that is adequate to run on a low-cost system. |

*(Continued)*

TABLE 12.1 A Brief Comparison between Some IoT-Based Systems Proposed in Related Work *(Continued)*

| References | Year | Data Processing and Transmission Modules | Communication Protocols | IoT Platforms | Achievements |
|---|---|---|---|---|---|
| Rouibah et al. (2019) | 2019 | Arduino Mega2560 + ESP8266-01 | HTTP | Web page | Designed a low-cost monitoring system for the maximum power point tracking (MPPT) in a PV system. |
| Pereira et al. (2019) | 2019 | ESP32 and ESP8266 | Not mentioned | Web page | Proposed IoT-embedded systems, which are based on free software, allowing online distribution, free usage, and communication with a server in the cloud wirelessly via WiFi. |
| Mudaliar and Sivakumar (2020) | 2020 | Raspberry Pi | HTTP/TCP | Node.js, Grafana and InfluxDB | Designed an IoT-based real-time energy monitoring system for controlling and monitoring of a switchgear industry. |
| Aghenta et al. (2020) | 2020 | Raspberry Pi + ESP32 | MQTT | Thingsboard | Presented a design and implementation of a low-cost, open-source SCADA system based on the Internet of Things (IoT) SCADA architecture. |
| Wedashwara et al. (2020) | 2020 | Arduino nano + ESP8266 + Raspberry Pi | Not mentioned | Web-based interfaces | Proposed a smart small-scale solar energy planning using the IoT by collaborating wireless sensor network as a data collector and evolutionary fuzzy association rule mining as a decision support system. |
| Cheddadi et al. (2020) | 2020 | ESP32 board | HTTP | Grafana, InfluxDB | Proposed a cost-effective and open-source IoT solution that collects intelligently and monitors in real-time the produced power and environmental conditions of solar stations. |
| De Arquer Fernández et al. (2021) | 2021 | Moxa UC-2112 PC with Kura installed as IoT Gateway | MQTT and MODBUS protocols | Eclipse Kura, Eclipse Kapua, Grafana platform | Proposed an IoT solution for photovoltaic plants monitoring based entirely on open-source software. |

*(Continued)*

TABLE 12.1 A Brief Comparison between Some IoT-Based Systems Proposed in Related Work *(Continued)*

| References | Year | Data Processing and Transmission Modules | Communication Protocols | IoT Platforms | Achievements |
|---|---|---|---|---|---|
| Gupta et al. (2021) | 2021 | NodeMCU ESP8266 modules | Not mentioned | ThingSpeak | Proposed a data acquisition system for sensing and monitoring of the PV system. |
| Boubakr et al. (2022) | 2022 | ESP32 Dev board | Not mentioned | ThingSpeak | Presented an IoT-based intelligent real-time monitoring system for PV power systems, including an MPPT controller. |
| Mallahi et al. (2022) | 2022 | Arduino Mega2560 + HC-05 Bluetooth module | Bluetooth | Mobile App | Developed a photovoltaic tracking system and also a cleaning system connected to a smart board. |
| Jabbar et al. (2022) | 2022 | Arduino Uno R3 + SX1278-LoRa shield + SX1301 LoRaWAN gateway | LoRa | The Things Network and AllThingsTalk Maker | Designed of a smart energy meter based on LoRaWAN for monitoring PV systems over IoT. |
| Sheba et al. (2023) | 2023 | ESP8266 board | MQTT and Modbus protocols | ThingSpeak | Proposed an IoT infrastructure solution based on a newly designed low-cost microcontroller-based IoT remote terminal unit to integrate new, old, and conventional sites of existing grids with smart grids. |
| Ul Mehmood et al. (2023) | 2023 | NodeMCU | MQTT protocol | NET Framework using the C# programming language | Suggested a cloud-based IoT solution for remotely monitoring the amount of soiling on PV panels using an ANN. |
| Radia et al. (2023) | 2023 | Raspberry Pi + ESP8266 | MQTT | Node-Red, Mosquitto broker, Grafana, and InfluxDB | Introduce a low-cost wireless monitoring system for PV modules that employs NodeMCU boards, Raspberry Pi, and IoT technologies, using free and open-source software packages including MQTT Mosquitto broker, InfluxDB, Node-RED, and Grafana. |

*(Continued)*

TABLE 12.1 A Brief Comparison between Some IoT-Based Systems Proposed in Related Work *(Continued)*

| References | Year | Data Processing and Transmission Modules | Communication Protocols | IoT Platforms | Achievements |
|---|---|---|---|---|---|
| Asnil et al. (2023) | 2023 | Arduino UNO + ESP8266 | Not mentioned | Thinger.io platform | Discussed the design of a photovoltaic parameter monitoring system that uses the IoT technology to monitor in real-time. |
| Supo et al. (2023) | 2023 | Raspberry Pi | MQTT | Node-RED and Azure IoT Central | Designed and implemented a monitoring system for a grid-connected photovoltaic system of 12.7 kW of installed power based on an IoT solution. |
| Muñiz et al. (2024) | 2024 | Raspberry Pi | HTTP, MQTT, and Modbus | Amazon Web Services (AWS) | Presented an integrated energy management solution for solar-powered smart buildings, combining a multifaceted physical system with advanced IoT and cloud-based control systems. |

properties and limitations, such as processing power, memory capacity, and power requirements (Ansari et al., 2020; Islam et al., 2023).

12.3.1.1.1 Arduino Arduino is an open-source, low-cost, and low-power electronics platform based on simple hardware design and software programming codes. It offers several versions of boards such as UNO (Figure 12.4), Mega, Micro, Pro, and Nano as hardware platforms, which include a microcontroller (Atmel AVR series), digital and analog input/output pins USB interface for programming and communication, power jack, and more. It also provides an Arduino Integrated Development Environment (IDE) for programming, compiling, and uploading code to the board, which provides a simplified programming language (C/C++). Arduino is currently a popular open-source platform used in electronics, robotics, and IoT projects. Arduino boards can be expanded using shields, which are additional circuit boards that plug into the main Arduino board to add specific functionality such as Ethernet Shield, Relay Shield, CAN-BUS Shield, GSM/GPRS Shield, WiFi Shield (*Arduino.cc*, 2024).

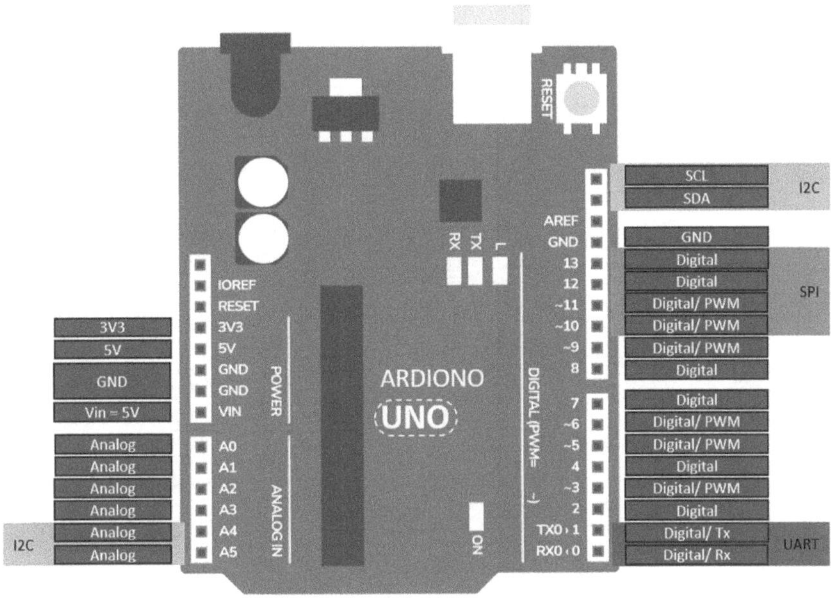


FIGURE 12.4 Pinout diagram of Arduino UNO (Arduino.cc).

An IoT application for real-time monitoring of Solar Home Systems based on Arduino UNO with an Ethernet shield, which is connected to the internet using 3G, from the TLMR3020 router was reported by Lopez-Vargas et al. (2019). Rouibah et al. (2019) designed a low-cost monitoring system for MPPT in a PV system. The authors used an Arduino Mega2560 and a WiFi module ESP8266-01 as data acquisition and transmission modules. Muthukumar et al. (2023) proposed a dual-axis solar tracker using Arduino Nano to control the solar panel's direction based on the sun's position, while the performances of the system are monitored using Node MCU. Omidi et al. (2023) presented a low-cost, low-power, open-source SCADA system for hybrid renewable energy systems. This system is based on Arduino Mega2560 and a Wio terminal as remote terminal units.

12.3.1.1.2 ESP8266 and ESP32 ESP8266 is a system-on-chip (SOC) with built-in WiFi capabilities, produced by Espressif Systems. it integrates a 32-bit Tensilica microcontroller, digital peripheral interfaces, and various components for wireless connectivity and power management into a compact package (Mesquita et al., 2018). It is often used to enable internet connectivity in various electronic projects, in particular in IoT applications. ESP32 (Figure 12.5) is

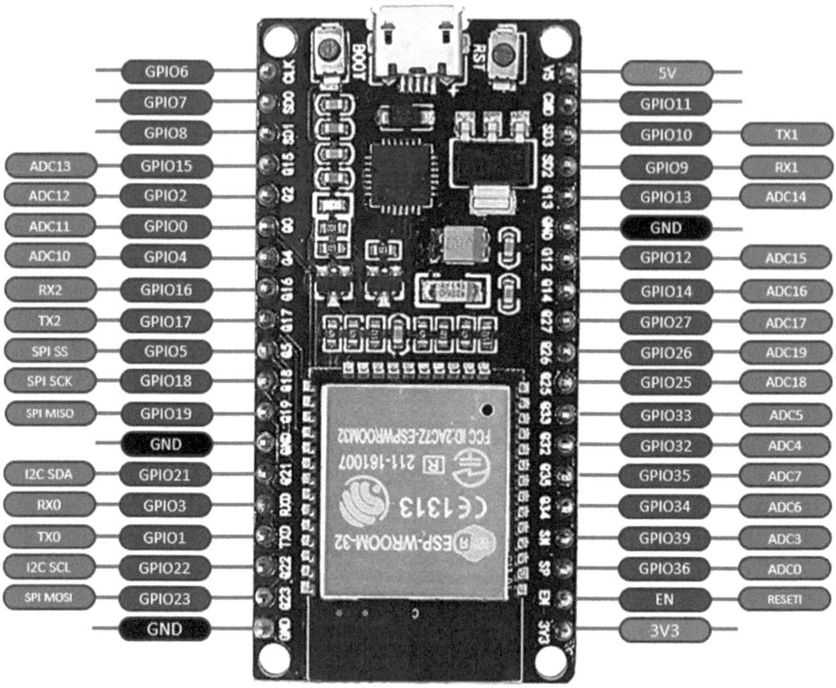


FIGURE 12.5 Pinout diagram for ESP32.

the successor of ESP8266 (Figure 12.6); it comes with another CPU core, faster WiFi, more GPIOs, and supports Bluetooth 4.2 and Bluetooth low energy, higher performance and more memory (RAM and flash), and more (2024c).

Both the ESP8266 and ESP32 are compatible with the Arduino IDE, as well as several other development platforms like PlatformIO and Espressif's own ESP32 IoT Development Framework (ESP-IDF). ESP8266 and ESP32 are widely used in IoT projects to build connected and smart devices due to their low-cost, small size, powerful features, and wireless capabilities; they are most popular in IoT-based PV systems. In this context, Pereira et al. (2019) proposed an IoT embedded systems for three grid-connected PV plants, using ESP8266 and ESP32 as data processing and transmission modules. Baig et al. (2021) proposed a system for real-time data acquisition, monitoring, and control of self-generated energy at a remote location is presented. The authors used an ESP32-S2 microcontroller to collect and process energy data. Lahabar et al. (2023) presented an IoT PV monitoring system that collects and analyzes solar energy parameters to predict performance and produce sustainable energy. This

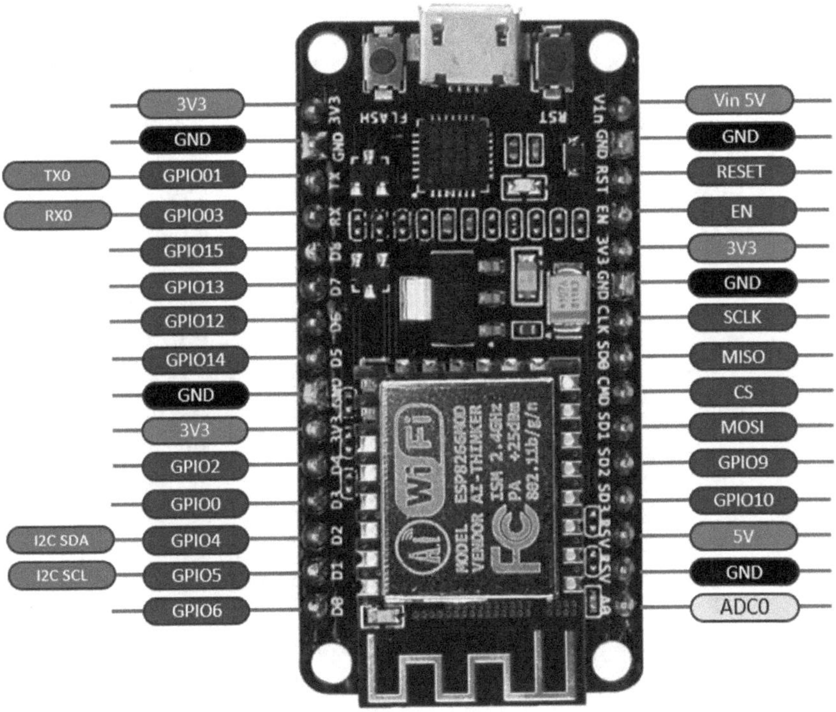


FIGURE 12.6 Pinout diagram for ESP8266.

solar monitoring system comprises a small WiFi module ESP32 with an embedded Arduino Nano.

12.3.1.1.3 Raspberry Pi The Raspberry Pi is a low-power, miniature computer developed by the Raspberry Pi Foundation. It serves as a cost-effective and robust platform for constructing a diverse array of IoT projects and smart devices. There are three series of Raspberry Pi and several generations of each, with varying specifications and capabilities. Setting up the Raspberry Pi involves inserting an SD card into the provided slot. This SD card should contain the necessary operating system for booting the computer. While there are numerous operating systems compatible with the Raspberry Pi, it is recommended to choose the official one, known as Raspberry Pi OS or Raspbian, which is based on Linux. Raspberry Pi is widely used in IoT systems due to its ease of use, high performance, and various benefits and capabilities. Pinout diagram for a Raspberry Pi 4 is presented in Figure 12.7.

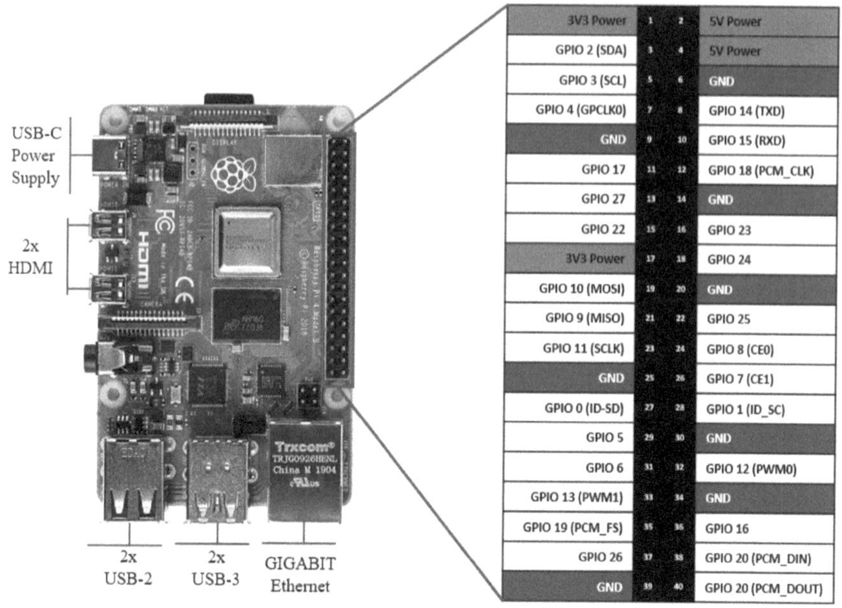


FIGURE 12.7 Pinout diagram for Raspberry Pi 4 (Raspberrypi.com).

Ranjit et al. (2018) proposed an IoT-based PV system health monitoring system for an individual PV panel. This system comprises of a Raspberry Pi Zero, four thermocouples (K-Type and Adafruit MAX31855), and a voltage/current sensor (INA219). Gimeno-Sales et al. (2020) have used a Raspberry Pi as the data gateway stage of a monitoring system applied to a small-scale prototype of an autonomous solar water pumping system. A Raspberry Pi 4 module is used as a computer board to control an intelligent energy management system for conventional grid-connected solar system in (Tajjour and Chandel, 2023). Supo et al. (2023) designed and implemented a monitoring system for a grid-connected PV system based on an IoT solution and Raspberry Pi 4.

After a brief overview of the previous data processing and transmission modules based on their manufacturer descriptions and applications in the literature. Table 12.2 shows a comparison of some different characteristics and features of the previous data processing and transmission modules, that are used in IoT-based PV systems, such as processors, memory, and RAM. As a result of this comparison, the choice of the data processing and transmission modules for IoT-based PV systems depends on the specific needs and constraints of the project and applications in

TABLE 12.2 Comparison of Some Different Characteristics and Features of Various Data Processing and Transmission Modules

| **Features** | **Arduino (UNO)** | **ESP8266** | **ESP32** | **Raspberry Pi (4)** |
|---|---|---|---|---|
| Processor | ATmega328P | Tensilica Xtensa Single-Core 32-bit LX106 | Tensilica Xtensa Dual-Core 32-bit LX6 | 64-bit quad-core ARM Cortex-A72 |
| Storage memory | 32 KB | 16 Mb | Up to 16 MB | SD card |
| RAM | 2 KB | 160 KB | 512 KB | 8 GB LPDDR4 |
| Clock frequency | 16 MHz | 80 MHz | 240 MHz | 1.5 GHz |
| Operating voltage | 5V | 12.3V | 12.3 V | 5 V |
| Portes | • 14 digital I/O pins (6 provide PWM output)<br>• 6 analog input pins<br>• USB B Connector<br>• Power jack 2.1 × 5.5 mm | • 16 GPIOs (4 provide PWM output)<br>• 1 analog input pin<br>• USB-TTL based on CP2102 | • 34 GPIOs (28 PWM channels)<br>• 10 capacitive-sensing GPIOs<br>• 1 analog input pin<br>• 1 host (SD/eMMC/SDIO)<br>• USB-TTL based on CP2102 | • 40 GPIOs<br>• 2 micro-HDMI (supports up to 4 Kp60)<br>• 2 USB 12.0<br>• 2 USB 2.0<br>• Gigabit Ethernet<br>• Camera Serial Interface (CSI)<br>• Display Serial Interface (DSI)<br>• 12.5 mm analog audio-video jack<br>• USB-C |
| Communication protocols | 1 × UART, 1 × SPI and 1 × I2C | 2 × UART, 3 × SPI, 1 × I2C and 1 × I2S | 3 × UART, 4 × SPI, 2 × I2 C and 2 ×I2 S | 6 × UART, 6 × I2C, 5 × SPI, 1 × SDIO, 1 × DPI, 1× PCM, 2 × PWM channels, and 3 × GPCLK outputs |
| ADC | 6 | 1 | 18 × 12 bits ADC channels | 0 |
| DAC | 0 | 0 | 2 × 8 bits DAC channels | 0 |
| WiFi | No | 802.11 b and 802.11 g | • 802.11 b/g/n,<br>• 802.11 n (2.4 GHz) up to 150 Mbps | 802.11b/g/n/ac |
| Bluetooth | No | No | Bluetooth v 4.2 and BLE | Bluetooth 5.0 and BLE |
| Programing language | C/C++ | C/C++ | C/C++, MicroPython, JavaScript | Java, Python, C, HTML, C++, etc. |
| Cost | Very Low-cost | Low-cost | Low-cost | Expensive |

terms of complexity and functionality. Additionally, these modules can be classified according to their processing power, memory capacity, cost, connectivity capabilities (WiFi or Bluetooth), power consumption, available communication interfaces, analog-digital converters (ADC), digital-analog converters (DAC), programming languages, and General-Purpose Input/Output (GPIO) pins.

#### *12.3.1.2 Sensors*

According to the IEC 61724 standard (IEC 61724 standard 1998), a comparison is carried out between different PV installations of varying sizes, operating in different climates, and serving different energy needs. This study involves examining monitoring parameters customized for each system to evaluate their performance and efficiency. A list of required parameters for each class is given in Table 12.3. There are several types of sensors that collect and measure these parameters, each on is based on different techniques and physical principles. Table 12.4 presents the various monitoring parameters and sensors used in different related applications of IoT-based PV systems.

### 12.3.2 Software

In this section, we present some of the most popular IoT software platforms used to monitor and control PV systems, which are divided into two categories: commercial platforms including Microsoft Azure platform, Amazon Web Services (AWS) platform, IBM Cloud platform, etc. These platforms are designed for professional and industrial use, which offer advanced options and functionalities, and professional support services; they require a subscription or license the use. However, free and open-source ones are usually available for free or under open-source licenses

TABLE 12.3 Monitoring Parameters Required for Different Classes

| Class of Monitoring Parameters | Required Parameters |
|---|---|
| Utility grid | Grid voltage, current from grid, current to grid, power from grid, power to grid, and grid impedance. |
| Photovoltaic array | Output voltage, output current, output power, and output energy. |
| Energy storage | Operating voltage, current from storage, current to storage, power from storage, and power to storage. |
| Load | Output voltage, output current, and output power. |
| Meteorology | Total irradiance, ambient temperature, module temperature, wind speed and direction (optional), humidity (optional), and pressure (optional). |

TABLE 12.4 Various Monitoring Parameters and Sensors Used in Different Related Applications of IoT-Based PV Systems

| Reference | Monitoring Parameters | Sensors Used | Application |
|---|---|---|---|
| Rouibah et al. (2019) | • PV current,<br>• PV voltage. | • ACS712<br>• Voltage sensor LM24 | Designed a low-cost monitoring system for the MPPT in a PV system. |
| Gupta et al. (2021) | • PV current and voltage,<br>• Ambient temperature and humidity,<br>• Wind speed,<br>• Dust density,<br>• Solar irradiance. | • INA219<br>• DHT22 & LM35<br>• INSPEED VORTEX sensor<br>• GP2Y1010AU0F sensor<br>• Pyranometer | Designed a data acquisition system for sensing and monitoring of the PV system. |
| Samuel and Rajagopal (2022) | • Dust density,<br>• Voltage and Current of solar panel,<br>• Temperature and Humidity | • Dust sensors (GP2Y1010AU0F),<br>• INA226,<br>• DHT11 | Proposed a solar panel cleaning system based on dust monitoring using IoT. |
| Boubakr et al. (2022) | • PV current,<br>• PV voltage,<br>• Ambient temperature and humidity,<br>• Irradiance incident on PV. | • ACS712 40/30<br>• Voltage divider circuit<br>• DHT11 BH1750 | Presented an IoT-based intelligent real-time monitoring system for PV power systems, including an MPPT controller. |
| Jabbar et al. (2022) | • Load current,<br>• PV voltage,<br>• Ambient temperature and humidity,<br>• Light intensity. | • ACS712<br>• B25<br>• DHT11<br>• LDR | Designed a smart energy meter based on LoRaWAN for monitoring stand-alone PV systems using IoT. |
| Ul Mehmood et al. (2023) | • PV current,<br>• PV voltage,<br>• Ambient temperature,<br>• Light intensity. | • ACS712<br>• Voltage divider circuit<br>• DS18B20<br>• MAX44009 | Suggested a cloud-based IoT solution for remotely monitoring the amount of soiling on PV panels using an ANN. |
| Radia et al. (2023) | • PV voltage,<br>• PV current,<br>• PV module Temperature,<br>• Irradiance,<br>• Ambient temperature and Relative humidity | • Resistive divider by 4.6<br>• ACS712 – 5A<br>• LM35<br>• Pyranometer LP02<br>• DHT11 | Proposed a monitoring technique for hybrid renewable energy-based power sources through WiFi. |

such as Thingsboard, ThingSpeak, Eclipse, Node-Red, Thinger.io, and others (Domínguez-Bolaño et al., 2022); they may have limited options and functionalities. These open-source IoT platforms offer flexibility, scalability, and low cost for deploying PV monitoring and management IoT solutions.

- **Microsoft Azure platform:** Microsoft Azure IoT platform is a combination of Microsoft-managed cloud services, edge components, and SDKs that enable you to connect, monitor, and control your IoT systems at scale (https://microsoft.azure, 2024). Supo et al. (2023) proposed an IoT-based monitoring system using the Microsoft cloud platform for visualize in real-time and store the data of a grid-connected PV system of 12.7 kW.
- **AWS platform:** AWS platform is another powerful, extensive and continually developing cloud computing platform. It enables secure connection, data storage, web visualization, data analysis, and remote control of the IoT systems, allowing users the ability to monitor and manage their IoT systems (https://aws.amazon.com, 2024). An integrated energy management solution for solar-powered smart buildings, incorporating advanced IoT and cloud-based control systems alongside AWS's IoT platforms, is presented in Muñiz et al., (2024).
- **Thingsboard platform:** Thingsboard is an open-source IoT platform, that provides data collection, processing, visualization, rule engine capabilities, and device management (https://*thingsboard.io*, 2024). The Thingsboard IoT server is used as the master terminal unit in an IoT-based SCADA system for PV systems designed by Aghenta et al., (2020).
- **Thinger.io IoT platform:** Thinger.io is a cloud IoT open-source platform that offers all the tools you need to easily prototype, scale, and manage connected devices. It provides features such as device management, data storage, visualization, and analysis, real-time monitoring, and remote control of connected devices (https://thinger.io/, 2024). In addition, a monitoring system based on the Thinger.io platform was used to monitor in real-time the PV parameters proposed in this work (Asnil et al., 2023).
- **Node-Red platform:** Node-Red is an open-source flow-based, low-code development tool for visual IoT programming. It allows users to connect together hardware devices, APIs, and online services in new

and interesting ways (https://nodered.org/, 2024). Radia et al. (2023) and Supo et al. (2023) both proposed a Node-Red-based monitoring system for a grid-connected PV system and PV modules, respectively.

- **ThingSpeak platform:** ThingSpeak is an open-source IoT platform with MATLAB analytics that allows users to collect, visualize, and analyze real-time data streams in the cloud (https://thingspeak.com/, 2024). Gupta et al. (2021) proposed a ThingSpeak-based low-cost data acquisition system for monitoring and gathering operational information of the PV system. Boubakr et al. (2022) presented an IoT-based intelligent real-time monitoring system for PV power systems, including an MPPT controller. Indeed, the authors used ThingSpeak as a cloud server to visualize, analyze, and store the collected data.

## 12.4 COMMUNICATION NETWORK TECHNOLOGIES FOR IoT PV SYSTEMS

### 12.4.1 Infrastructure Communication Technologies

Infrastructure communication technologies are essential for connecting networks and systems, ensuring consistency and standardization in message transmission. These technologies vary widely in terms of range, memory usage, energy requirements, installation costs, etc. Such as WiFi, Bluetooth/BLE, Zigbee, LoRa/LoRaWAN, GSM/GPRS, and Zifox (Kuzlu et al., 2015). A comparison of the main characteristics of some IoT communication technologies used in IoT-based PV systems is presented in Table 12.5 (Patchmuthu et al., 2022; Taghizad-Tavana et al., 2022).

TABLE 12.5 Comparison of IoT Communication Technologies Used in IoT-based PV Systems

| Communication Technologies | Network Type | Frequency | Range | Data Rate | Power Consumption |
|---|---|---|---|---|---|
| WiFi | LAN | 2.4, 5 GHz | 100–250 m | 100–250 Kbps | High |
| Bluetooth/BLE | PAN | 2.4 GHz | 100 m | 1 Mbps | Low |
| Zigbee | PAN | 2.4 GHz | 75 m | 250 Kbps | Very low |
| LoRa/LoRaWAN | LPWAN | 433/868 MHz<br>915 MHz<br>43 MHz | 2–15 km | 50 Kbps | Low |
| GSM/GPRS | WAN | 850 MHz–1.5 GHz | +35 km | 168 Kbps | Very high |

### 12.4.2 Messaging Protocols

There are several messaging protocols that address a variety of application requirements such as Hypertext Transfer Protocol (HTTP), Message Queuing Telemetry Transport (MQTT), Constrained Application Protocol (CoAP), Advanced Message Queuing Protocol (AMQP), and Extensible Messaging and Presence Protocol (XMPP) (Al-Masri et al., 2020; Qays et al., 2023). These protocols differ in terms of performance, functionality, scalability, security, and quality of service (QoS), among other aspects. A basic comparison of some of the most popular messaging protocols used in IoT-based PV systems is presented in Table 12.6 (Al-Fuqaha et al., 2015; Naik, 2017; Nikolov, 2020; Islam et al., 2023).

#### *12.4.2.1 HTTP (Hypertext Transfer Protocol)*

HTTP is an application layer protocol that is widely used for data transmission across the World Wide Web (WWW). It supports the RESTful Web architecture of request/response (Figure 12.8(a)), where a client delivers a request message to a server that generates a response message. HTTP uses Transmission Control Protocol (TCP) as the default transport protocol, which demands a lot of computing resources, making it unsuitable for battery-powered applications. HTTP can be secured using Transport Layer Security (TLS) or its predecessor, Secure Sockets Layer (SSL), to provide encryption, data integrity, and authentication.

TABLE 12.6 Comparison of Some IoT Messaging Protocols

| **Protocols** | **HTTP** | **MQTT** | **CoAP** | **AMQP** | **XMPP** |
|---|---|---|---|---|---|
| Transport | TCP | TCP | UDP | TCP | TCP |
| Transport port | 80/443 | 1883/8883 | 5683/5684 | 5671/5672 | 5222 |
| Message pattern | Request/response | Publish/subscribe | Publish/subscribe and request/response | Publish/subscribe and request/response | Publish/subscribe and request/response |
| Architecture | Client/server | Tree | Tree | Star | Client/server |
| Security | TLS/SSL | TLS | DTLS | SASL, TLS/SSL | SASL/TLS |
| QoS | Based on TCP | 3 level | 2 level | 3 level | None |
| Header size | Undefined | 2 byte | 4 byte | 8 byte | Variable |
| Message size | Undefined | Up to 256 MB | 64 KB | Undefined | Undefined |

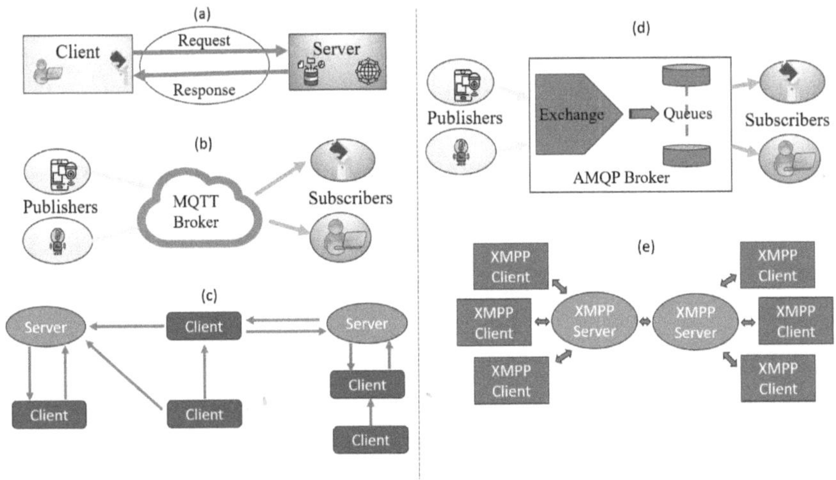


FIGURE 12.8 Messaging protocol architectures: (a) HTTP, (b) MQTT, (c) CoAP, (d) AMQP, and (e) XMPP.

#### *12.4.2.2 MQTT (Message Queuing Telemetry Transport)*

MQTT is a publish/subscribe messaging protocol developed by IBM in 1999, designed for lightweight Machine-to-Machine (M2M) communications and IoT applications. The MQTT architecture is based on the publish/subscribe model (Figure 12.8(b)), consisting of a broker (server) and clients that can be publishers or subscribers. The publisher clients publish the messages on a topic to the broker, while the subscriber clients subscribe to a specific topic to receive published messages. MQTT operates on top of TCP protocol. It delivers messages with three levels of QoS. It supports easy encryption of messages using TLS and authenticate clients using modern authentication protocols.

#### *12.4.2.3 CoAP (Constrained Application Protocol)*

CoAP is a functional and lightweight application layer protocol based on HTTP. The CoAP protocol expands the capabilities of HTTP to satisfy the needs of IoT devices without imposing significant complexity, as most IoT devices have limited power and storage. It is designed in the REST architectural model as shown in Figure 12.8(c) and uses User Datagram Protocol (UDP) or Datagram Transport Layer Security (DTLS) as a transport protocol.

#### *12.4.2.4 AMQP (Advanced Message Queuing Protocol)*

AMQP is another lightweight M2M protocol that supports message-oriented middleware. AMQP is based on both request/response and publish/subscribe architectures (Figure 12.8(d)). Similar to MQTT, it uses topic-based publish-subscribe messaging. Additionally, it provides message queuing, routing, reliability, and security. AMQP supports authentication and encryption using Simple Authentication and Security Layer (SASL) and TLS over TCP as the transport protocol.

#### *12.4.2.5 XMPP (Extensible Messaging and Presence Protocol)*

XMPP is an open standard protocol that allows a wide range of service communication technologies including instant messaging, voice/video calls, multi-party chat, lightweight middleware, collaboration, and generalized routing of Extensible Markup Language (XML) data, to enable the development of real-time applications. XMPP has a decentralized client-server architecture (Figure 12.8(e)). XMPP is based on XML and operates over TCP/IP. Additionally, XMPP provides point-to-point encryption (TLS).

## 12.5 CHALLENGES AND ISSUES OF IoT-BASED PV SYSTEMS

The integration of the IoT into PV systems has made them intelligent, more efficient, accessible, and reliable due to its numerous advantages such as real-time monitoring and control, total system management, energy optimization, and more. Despite these benefits, there are also various challenges and issues that affect the performance of IoT-based PV systems in terms of efficiency, compatibility, scalability, reliability, resilience, complexity, security, and privacy. We provide in this section a brief discussion of some of these challenges faced in designing and implementing IoT in PV systems (Madeti & Singh, 2017; Orumwense & Abo-Al-Ez, 2022; Qays et al., 2023).

### 12.5.1 Compatibility and Interoperability

An IoT-based PV system can be divided into two main systems: (1) a PV system and (2) an IoT system. The PV system comprises PV panels, inverters, converters, batteries, and other essential components. Meanwhile, IoT system consist of multiple sensors with different communication protocols and operating voltages. They also include data processing and transmission modules developed by various manufacturers with different characteristics and capabilities including architectures, storage and processing capacities, communication protocols, and more. The varying characteristics and capabilities of these components and the connected heterogeneous devices exhibit a challenge in ensuring compatibility and interoperability (Al-Masri et al., 2020).

### 12.5.2 Scalability

The increased implementation of IoT systems in PV systems poses scalability challenges, particularly in terms of data handling, transmission, and storage, as well as energy management. Therefore, as the number of IoT devices in a PV system grows, the volume of data generated increases exponentially, which involves difficulty in the collection, processing, transmission, and storage of this enormous amount of data. Additionally, more devices require a lot of energy to power and operate, leading to higher overall energy consumption. Moreover, as the scale increases integration of new IoT devices and technologies with existing PV systems can be complex (Saleem et al., 2019).

### 12.5.3 Security and Privacy

IoT-based PV systems use various communication technologies to establish connections between devices, gateways, and servers, enabling data transmission, storage, and system control. The lack of common communication standards leads to significant security and privacy challenges. Additionally, the heterogeneity of IoT devices complicates efforts to ensure the security and privacy of the system. Security is a major challenge in any IoT-networked system, as it is vulnerable to numerous attacks and threats. These include replay attacks, where an attacker intercepts and retransmits data to create unauthorized actions; Denning-Sacco attacks, which involve compromising session keys to decrypt communications; denial-of-service (DoS) attacks, which overload the system to make it unavailable where the attackers often employ flooding attacks to deplete a system's resources, such as memory, CPU, and bandwidth; password-guessing attacks, where attackers attempt to gain access by systematically guessing passwords; and other (Bedi et al., 2018; Azrour et al., 2021; Babun et al., 2021).

Robust security protocols and frequent updates are essential to limit these risks. IoT-based PV systems can be protected by implementing end-to-end encryption, cryptosecurity algorithms, strong authentication techniques, intrusion detection systems, and more secure firmware updates can help protect. Moreover, using software and hardware security modules can improve the overall security status of the system. Continuous monitoring and proactive security measures are crucial to defend against evolving threats and ensure the reliable operation of IoT-based PV systems.

## 12.6 CONCLUSION

The integration of IoT with PV energy systems has introduced several advancements to the solar field, such as remote monitoring, control and optimization, data collection and analysis, and fault detection. These features enable optimal performance, improved efficiency, ensured reliability, and reduced maintenance costs. This chapter provided a comprehensive overview of IoT systems for solar energy applications. It highlighted the role of IoT in solar energy systems and the particular architectures of IoT-based PV systems. Additionally, it discussed and compared potential hardware and software platforms suitable for IoT-based PV systems, focusing on data processing and transmission modules as well as IoT software platforms in terms of various characteristics and features. Moreover, an investigation is conducted on communication network technologies, including both infrastructure communication technologies and IoT messaging protocols. Despite the numerous advantages of IoT-based PV systems, various challenges and issues affect their performance, including efficiency, compatibility, scalability, reliability, resilience, complexity, security, and privacy. A brief discussion of some of these challenges faced in designing and implementing IoT in PV systems was provided.

## REFERENCES

Abir, S. M. A. A., Anwar, A., Choi, J., & Kayes, A. S. M. (2021). IoT-Enabled Smart Energy Grid: Applications and Challenges. *IEEE Access*, *9*, 50961–50981. https://doi.org/10.1109/ACCESS.2021.3067331

Adhya, S., Saha, D., Das, A., Jana, J., & Saha, H. (2016). An IoT based smart solar photovoltaic remote monitoring and control unit. *2016 2nd International Conference on Control, Instrumentation, Energy & Communication (CIEC)*, 432–436. https://doi.org/10.1109/CIEC.2016.7513793

Aghenta, L. O., & Tariq Iqbal, M. (2020). Design and Implementation of a Low-Cost, Open Source IoT-Based SCADA System Using ESP32 with OLED, ThingsBoard and MQTT Protocol. *AIMS Electronics and Electrical Engineering*, *4*(1), 57–86. https://doi.org/10.3934/ElectrEng.2020.1.57

Ahmad, F. F., Ghenai, C., & Bettayeb, M. (2021). Maximum Power Point Tracking and Photovoltaic Energy Harvesting for Internet of Things: A Comprehensive Review. *Sustainable Energy Technologies and Assessments*, *47*, 101430. https://doi.org/10.1016/j.seta.2021.101430

Al-Fuqaha, A., Guizani, M., Mohammadi, M., Aledhari, M., & Ayyash, M. (2015). Internet of Things: A Survey on Enabling Technologies, Protocols, and Applications. *IEEE Communications Surveys & Tutorials*, *17*(4), 2347–2376. https://doi.org/10.1109/COMST.2015.2444095

Al-Masri, E., Kalyanam, K. R., Batts, J., Kim, J., Singh, S., Vo, T., & Yan, C. (2020). Investigating Messaging Protocols for the Internet of Things (IoT). *IEEE Access*, *8*, 94880–94911. https://doi.org/10.1109/ACCESS.2020.2993363

Amazon. (2024). *Amazon Web Services (AWS)*. Retrieved May 20, 2024, From https://aws.amazon.com/

Andreoni Lopez, M. E., Galdeano Mantinan, F. J., & Molina, M. G. (2012). Implementation of wireless remote monitoring and control of solar photovoltaic (PV) system. *2012 Sixth IEEE/PES Transmission and Distribution: Latin America Conference and Exposition (T&D-LA)*, 1–6. https://doi.org/10.1109/TDC-LA.2012.6319050

Ansari, S., Ayob, A., Hossain Lipu, M. S., Saad, M. H. M., & Hussain, A. (2020). Comparison of the IoT based modules for solar PV environment: A review. *2020 IEEE Student Conference on Research and Development (SCOReD)*, 401–405. https://doi.org/10.1109/SCOReD50371.2020.9250946

Ansari, S., Ayob, A., Lipu, M. S. H., Saad, M. H. M., & Hussain, A. (2021). A Review of Monitoring Technologies for Solar PV Systems Using Data Processing Modules and Transmission Protocols: Progress, Challenges and Prospects. *Sustainability*, *13*(15), 8120. https://doi.org/10.3390/su13158120

Arduino. (2024). *Arduino.cc*. Retrieved May 02, 2024, From https://www.arduino.cc/

Argyrou, M. C., Spanias, C., Marouchos, C. C., Kalogirou, S. A., & Christodoulides, P. (2019). Energy management and modeling of a grid-connected BIPV system with battery energy storage. *2019 54th International Universities Power Engineering Conference (UPEC)*, 1–6. https://doi.org/10.1109/UPEC.2019.8893495

Asnil, A., Krismadinata, K., Husnaini, I., Hazman, H., & Astrid, E. (2023). Real-Time Monitoring System Using IoT for Photovoltaic Parameters. *TEM Journal*, 1316–1322. https://doi.org/10.18421/TEM123-11

Azrour, M., Mabrouki, J., Guezzaz, A., & Kanwal, A. (2021). Internet of Things Security: Challenges and Key Issues. *Security and Communication Networks*, *2021*, 1–11. https://doi.org/10.1155/2021/5533843

Babun, L., Denney, K., Celik, Z. B., McDaniel, P., & Uluagac, A. S. (2021). A Survey on IoT Platforms: Communication, Security, and Privacy Perspectives. *Computer Networks*, *192*, 108040. https://doi.org/10.1016/j.comnet.2021.108040

Baig, M. J. A., Iqbal, M. T., Jamil, M., & Khan, J. (2021). Design and Implementation of an Open-Source IoT and Blockchain-Based Peer-to-Peer Energy Trading Platform Using ESP32-S2, Node-Red and, MQTT Protocol. *Energy Reports*, *7*, 5733–5746. https://doi.org/10.1016/j.egyr.2021.08.190

Bedi, G., Venayagamoorthy, G. K., Singh, R., Brooks, R. R., & Wang, K.-C. (2018). Review of Internet of Things (IoT) in Electric Power and Energy Systems. *IEEE Internet of Things Journal*, *5*(2), 847–870. https://doi.org/10.1109/JIOT.2018.2802704

Boubakr, G., Gu, F., Farhan, L., & Ball, A. (2022). Enhancing Virtual Real-Time Monitoring of Photovoltaic Power Systems Based on the Internet of Things. *Electronics*, *11*(15), 2469. https://doi.org/10.3390/electronics11152469

Caraveo-Cacep, M. A., Vázquez-Medina, R., & Hernández Zavala, A. (2023). A Survey on Low-Cost Development Boards for Applying Cryptography in IoT Systems. *Internet of Things, 22*, 1007412. https://doi.org/10.1016/j.iot.20212.100743

Cheddadi, Y., Cheddadi, H., Cheddadi, F., Errahimi, F., & Es-sbai, N. (2020). Design and Implementation of an Intelligent Low-Cost IoT Solution for Energy Monitoring of Photovoltaic Stations. *SN Applied Sciences, 2*(7), 1165. https://doi.org/10.1007/s42452-020-2997-4

Chinnathambi, N. D., Nagappan, K., Samuel, C. R., & Tamilarasu, K. (2022). Internet of Things-Based Smart Residential Building Energy Management System for a Grid-Connected Solar Photovoltaic-Powered DC Residential Building. *International Journal of Energy Research, 46*(2), 1497–1517. https://doi.org/10.1002/er.7264

De Arquer Fernández, P., Fernández Fernández, M. Á., Carús Candás, J. L., & Arboleya Arboleya, P. (2021). An IoT Open Source Platform for Photovoltaic Plants Supervision. *International Journal of Electrical Power & Energy Systems, 125*, 106540. https://doi.org/10.1016/j.ijepes.2020.106540

Destek, M. A., & Sinha, A. (2020). Renewable, Non-Renewable Energy Consumption, Economic Growth, Trade Openness and Ecological Footprint: Evidence from Organisation for Economic Co-Operation and Development Countries. *Journal of Cleaner Production, 242*, 118537. https://doi.org/10.1016/j.jclepro.2019.118537

Domínguez-Bolaño, T., Campos, O., Barral, V., Escudero, C. J., & García-Naya, J. A. (2022). An Overview of IoT Architectures, Technologies, and Existing Open-Source Projects. *Internet of Things, 20*, 100626. https://doi.org/10.1016/j.iot.2022.100626

Espressif. (2024). *Espressif.* Retrieved May 03, 2024, From https://www.espressif.com/

Fathima, A. H., & Palanisamy, K. (2015). Optimization in Microgrids with Hybrid Energy Systems – A Review. *Renewable and Sustainable Energy Reviews, 45*, 431–446. https://doi.org/10.1016/j.rser.2015.01.059

Gimeno-Sales, F. J., Orts-Grau, S., Escribá-Aparisi, A., González-Altozano, P., Balbastre-Peralta, I., Martínez-Márquez, C. I., Gasque, M., & Seguí-Chilet, S. (2020). PV Monitoring System for a Water Pumping Scheme with a Lithium-Ion Battery Using Free Open-Source Software and IoT Technologies. *Sustainability, 12*(24), 10651. https://doi.org/10.3390/su122410651

Güney, T. (2019). Renewable Energy, Non-Renewable Energy and Sustainable Development. *International Journal of Sustainable Development & World Ecology, 26*(5), 389–397. https://doi.org/10.1080/13504509.2019.1595214

Gupta, V., Sharma, M., Pachauri, R. K., & Babu, K. N. D. (2021). A Low-Cost Real-Time IOT Enabled Data Acquisition System for Monitoring of PV System. *Energy Sources, Part A: Recovery, Utilization, and Environmental Effects, 43*(20), 2529–25412. https://doi.org/10.1080/15567036.2020.1844351

Han, J., Jeong, J.-D., Lee, I., & Kim, S.-H. (2017). Low-Cost Monitoring of Photovoltaic Systems at Panel Level in Residential Homes Based on Power Line Communication. *IEEE Transactions on Consumer Electronics, 63*(4), 435–441. https://doi.org/10.1109/TCE.2017.015074

International Electrotechnical Commission. (1998). *IEC 61724: Photovoltaic system performance monitoring-Guidelines for measurement, data exchange and analysis* (1st ed). Geneva: International Electrotechnical Commission.

Islam, Md. M., Nooruddin, S., Karray, F., & Muhammad, G. (2023). Internet of Things: Device Capabilities, Architectures, Protocols, and Smart Applications in Healthcare Domain. *IEEE Internet of Things Journal, 10*(4), 3611–3641. https://doi.org/10.1109/JIOT.2022.3228795

Jabbar, W. A., Annathurai, S., A. Rahim, T. A., & Mohd Fauzi, M. F. (2022). Smart Energy Meter Based on a Long-Range Wide-Area Network for a Stand-Alone Photovoltaic System. *Expert Systems with Applications, 197*, 1167012. https://doi.org/10.1016/j.eswa.2022.116703

Kabir, E., Kumar, P., Kumar, S., Adelodun, A. A., & Kim, K.-H. (2018). Solar Energy: Potential and Future Prospects. *Renewable and Sustainable Energy Reviews, 82*, 894–900. https://doi.org/10.1016/j.rser.2017.09.094

Khan, R., Khan, S. U., Zaheer, R., & Khan, S. (2012). Future internet: The Internet of Things architecture, possible applications and key challenges. *2012 10th International Conference on Frontiers of Information Technology*, 257–260. https://doi.org/10.1109/FIT.2012.53

Kirmani, S., Mazid, A., Khan, I. A., & Abid, M. (2022). A Survey on IoT-Enabled Smart Grids: Technologies, Architectures, Applications, and Challenges. *Sustainability, 15*(1), 717. https://doi.org/10.3390/su15010717

Kumar, N. M., Atluri, K., & Palaparthi, S. (2018). Internet of Things (IoT) in photovoltaic systems. *2018 National Power Engineering Conference (NPEC)*, 1–4. https://doi.org/10.1109/NPEC.2018.8476807

Kumar, N. M., Dash, A., & Singh, N. K. (2018). Internet of Things (IoT): An opportunity for energy-food-water Nexus. *2018 International Conference on Power Energy, Environment and Intelligent Control (PEEIC)*, 68–72. https://doi.org/10.1109/PEEIC.2018.8665632

Kuzlu, M., Pipattanasomporn, M., & Rahman, S. (2015). Review of communication technologies for smart homes/building applications. *2015 IEEE Innovative Smart Grid Technologies - Asia (ISGT ASIA)*, 1–6. https://doi.org/10.1109/ISGT-Asia.2015.7437036

Lahabar, S., Barapatre, A., & Heda, L. (2023). IOT – Powered Smart Photovoltaic Charge Controller. *International Journal of Electrical and Electronics Engineering, 10*(1), 219–225. https://doi.org/10.14445/23488379/IJEEE-V10I1P121

Lopez-Vargas, A., Fuentes, M., & Vivar, M. (2019). IoT Application for Real-Time Monitoring of Solar Home Systems Based on Arduino™ With 3G Connectivity. *IEEE Sensors Journal, 19*(2), 679–691. https://doi.org/10.1109/JSEN.2018.2876635

Madeti, S. R., & Singh, S. N. (2017). Monitoring System for Photovoltaic Plants: A Review. *Renewable and Sustainable Energy Reviews, 67*, 1180–1207. https://doi.org/10.1016/j.rser.2016.09.088

Mallahi, F., Mohamed, M., & Shaker, Y. (2022). Integration of Solar Energy Supply on Smart Distribution Board Based on IoT System. *Designs, 6*(6), 118. https://doi.org/10.3390/designs6060118

Mesquita, J., Guimaraes, D., Pereira, C., Santos, F., & Almeida, L. (2018). Assessing the ESP8266 WiFi module for the Internet of Things. *2018 IEEE 23rd International Conference on Emerging Technologies and Factory Automation (ETFA)*, 784–791. https://doi.org/10.1109/ETFA.2018.8502562

Microsoft. (2024). *Microsoft Azure IoT Platform*. Retrieved May 15, 2024, From https://azure.microsoft.com/,

Mokhtari, G., Anvari-Moghaddam, A., & Zhang, Q. (2019). A New Layered Architecture for Future Big Data-Driven Smart Homes. *IEEE Access, 7*, 19002–19012. https://doi.org/10.1109/ACCESS.2019.2896403

Mudaliar, M. D., & Sivakumar, N. (2020). IoT Based Real Time Energy Monitoring System Using Raspberry Pi. *Internet of Things, 12*, 100292. https://doi.org/10.1016/j.iot.2020.100292

Muñiz, R., Del Coso, R., Nuño, F., Villegas, P. J., Álvarez, D., & Martínez, J. A. (2024). Solar-Powered Smart Buildings: Integrated Energy Management Solution for IoT-Enabled Sustainability. *Electronics, 13*(2), 317. https://doi.org/10.3390/electronics13020317

Muthukumar, P., Manikandan, S., Muniraj, R., Jarin, T., & Sebi, A. (2023). Energy Efficient Dual Axis Solar Tracking System Using IOT. *Measurement: Sensors, 28*, 100825. https://doi.org/10.1016/j.measen.20212.100825

Naik, N. (2017). Choice of effective messaging protocols for IoT systems: MQTT, CoAP, AMQP and HTTP. *2017 IEEE International Systems Engineering Symposium (ISSE)*, 1–7. https://doi.org/10.1109/SysEng.2017.8088251

Nathaniel, S. P., & Iheonu, C. O. (2019). Carbon Dioxide Abatement in Africa: The Role of Renewable and Non-Renewable Energy Consumption. *Science of the Total Environment, 679*, 337–345. https://doi.org/10.1016/j.scitotenv.2019.05.011

Nath, D. C., Kundu, I., Sharma, A., Shivhare, P., Afzal, A., Soudagar, M. E. M., & Park, S. G. (2023). Internet of Things Integrated with Solar Energy Applications: A State-of-the-Art Review. *Environment, Development and Sustainability*. https://doi.org/10.1007/s10668-023-03691-2

Ngo, G. C., Floriza, J. K. I., Creayla, C. M. C., Garcia, F. C. C., & Macabebe, E. Q. B. (2015). Real-time energy monitoring system for grid-tied Photovoltaic installations. *TENCON 2015 - 2015 IEEE Region 10 Conference*, 1–4. https://doi.org/10.1109/TENCON.2015.7372784

Nikolov, N. (2020). Research of MQTT, CoAP, HTTP and XMPP IoT communication protocols for embedded systems. *2020 XXIX International Scientific Conference Electronics (ET)*, 1–4. https://doi.org/10.1109/ET50336.2020.9238208

Nižetić, S., Šolić, P., López-de-Ipiña González-de-Artaza, D., & Patrono, L. (2020). Internet of Things (IoT): Opportunities, Issues and Challenges towards a Smart and Sustainable Future. *Journal of Cleaner Production, 274*, 122877. https://doi.org/10.1016/j.jclepro.2020.122877

Node-Red. (2024). *Node-Red IoT tool*. Retrieved May 21, 2024, From https://nodered.org/

Omidi, S. A., Baig, M. J. A., & Iqbal, M. T. (2023). Design and Implementation of Node-Red Based Open-Source SCADA Architecture for a Hybrid Power System. *Energies, 16*(5), 2092. https://doi.org/10.3390/en16052092

Orumwense, E. F., & Abo-Al-Ez, K. (2022). Internet of Things for Smart Energy Systems: A Review on Its Applications, Challenges and Future Trends. *AIMS Electronics and Electrical Engineering*, *7*(1), 50–74. https://doi.org/10.3934/electreng.2023004

Østergaard, P. A., Duic, N., Noorollahi, Y., & Kalogirou, S. (2022). Renewable Energy for Sustainable Development. *Renewable Energy*, *199*, 1145–1152. https://doi.org/10.1016/j.renene.2022.09.065

Panagopoulos, O., & Argiriou, A. A. (2022). Low-Cost Data Acquisition System for Solar Thermal Collectors. *Electronics*, *11*(6), 934. https://doi.org/10.3390/electronics11060934

Patchmuthu, R. K., Wan, A. T., & Suhaili, W. S. (2022). Exploring Data Security and Privacy Issues in Internet of Things Based on Five-Layer Architecture. *International Journal of Communication Networks and Information Security (IJCNIS)*, *12*(1). https://doi.org/10.17762/ijcnis.v12i1.4345

Pereira, R. I. S., Dupont, I. M., Carvalho, P. C. M., & Jucá, S. C. S. (2018). IoT Embedded Linux System Based on Raspberry Pi Applied to Real-Time Cloud Monitoring of a Decentralized Photovoltaic Plant. *Measurement*, *114*, 286–297. https://doi.org/10.1016/j.measurement.2017.09.033

Pereira, R. I. S., Jucá, S. C. S., & Carvalho, P. C. M. (2019). IoT Embedded Systems Network and Sensors Signal Conditioning Applied to Decentralized Photovoltaic Plants. *Measurement*, *142*, 195–212. https://doi.org/10.1016/j.measurement.2019.04.085

Qays, Md. O., Ahmad, I., Abu-Siada, A., Hossain, Md. L., & Yasmin, F. (2023). Key Communication Technologies, Applications, Protocols and Future Guides for IoT-Assisted Smart Grid Systems: A Review. *Energy Reports*, *9*, 2440–2452. https://doi.org/10.1016/j.egyr.20212.01.085

Radia, M. A. A., Nimr, M. K. E., & Atlam, A. S. (2023). IoT-Based Wireless Data Acquisition and Control System for Photovoltaic Module Performance Analysis. *E-Prime - Advances in Electrical Engineering, Electronics and Energy*, *6*, 100348. https://doi.org/10.1016/j.prime.20212.100348

Rahman, M. M., Selvaraj, J., Rahim, N. A., & Hasanuzzaman, M. (2018). Global Modern Monitoring Systems for PV Based Power Generation: A Review. *Renewable and Sustainable Energy Reviews*, *82*, 4142–4158. https://doi.org/10.1016/j.rser.2017.10.111

Ranjit, S. S. S., Abbod, M., & Reader. (2018). Research and Integration of IoT based Solar Photovoltaic Panel Health Monitoring System. *Indian Journal of Public Health Research & Development*, *9*(12), 1678. https://doi.org/10.5958/0976-5506.2018.02232.5

Rao, C. K., Sahoo, S. K., & Yanine, F. F. (2024). A Literature Review on an IoT-Based Intelligent Smart Energy Management Systems for PV Power Generation. *Hybrid Advances*, *5*, 100136. https://doi.org/10.1016/j.hybadv.20212.100136

Rouibah, N., Barazane, L., Mellit, A., Hajji, B., & Rabhi, A. (2019). A low-cost monitoring system for maximum power point of a photovoltaic system using IoT technique. *2019 International Conference on Wireless Technologies, Embedded and Intelligent Systems (WITS)*, 1–5. https://doi.org/10.1109/WITS.2019.8723724

Saleem, Y., Crespi, N., Rehmani, M. H., & Copeland, R. (2019). Internet of Things-Aided Smart Grid: Technologies, Architectures, Applications, Prototypes, and Future Research Directions. *IEEE Access*, *7*, 62962–630012. https://doi.org/10.1109/ACCESS.2019.2913984

Saleem, M. U., Usman, M. R., Usman, M. A., & Politis, C. (2022). Design, Deployment and Performance Evaluation of an IoT Based Smart Energy Management System for Demand Side Management in Smart Grid. *IEEE Access*, *10*, 15261–15278. https://doi.org/10.1109/ACCESS.2022.3147484

Samara, S., & Natsheh, E. (2019). Intelligent Real-Time Photovoltaic Panel Monitoring System Using Artificial Neural Networks. *IEEE Access*, *7*, 50287–50299. https://doi.org/10.1109/ACCESS.2019.2911250

Samuel, J., & Rajagopal, Dr. B. (2022). IoT Based PV Panel Cleaning System. *International Journal for Research in Applied Science and Engineering Technology*, *10*(11), 775–779. https://doi.org/10.22214/ijraset.2022.47420

Sheba, M. A., Mansour, D. A., & Abbasy, N. H. (2023). A New Low-Cost and Low-Power Industrial Internet of Things Infrastructure for Effective Integration of Distributed and Isolated Systems with Smart Grids. *IET Generation, Transmission & Distribution*, *17*(20), 4554–45712. https://doi.org/10.1049/gtd2.12951

Srivastava, N., & Pandey, P. (2022). Internet of Things (IoT): Applications, Trends, Issues and Challenges. *Materials Today: Proceedings*, *69*, 587–591. https://doi.org/10.1016/j.matpr.2022.09.490

Supo, J. F., Puma, P. Y., Ramos, J. M., & Beltrán, N. J. (2023). Implementation of a Monitoring System for a Photovoltaic Installation Based on IoT Technology. *Journal of Physics: Conference Series*, *2538*(1), 0120012. https://doi.org/10.1088/1742-6596/2538/1/012003

Taghizad-Tavana, K., Ghanbari-Ghalehjoughi, M., Razzaghi-Asl, N., Nojavan, S., & Alizadeh, A. (2022). An Overview of the Architecture of Home Energy Management System as Microgrids, Automation Systems, Communication Protocols, Security, and Cyber Challenges. *Sustainability*, *14*(23), 15938. https://doi.org/10.3390/su142315938

Tajjour, S., & Chandel, S. S. (2023). Experimental Investigation of a Novel Smart Energy Management System for Performance Enhancement of Conventional Solar Photovoltaic Microgrids. *Discover Energy*, *3*(1), 8. https://doi.org/10.1007/s43937-023-00021-5

Thinger.io. (2024). *Thinger IoT Platform..* Retrieved May 20, 2024, From https://thinger.io/

Thingsboard.io. (2024). *Thingsboard IoT Platform..* Retrieved May 21, 2024, From https://thingsboard.io/

ThingSpeak. (2024). *ThingSpeak IoT Platform..* Retrieved May 21, 2024, From https://thingspeak.com/

Touati, F., Al-Hitmi, M. A., Chowdhury, N. A., Hamad, J. A., & San Pedro Gonzales, A. J. R. (2016). Investigation of Solar PV Performance Under Doha Weather Using a Customized Measurement and Monitoring System. *Renewable Energy*, *89*, 564–577. https://doi.org/10.1016/j.renene.2015.12.046

Ud Din, I., Guizani, M., Hassan, S., Kim, B.-S., Khurram Khan, M., Atiquzzaman, M., & Ahmed, S. H. (2019). The Internet of Things: A Review of Enabled Technologies and Future Challenges. *IEEE Access*, *7*, 7606–7640. https://doi.org/10.1109/ACCESS.2018.2886601

Ul Mehmood, M., Ulasyar, A., Ali, W., Zeb, K., Zad, H. S., Uddin, W., & Kim, H.-J. (2023). A New Cloud-Based IoT Solution for Soiling Ratio Measurement of PV Systems Using Artificial Neural Network. *Energies*, *16*(2), 996. https://doi.org/10.3390/en16020996

Wedashwara, W., Wayan Agus Arimbawa, I., Hidayat Jatmika, A., Zubaidi, A., & Mulyana, T. (2020). IoT based smart small scale solar energy planning using evolutionary fuzzy association rule mining. *2020 International Conference on Advancement in Data Science, E-Learning and Information Systems (ICADEIS)*, 1–6. https://doi.org/10.1109/ICADEIS49811.2020.9276905

CHAPTER 13

# IoT in Energy Systems

## *Innovations, Challenges, and Future Outlooks*

Raja Mahmou

## 13.1 INTRODUCTION

The integration of the Internet of Things (IoT) into embedded systems has revolutionized energy management, driving significant advancements and efficiencies across various sectors. IoT serves as a powerful innovation catalyst, enhancing the functionality and connectivity of embedded systems. This chapter explores the role of IoT in energy systems, focusing on its impact on operational efficiency and sustainability. It begins with a brief overview of IoT principles and their application in enhancing embedded systems. The chapter then delves into the integration of IoT in energy management, highlighting innovations such as renewable energy sources and optimization of energy production processes. Additionally, it examines the role of IoT in traditional power plants and the integration of renewable energies such as wind turbines and solar panels. Emerging trends and future outlooks, including drone technology, digital twins (DT), and green IoT (G-IoT), are also discussed to provide a comprehensive understanding of IoT's potential in transforming energy systems.

## 13.2 INTRODUCTION TO IoT IN EMBEDDED SYSTEMS AND ENERGY MANAGEMENT

The integration of the IoT into embedded systems has revolutionized the way we manage and utilize energy. IoT serves as a powerful innovation catalyst in embedded systems, driving advancements and efficiencies across various sectors. This section explores the role of IoT in enhancing


DOI: 10.1201/9781003499732-13

embedded systems and its significant impact on energy management, particularly in terms of operational efficiency and sustainability.

### 13.2.1 IoT's Role as an Innovation Catalyst in Embedded Systems

#### *13.2.1.1 A Brief Overview and Principles of IoT Operation*

The IoT is a network of physical objects equipped with sensors, actuators, processors, and software that enable data collection, processing, and exchange. These objects can communicate with each other via local networks or the Internet for remote analysis and control. IoT devices include household appliances, vehicles, and industrial equipment.

IoT devices operate through interconnected steps involving key components: sensors collect environmental data (e.g., temperature, humidity, motion), while actuators perform actions based on received data (e.g., turning on a light, adjusting a thermostat). Embedded processors analyze the data and make decisions using pre-defined algorithms or rules.

Various communication protocols (e.g., Zigbee, Bluetooth, Wi-Fi) facilitate data exchange between devices or central systems, locally or via the Internet. Data is often sent to cloud servers for storage, analysis, and insight generation using advanced analytics, machine learning, and big data techniques. Users interact with IoT systems through interfaces like mobile apps, web dashboards, or voice commands, enabling monitoring, control, and customization of IoT device behavior. The following diagram (Figure 13.1) illustrates the fundamental components and operational flow of IoT systems, providing a visual overview of how these interconnected devices function.

#### *13.2.1.2 Overview of Embedded Systems*

Embedded systems are specialized computer systems integrated into electronic devices to perform specific, real-time application. They consist of a processor for executing instructions, memory (ROM for programs and RAM for temporary data), communication interfaces (e.g., UART, SPI, I2C), and input/output peripherals like sensors and actuators. ADC and DAC converters, often part of the processor, are used to convert signals, and a power supply provides necessary energy.

The operation of an embedded system starts with sensors collecting data, which the processor then processes based on embedded software instructions. Real-time decisions are made and executed by actuators, often with feedback loops for continuous optimization. By combining hardware and software, embedded systems achieve high efficiency in dedicated functions, making them essential in various applications. The following diagram (Figure 13.2)

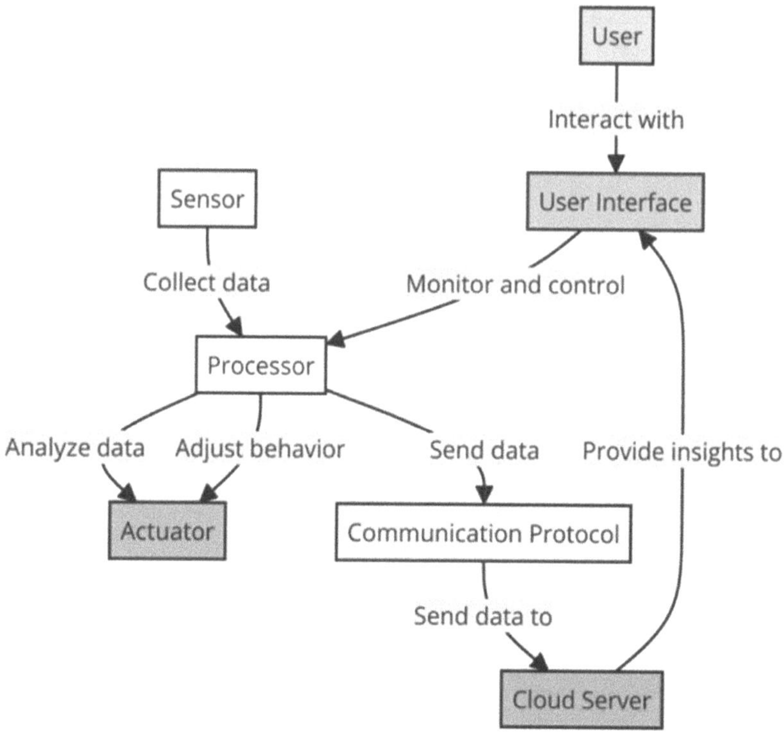


FIGURE 13.1 Data flow diagram of IoT system components and interactions.

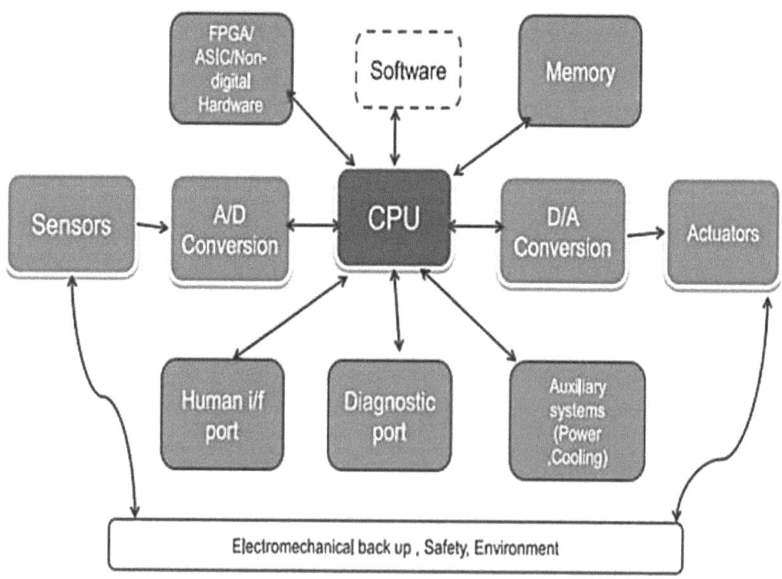


FIGURE 13.2 Typical embedded system architecture.

illustrates the typical architecture of an embedded system, highlighting the interaction between its components.

### *13.2.1.3 Enhancing Embedded Systems through IoT Integration*

The IoT enhances embedded systems, which already incorporate advanced sensors, real-time data processing, and communication technologies. IoT amplifies these capabilities by providing greater connectivity, broader data collection, and more adaptive responses, leading to smarter, more efficient systems that can operate autonomously and adapt to changing conditions [1].

Numerous innovations have emerged by combining IoT and embedded systems, particularly in energy management (Figure 13.3). In smart grids,

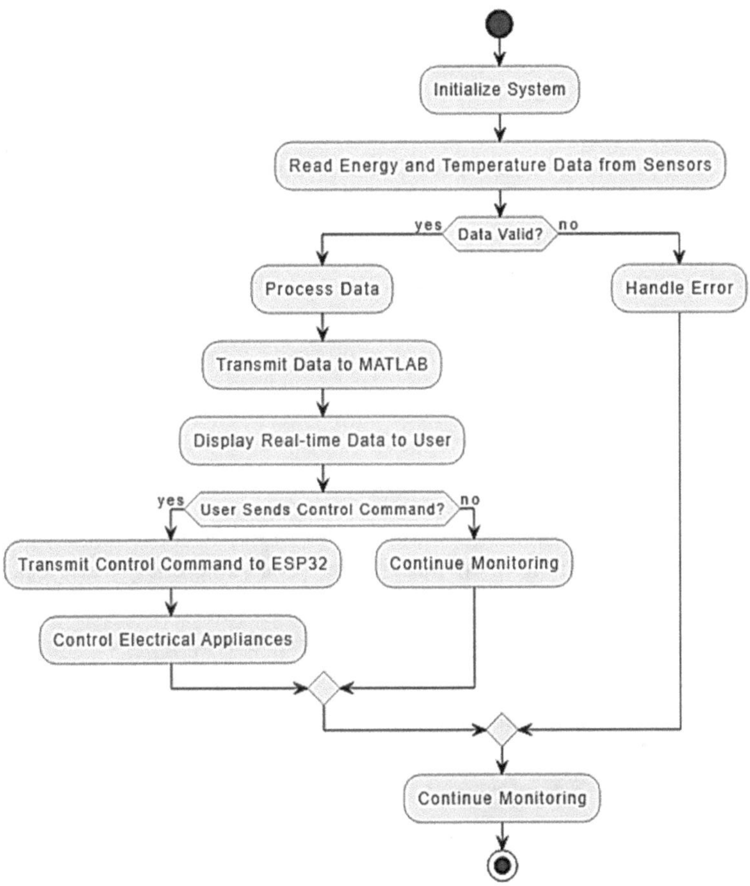


FIGURE 13.3 Proposed complementary diagram for IoT-based energy management system inspired by [3].

IoT devices manage energy distribution, reducing peak load and improving efficiency. IoT sensors collect data on energy consumption patterns, enabling better demand response strategies and optimizing energy use [2].

By combining the strengths of IoT and embedded systems, numerous innovations have emerged across various industries, including manufacturing and healthcare, where IoT-enabled embedded systems optimize performance and reduce energy consumption.

### 13.2.2 Impact of IoT on Energy Management: Operational Efficiency and Sustainability

The integration of the IoT into energy management has significantly enhanced operational efficiency and sustainability. IoT facilitates real-time monitoring, control, and optimization of energy usage, contributing to reduced energy consumption and improved sustainability practices.

#### *13.2.2.1 Operational Efficiency*

For example, IoT-based energy management systems enable real-time monitoring and control of household energy consumption. A system using the ESP32 microcontroller can process energy consumption and temperature readings, allowing for remote control of electric appliances via a MATLAB application. This system helps reduce electrical wastage through proper scheduling and monitoring of appliances [3]. An illustration of this system could include an activity diagram of the ESP32 microcontroller setup for energy and temperature monitoring, along with a flowchart describing the process of real-time monitoring, data transmission, and remote control, as shown in Figure 13.3.

#### *13.2.2.2 Sustainability*

Integration with Renewable Energy Sources: IoT applications in energy management systems help in integrating renewable energy sources and managing energy consumption more sustainably. For example, IoT technologies in wind energy generation and electric drives improve monitoring and management, leading to energy savings [4] (section II). To illustrate this concept more generally, a proposed diagram of an IoT-enabled wind energy generation system is provided in Figure 13.4.

In summary, IoT has proven to be a transformative force in embedded systems and energy management. By enabling real-time monitoring, control, and optimization, IoT enhances operational efficiency and promotes sustainable energy practices. These advancements not only reduce energy consumption but also pave the way for more intelligent and responsive energy management systems.

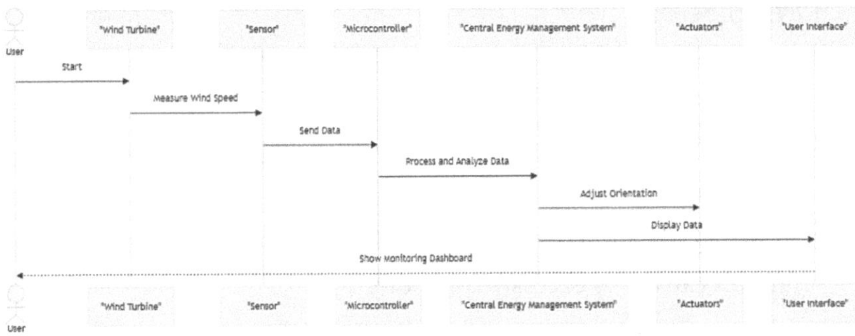


FIGURE 13.4 Wind turbine energy management system.

## 13.3 OPTIMIZATION OF ENERGY PRODUCTION THROUGH IoT

This section examines how IoT is applied to optimize electricity production processes, focusing on traditional power plants and the integration of renewable energies such as wind turbines and solar panels. Real-time data collection and advanced analysis enable smarter management of energy resources.

### 13.3.1 IoT Applications for Control and Monitoring of Electricity Production Processes

Traditional power plants benefit from the integration of IoT sensors, which continuously monitor various parameters such as temperature, pressure, and turbine performance. These sensors collect real-time data on equipment operating conditions. The data is then analyzed to optimize performance and prevent failures. The following illustration (Figure 13.5) highlights this principle.

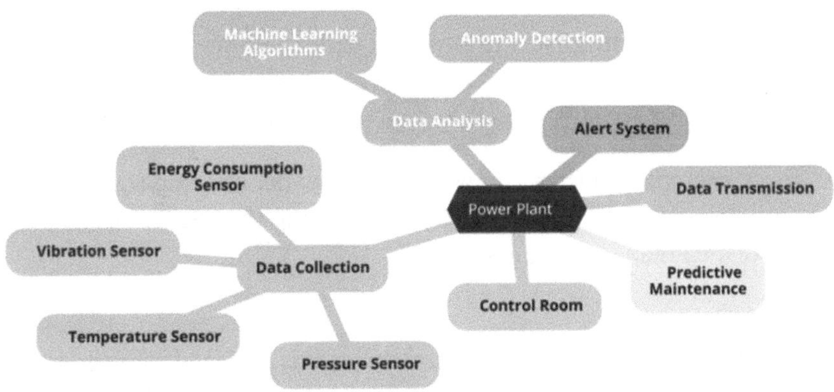


FIGURE 13.5 Comprehensive overview of IoT integration in power plant operations.

#### *13.3.1.1 IoT Sensors*

IoT sensors are strategically placed on critical equipment in the power plant, such as (Figure 13.5):

- **Temperature Sensor:** Continuously measures the temperature of key components to prevent overheating.
- **Pressure Sensor:** Monitors pressure levels in the systems to prevent failures caused by excessive pressure.
- **Vibration Sensor:** Detects abnormal vibrations that may indicate imbalance or wear of parts.
- **Energy Consumption Sensor:** Tracks the energy consumption of various components to optimize energy efficiency.

The data collected by these sensors are wirelessly transmitted to a control center.

#### *13.3.1.2 Data Transmission*

The data transmission utilizes secure networks to ensure the confidentiality and integrity of the information. Here are some examples of secure networks used for data transmission in power plants:

- **Power Line Communication (PLC)**: It utilizes the existing power grid infrastructure for data transmission, connecting numerous sensors and IoT devices without needing a dedicated network [5].
- **Wireless Sensor Networks (WSN)**: It uses wireless connections to transmit environmental and operational data. The nodes in this network can be energy-autonomous, often powered by photovoltaic panels (PVs) [6].
- **Edge Computing Network over PLC**: It combines PLC technology with edge computing to reduce transmission latency and offload some computations from the cloud to the network edges [7].

These examples illustrate how secure networks are employed for the transmission of data in power plants, enhancing the efficiency and reliability of data management and analysis.

#### *13.3.1.3 Data Analysis*

Once the data is received at the control center, it is analyzed using machine learning algorithms and anomaly detection techniques. The machine learning algorithms are used to identify patterns in the data that could indicate potential problems. For example, a gradual increase in temperature could signal a cooling issue.

Here is a simple linear regression model used to predict the state of equipment based on sensor readings (Equation (1)):

$$y = f(x) = \beta_0 + \beta_1 x_1 + \beta_2 x_2 + \cdots + \beta_n x_n + \varepsilon \tag{13.1}$$

where

- $y$ is the dependent variable (e.g., the state of the equipment)
- $x_i$ are the independent variables (e.g., sensor readings)
- $\beta_i$ are the coefficients of the independent variables
- $\varepsilon$ epsilon $\epsilon$ is the residual error

#### *13.3.1.4 Anomaly Detection*

Anomaly detection algorithms identify data points that deviate from expected normal behavior, which can indicate imminent failures or inefficiencies. A common algorithm for anomaly detection is the Isolation Forest. Here is an example of using the Isolation Forest algorithm in Python (Figure 13.6).

```
from sklearn.ensemble import IsolationForest

# Create an IsolationForest model with 100 trees and contamination set to 10%
# n_estimators defines the number of base estimators in the ensemble
# contamination defines the proportion of outliers in the data set
model = IsolationForest(n_estimators=100, contamination=0.1)

# Train the model on the data
# The fit method trains the Isolation Forest model using the input data
model.fit(data)

# Predict anomalies in the data
# The predict method returns -1 for anomalies and 1 for normal observations
anomalies = model.predict(data)
```

FIGURE 13.6 Example of Python code for anomaly detection using isolation forest algorithm.

#### *13.3.1.5 Alert Systems and Predictive Maintenance*

13.3.1.5.1 Alert System When an anomaly is detected, an alert system is activated to notify the plant operators, enabling a quick response. This system is essential for minimizing reaction times in case of potential failures, thus ensuring the continuity of operations and the safety of the installations.

13.3.1.5.2 Predictive Maintenance The analyzed data allows for the prediction of failures before they occur, enabling proactive maintenance planning. This reduces downtime and repair costs. Predictive maintenance uses advanced algorithms to analyze data and predict failures.

A predictive maintenance model could use "logistic regression" (Equation (13.2)) to predict the probability of failure [8]:

$$logit(P) = \log\left(\frac{P}{1-P}\right) = \beta_0 + \beta_1 X_1 + \beta_2 X_2 + \cdots + \beta_n X_n \tag{13.2}$$

- $P$ is the probability of failure
- $X_i$ are the independent variables (e.g., sensor readings)
- $\beta_i$ are the coefficients of the independent variables

Here is an example of Python code to implement a logistic regression model (Figure 13.7):

13.3.1.5.3 The Control Center Uses this information to adjust the operational parameters of the equipment in real time, thus maximizing operational efficiency and sustainability. This optimization not only improves energy production but also enhances resource management and reduces operational costs.

### 13.3.2 IoT for the Exploration and Exploitation of Renewable Energy Sources

The utilization of IoT technologies is essential for the effective exploration and exploitation of renewable energy sources. The following image illustrates how IoT systems contribute to optimizing these processes (Figure 13.8).

Such as wind turbines and solar panels. The following sections provide detailed illustrations and theoretical approaches to how IoT systems optimize these processes.

```python
from sklearn.model_selection import train_test_split
from sklearn.linear_model import LogisticRegression
from sklearn.metrics import classification_report

# Assume 'data' is your DataFrame containing the collected data
# 'features' are the independent variables (e.g., sensor readings)
# 'target' is the dependent variable (e.g., equipment failure or success)

X = data[features]
y = data[target]
```

```python
# Split the data into training and test sets
X_train, X_test, y_train, y_test = train_test_split(X, y, test_size=0.2, random_state=42)

# Create and train the logistic regression model
model = LogisticRegression()
model.fit(X_train, y_train)

# Predict on the test set
y_pred = model.predict(X_test)

# Display the classification report
print(classification_report(y_test, y_pred))
```

FIGURE 13.7 Implementation of logistic regression for predictive maintenance using Python.

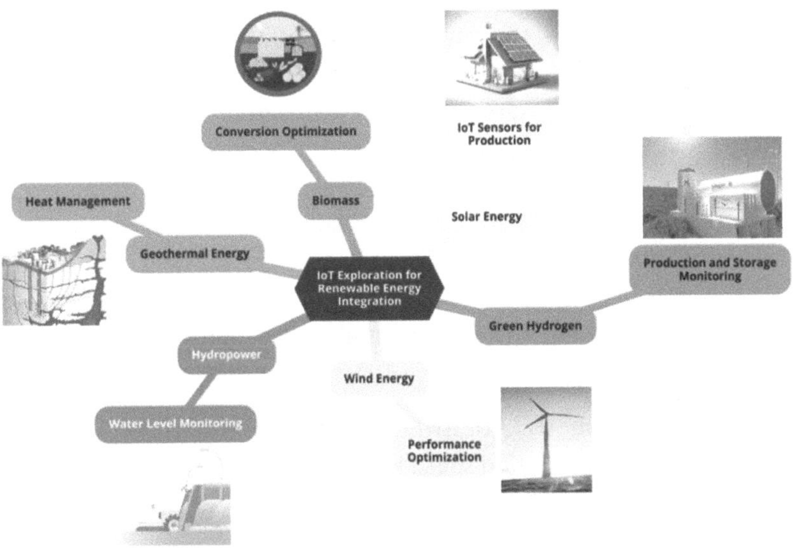


FIGURE 13.8 IoT exploration for renewable energy integration.

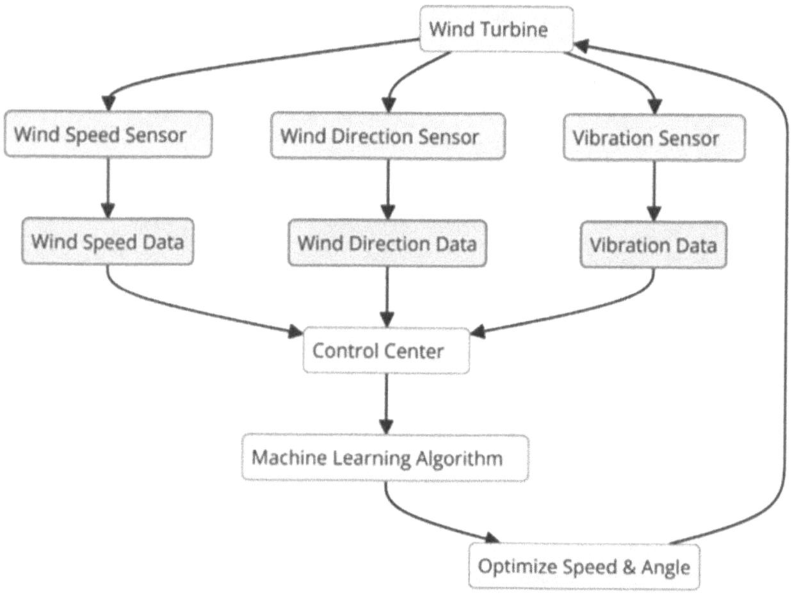


FIGURE 13.9 IoT sensors on wind turbines with data transmission and optimization.

### *13.3.2.1 Wind Turbines*

The following diagram illustrates how IoT sensors on a wind turbine collect various types of data, which are then processed by a machine learning algorithm to optimize the turbine's operational parameters (Figure 13.9).

The collected data is transmitted to a control center where machine learning algorithms analyze the information to optimize blade angle, rotation speed, and other operational parameters to maximize energy production. For instance, if wind speed increases, the system can adjust blade angle to optimize energy capture without overloading the turbine.

Here is an example of Python pseudocode for a real-time control algorithm that adjusts wind turbine parameters based on IoT data (Figure 13.10):

- **Parameter Initialization**: The initial parameters of the wind turbine are defined.
- **Turbine Parameter Adjustment Function**: This function adjusts blade angle and rotation speed based on wind speed, wind direction, and temperature.
- **Main Control Loop**: This loop collects data from IoT sensors at regular intervals and adjusts turbine parameters accordingly.

```python
# Pseudocode for real-time wind turbine control

# Initialize parameters
current_blade_angle = initial_blade_angle
current_rotation_speed = initial_rotation_speed

# Function to adjust turbine parameters based on sensor data
def adjust_turbine_parameters(wind_speed, wind_direction, vibration):
    # Adjust blade angle based on wind speed
    if wind_speed > optimal_wind_speed:
        current_blade_angle = decrease_blade_angle()
    elif wind_speed < optimal_wind_speed:
        current_blade_angle = increase_blade_angle()

    # Adjust rotation speed based on wind direction and vibration
    if wind_direction_changes():
        current_rotation_speed = adjust_rotation_speed()

    if vibration_exceeds_threshold():
        current_rotation_speed = reduce_rotation_speed()

    # Send new parameters to turbine
    update_turbine_parameters(current_blade_angle, current_rotation_speed)

# Main control loop
while True:
    # Collect data from IoT sensors
    wind_speed = read_sensor("wind_speed")
    wind_direction = read_sensor("wind_direction")
    vibration = read_sensor("vibration")

    # Adjust turbine parameters based on collected data
    adjust_turbine_parameters(wind_speed, wind_direction, vibration)

    # Wait for the next measurement cycle
    sleep(measurement_interval)
```

FIGURE 13.10 Real-time control Python algorithm for wind turbine optimization using IoT data.

For example, this study [9] innovatively integrated IoT into wind energy conversion systems by dynamically adjusting blade angle, rotation speed, and other operational parameters based on real-time environmental data. It not only improved energy production efficiency but also reduced operational risks. This approach holds transformative potential for wind farm management, promising to maximize energy yield while enhancing overall operational efficiency and reliability.

#### *13.3.2.2 Solar Panel*

The following diagram illustrates an IoT system monitoring solar panels and adjusting their orientation based on weather conditions (Figure 13.11):

The system collect data on environmental conditions such as sunlight intensity and temperature. The machine learning algorithms analyze the cited data to determine the optimal tilt angle for the solar panel. The system adjusts the tilt angle in real time to maximize light absorption and energy production.

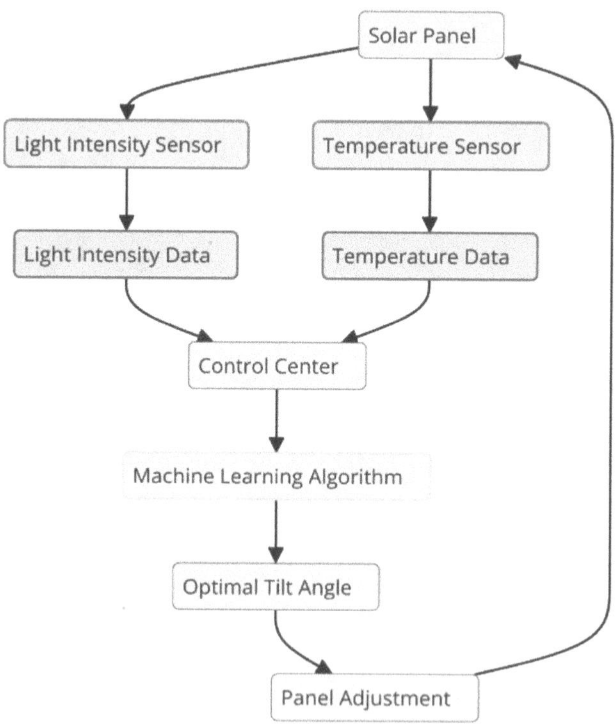


FIGURE 13.11 IoT sensors on solar panel with data transmission and optimization.

13.3.2.2.1 Sunlight Intensity

$$E = I.\cos\theta \tag{13.3}$$

where $E$ is the solar energy received per unit area, $I$ is the sunlight intensity, and $\theta$ is the angle of incidence of the solar rays.

13.3.2.2.2 Energy Conversion Efficiency

$$\theta = \frac{Pout}{Pin} \tag{13.4}$$

where $\boldsymbol{\theta}$ is the energy conversion efficiency, Pout is the output power, and Pin is the input power.

13.3.2.2.3 Tilt Angle Optimization

$$\boldsymbol{\theta}_{opt} = \mathbf{arctan}\left(\frac{\boldsymbol{h}}{\boldsymbol{d}}\right) \tag{13.5}$$

where $\theta_{opt}$ is the optimal tilt angle, $h$ is the height of the sun above the horizon, and $d$ is the horizontal distance from the measurement point to the solar panel.

13.3.2.2.4 Real-Time Adjustment

$$\boldsymbol{\theta}(\boldsymbol{t}) = \boldsymbol{f}\left(\boldsymbol{I}(\boldsymbol{t}), \boldsymbol{T}(\boldsymbol{t})\right) \tag{13.6}$$

where $\theta(t)$ is the tilt angle at time t, $I(t)$ is the sunlight intensity at time $t$, and $T(t)$ is the temperature at time $t$.

A recent study by [10] presents an IoT-based solar energy monitoring and management system that optimizes solar energy utilization. The system integrates IoT sensors to continuously monitor and adjust the orientation of solar panels in real time, resulting in improved system performance and energy efficiency. The study demonstrates a significant improvement in solar panel efficiency by using IoT systems to dynamically adjust the panels' orientation based on real-time environmental conditions.

In conclusion, the application of IoT technologies significantly enhances the efficiency and reliability of energy production. By leveraging real-time data and advanced analytics, both traditional power plants and renewable energy sources can achieve optimal performance. This integration paves the way for smarter, more sustainable energy management.

## 13.4 EMERGING TRENDS AND FUTURE OUTLOOKS OF IoT IN THE ENERGY SECTOR

Technological advancements in the field of the IoT are transforming energy management by improving efficiency and sustainability. These advancements not only enhance energy efficiency but also enable the exploitation of renewable energy sources to power embedded systems and IoT components. However, these innovations also bring their own set of challenges. Innovations and future projects show tremendous potential for the adoption and development of IoT in the energy sector. In this section, we will examine the emerging trends, and the future opportunities associated with the adoption of IoT in the energy sector.

### 13.4.1 Emerging Trends

#### *13.4.1.1 Drone Technology for Energy Management*

Drones equipped with IoT sensors play a crucial role in energy management Figures 13.12, facilitating the monitoring and maintenance of energy installations. For example, drones can be used for wireless charging using

FIGURE 13.12 Drone charging with building-integrated photovoltaics (BIPV).

TABLE 13.1 Photovoltaic Panel Charging Details and Energy Efficiency Result by [11]

| Photovoltaic Panel | Charging Details | Energy Percentage |
|---|---|---|
| Roof photovoltaic panel | Standard | 100 |
| Vertical photovoltaic panel | 12.6 V/5.2 Ah in 36.9 min | 85 |
| Inclined photovoltaic panel | 12.6 V/5.2 Ah in 31.29 min | 95 |
| Inclined PV with white reflective surface | Not specified | 90 |
| Tilted PV with black reflective surface | Produces 10.09% more energy compared to white reflective surface | 99 |

building-integrated photovoltaic (BIPV) structures. According to [11], these BIPV structures enable efficient wireless charging for drones by utilizing both inclined and vertical PVs.

To illustrate the charging details and energy efficiency of different photovoltaic panel configurations, Table 13.1 is based on the findings by [11].

#### *13.4.1.2 Roles of Drones to Explore Solar Energy Sources*

To further exploit solar energy sources, the construction of solar panels must move forward. However, the discovery, maintenance, and exploitation of geographically dispersed and isolated solar zones can be very difficult. A promising solution involves using autonomous drones equipped with thermographic imaging capabilities (UAVs) to detect high-temperature areas by measuring temperature changes Figures 13.13. These drones can also assist in energy maintenance after the installation of PVs.

For example, [12] focuses on the use of drones for the unattended maintenance of PV plants. This study applies rigorous drone photogrammetry using optical Red, Green, and Blue (RGB) and infrared thermography (IRT) images to detect hot spots, one of the most common faults in PV plants. The research introduces an in-house software application, SunMap, which allows automatic, accurate, and reliable detection of hot spots on PV panels. The validation performed in PV plants located in Spain showed positive results, with an error incidence lower than 15% as validated by the manufacturer's standard electrical tests.

This innovative approach using drones and advanced imaging techniques significantly enhances the efficiency and reliability of PV plant maintenance, addressing the challenges of maintaining solar zones in remote locations.

Table 13.2 summarizes the key points and quantitative results of the study.

FIGURE 13.13 Thermal zone detection by autonomous drones.

TABLE 13.2 Key Results of the SunMap Study [12] on Photovoltaic Plant Maintenance

| Key Results | Description |
|---|---|
| **Hot spot detection accuracy** | The system detects hot spots with an error incidence lower than 15%. |
| **Automatic detection** | The SunMap software enables automatic and reliable detection of hot spots on PV panels. |
| **Geolocation of malfunctions** | Accurate identification and geolocation of malfunctioning PV panels. |
| **High-quality cartographic products** | Generation of 3D models and true orthophotos for maintenance support. |
| **Validation sites** | Successfully validated in two different PV plants in Spain. |

### *13.4.1.3 Wind Energy Harvesting for Self-Powered IoT Applications*

13.4.1.3.1 Principle of Operation The piezoelectric wind energy harvesting (PWEH) system leverages the piezoelectric effect, wherein specific materials generate an electric charge in response to applied vibrations induced by wind. As wind causes the piezoelectric material to deform, it generates an alternating current. This current can be harvested and stored in energy storage devices such as batteries or capacitors. PWEH is a promising technology for powering

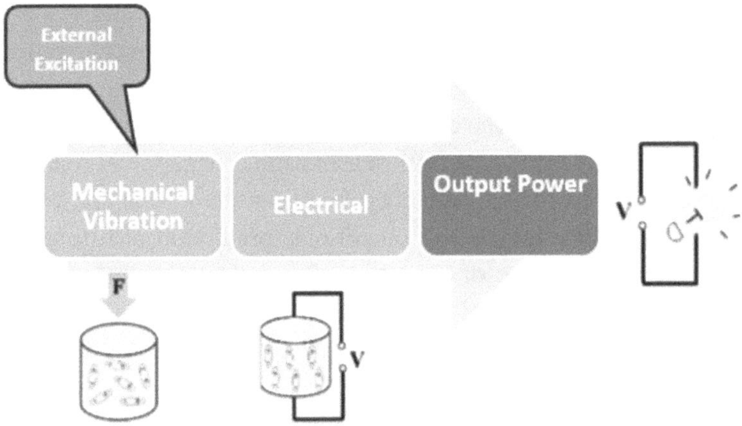


FIGURE 13.14 Schematic of piezoelectric energy harvesting system.

low-consumption IoT devices, especially in remote or hard-to-access areas where traditional power sources are unavailable. The harvested energy can be used to power various IoT devices, such as sensors for remote fire monitoring systems, environmental monitoring, and infrastructure health monitoring. The process is illustrated in Figure 13.14.

13.4.1.3.2 Research Example According to [13] work, the PWEH method has been successfully demonstrated to power low-consumption IoT devices such as sensors used in remote fire monitoring systems. In their study, the PWEH system was designed to harvest energy from wind vibrations in a tunnel environment, utilizing both electromagnetic and piezoelectric modules to maximize energy conversion efficiency.

Table 13.3 succinctly highlights the key components and performance metrics of the wind energy harvesting system developed by [13].

TABLE 13.3 Summary Table for Wind Energy Harvesting System [13]

| Component | Performance Details | Energy Output |
|---|---|---|
| Hybrid wind energy system | Combines electromagnetic and piezoelectric modules for enhanced energy harvesting. | Output power at a wind speed of 7 m/s is **59.31 mW.** |
| Energy storage module | Converts AC power into DC, stores it in a supercapacitor, powers WSN[a] nodes. | WSN[a] nodes consume 49.4%–59.8% of the harvested energy. |

[a] Wireless sensor network.

*13.4.1.4 Digital Twins for Energy Storage*

13.4.1.4.1 General Principle DT are accurate virtual representations of physical systems, enabling real-time simulation, analysis, and optimization of these systems. In the context of electrical energy storage, a digital twin integrates real-time data from sensors and other sources to faithfully reproduce the behavior of an energy storage system. This allows for the monitoring and management of operations, prediction of future performance, and identification of potential issues before they arise. DTs also facilitate the development of more efficient and reliable energy management strategies while reducing operational costs and extending the lifespan of storage systems.

The following diagram illustrates the interaction between different components involved in the application of DT for energy storage. It highlights how DTs simulate scenarios, generate artificial intelligence (AI) strategies, and optimize control systems to adjust operations and provide insights to operators (Figure 13.15).

13.4.1.4.2 Research Example The article by [14] examines the trends and challenges of applying DT in the field of energy storage. The authors highlight several key innovations brought by DTs in this domain. They emphasize the importance of DTs in enhancing real-time monitoring and management of energy storage systems, allowing for better performance forecasting and early fault detection. The study also identifies the main obstacles to DT adoption, such as data requirements and technical challenges related to integrating various technologies. Additionally, the authors discuss future prospects, including

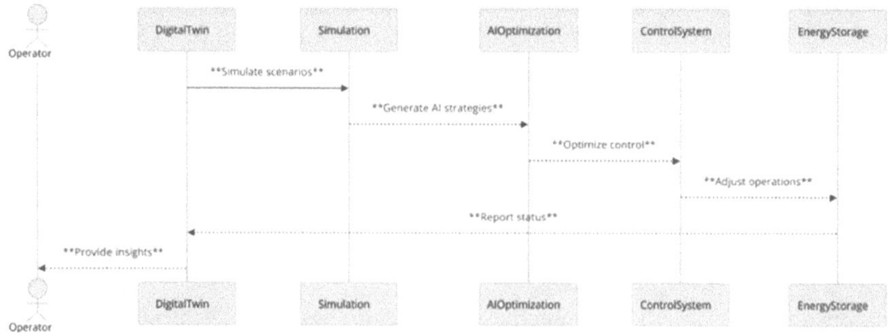


FIGURE 13.15 Sequential diagram of digital twin process for energy storage management.

TABLE 13.4 Summary Table of the Article's [14] Key Points

| Key Points of the Article | Description |
|---|---|
| **Improvement of real-time monitoring** | Use of DTs for continuous and detailed monitoring of energy storage systems. |
| **Performance prediction** | Enhanced capabilities to predict performance and anticipate maintenance needs. |
| **Early fault detection** | Rapid identification of potential problems before they affect the system. |
| **Optimization of energy management** | Development of optimized strategies for managing energy resources. |
| **Technical challenges and data requirements** | Analysis of the main obstacles to DT adoption, including data collection and integration challenges. |
| **Future prospects** | Exploration of possible improvements in modeling algorithms and AI integration. |

the improvement of modeling algorithms and increased integration of AI technologies to optimize the performance of energy storage systems.

Table 13.4 summarizes the article's key points [14], highlighting the main benefits and challenges of using DT in energy storage systems as well as the future directions proposed by the authors.

### 13.4.2 Future Outlook for IoT in Electrical Energy Management and Sustainability

The integration of IoT in energy management and sustainability holds significant potential for future advancements. The following examples highlight some of the most recent and relevant developments in this field.

#### *13.4.2.1 Energy-Sustainable IoT Ecosystems*

A recent study focuses on the development of self-sustainable IoT ecosystems that require minimal external resources and intervention. Key components include the use of renewable energy sources and recycling materials to optimize energy provision, transfer, and efficiency [15].

#### *13.4.2.2 Green IoT for Sustainable Smart Cities*

G-IoT aims to enhance urban sustainability by integrating eco-friendly technologies into IoT applications. This includes energy-efficient sensor networks, smart waste management, and optimized energy consumption in buildings. AI is used to enhance data analysis and decision-making processes, improving resource allocation and reducing carbon footprints in smart cities [16].

#### *13.4.2.3 Integration of Electric Vehicles, Renewable Energy Sources, and IoT*

The innovative aspect of this research is the optimization of energy use and the reduction of environmental impact through the intelligent coordination of electric vehicle (EV) charging and renewable energy generation. This approach can enhance grid stability, maximize the use of renewable energy, and provide flexible energy solutions [17].

#### *13.4.2.4 IoT-Based Smart Energy Meter (SEM)*

The future perspective in IoT-based smart energy meters (SEMs) lies in their ability to optimize energy consumption and reduce costs by using real-time data collected from various sensors. These systems can identify opportunities for energy savings and predict energy demands [18].

In this last section, we explored emerging trends and future perspectives in IoT for energy management and sustainability. We discussed four main emerging trends and cited four promising perspectives. These innovations demonstrate significant potential to improve energy efficiency and reduce environmental impact, contributing to the development of smart, sustainable cities. It is important to note that these are not the only emerging trends and perspectives in the field.

## 13.5 CONCLUSION

In conclusion, the application of IoT technologies has significantly enhanced the efficiency and reliability of energy systems. By leveraging real-time data and advanced analytics, both traditional power plants and renewable energy sources can achieve optimal performance. The integration of IoT in energy management not only reduces energy consumption but also promotes sustainable practices. Emerging trends such as drone technology, DT, and G-IoT demonstrate the potential for further advancements in this field. As IoT continues to evolve, its role in driving smarter, more sustainable energy management solutions will become increasingly crucial, paving the way for a more efficient and environmentally friendly energy future.

## REFERENCES

1. Wu Rong, Gobinath Tamil Vanan and Mark Phillips, "The Internet of Things (IoT) and Transformation of the Smart Factory," *2016 International Electronics Symposium (IES)*, 2016.

2. Francisco-Javier Ferrández-Pastor, Higinio Mora, Antonio Jimeno-Morenilla and Bruno Volckaert, "Deployment of IoT Edge and Fog Computing Technologies to Develop Smart Building Services," *Sustainability* vol. 10, no. 11, p. 3832, 2018.
3. Abubakar Salisu, Aminu Bugaje and A. Z. Loko, "IoT Based Household Electricity Energy Monitoring and Control," *FUDMA Journal of Sciences (FJS)*, vol. 4, no. 4, pp. 77–84, 2020.
4. Maria G. Ioannides, Anastasios Stamelos, Stylianos A. Papazis, Athanasios Papoutsidakis, Vasilios Vikentios and Nikolaos Apostolakis, "IoT Monitoring System for Applications with Renewable Energy Generation and Electric Drives," *19th International Conference on Renewable Energies and Power Quality (ICREPQ'21)*, vol. 19, pp. 565–570, 2021.
5. Abdellah Chehri and Alfred Zimmermann, "Spectrum Management of Power Line Communications Networks for Industrial Applications," in *Smart Innovation, Systems and Technologies*, Springer, 2020, pp. 173–182, volume 189.
6. Abella Salvo Crispino, S. C., Bonina Salvo, Cucuccio Antonino, D'Angelo Salvatore, Giustolisi Gianluca, Grasso Alfio D., Imbruglia Antonio, Mauro Giorgio S., Nastasi Giuseppe A. M., Palumbo Gaetano, Pennisi Salvatore, Sorbello Gino, & Scuderi Antonino. "Autonomous Energy-Efficient Wireless Sensor Network Platform for Home/Office Automation," *IEEE Sensors Journal*, vol. 19, no. 9, pp. 3501–3512, 2019.
7. Yuwen Qian, Long Shi, Jun Li, Xiangwei Zhou, Feng Shu and Jiangzhou Wang, "An Edge-Computing Paradigm for Internet of Things over Power Line Communication Networks," *IEEE Network*, vol. 34, no. 2, pp. 262–269, 2020.
8. O. Torres-Reyna, *Getting Started in Logit and Ordered Logit Regression*, Princeton University, 2008.
9. Lina Alhmoud and Hussein Al-Zoubi, "IoT Applications in Wind Energy Conversion Systems," *Open Engineering*, vol. 9, no. 1, pp. 490–499, 2019.
10. N. V. Phanendra Babu, A. Vivek Chandra, Ch. Rishikesh, M. Pooja, T. Murali Krishna, A. Anil Kumar, Saptarshi Roy, and Dipankar Chatterjee, "Solar Energy Monitoring and Management System Using IoT," *2023 IEEE 2nd International Conference on Industrial Electronics: Developments & Applications (ICIDeA)*, pp. 493–498, 2023.
11. Prithvi Krishna Chittoor and C. Bharatiraja, "Building Integrated Photovoltaic Powered Wireless Drone Charging System," *Solar Energy*, vol. 252, pp. 163–175, mars 2023.
12. David Hernández-López, Esteban Ruíz de Oña, Miguel A. Moreno and Diego González-Aguilera, "SunMap: Towards Unattended Maintenance of Photovoltaic Plants Using Drone Photogrammetry," *Drones*, vol. 7, no. 2, pp. 129, 2023.
13. Peng Zheng, Lingfei Qi, Mengdie Sun, Dabing Luo and Zutao Zhang, "A Novel Wind Energy Harvesting System with Hybrid Mechanism for Self-Powered Applications in Subway Tunnels," *Energy*, vol. 227, pp. 120446, 2021.

14. Semeraro Concetta, A.G Olabi, Haya Aljaghoub, Abdul Hai Alami, Muaz Al Radi, Michele Dassisti and Mohammad Ali Abdelkareem, "Digital Twin Application in Energy Storage: Trends and Challenges," *Journal of Energy Storage*, vol. 58, pp. 106347, 2023.
15. López Onel L. A., Rosabal Osmel M., Ruiz-Guirola David E., Raghuwanshi Prasoon, Mikhaylov Konstantin, Lovén Lauri, & Iyer Sridhar. "Energy-Sustainable IoT Connectivity: Vision, Technological Enablers, Challenges, and Future Directions," *IEEE Open Journal of the Communications Society*, vol. 4, pp. 2609–2666, 2023.
16. Martin Jacob, Abhay P. Aneesh, Rishikesh S., Joel Jogesh Varghese and Akhil Thomas, "Green IoT for Sustainable Smart Cities: Innovations and Challenges," *International Journal of Engineering Technology and Management Sciences*, vol. 7, no. 4, pp. 466–478, 2023.
17. N. V. A. Ravikumar, Ramakrishna S S Nuvvula, Polamarasetty P Kumar, Noor Hanoon Haroon, Umakant Dinkar Butkar and Alighazi Siddiqui, "Integration of Electric Vehicles, Renewable Energy Sources, and IoT for Sustainable Transportation and Energy Management: A Comprehensive Review and Future Prospects," *2023 12th International Conference on Renewable Energy Research and Applications (ICRERA)*, 1 September 2023.
18. Ms. B. Tulasi. Sowjanya, Haarini Tadiparthi, Kathyaeni Vidmahe Pidiyar, Sai Sri Gaddala and Deekshitha Sabavath, "IoT-Based Smart Energy Meter," *International Journal of Research Publication and Reviews*, vol. 4, no. 11, pp. 3175–3183, 2023.

CHAPTER 14

# Machine Learning-User Association Approach for Energy-Efficient and Mitigation Interference in HetNets

Samira Achki, Layla Aziz, Sanaa Kabil, Abdelali El Gorari, and Ait Lahcen Yassine

## 14.1 INTRODUCTION

One of the important key approaches that advance the success of 5G is the deployment of heterogeneous networks (HetNets) [1] that contain high transmission power base stations (BSs) such as macro-BSs and low transmission power BSs (pico and femto BSs). The deployment of heterogeneous networks (HetNets) is one of the important keys which allows us to evaluate and advance the success of 5G [2]. The improved capacity and efficient spectral of the network with a better link quality than users who are nearer to BSs considered the main advantages of HetNet. However, that necessitates the high energy consumption and impacts the problem of interference between different equipment in HetNets [3]; to overcome these problems, we proposed a strategy of user association using deep learning to improve the energy efficiency and mitigate the interference in HetNet. It is known that HetNet with small cells (pico and femto BSs) can resolve the physical deficiency of radiofrequency and develop the system capacity to some level, but at the purse of the extent and deploying

DOI: 10.1201/9781003499732-14

more hardware infrastructures, that means higher energy consumption consequently.

The HetNets [4] have enabled researchers to develop new concepts, protocols, and algorithms to improve the performance of the HetNets. They have also enabled the development of innovative solutions for a wide range of wireless networking and communication applications. The authors in [5] discussed the potential and challenges of combining various wireless technologies in HetNets; on the other hand, the authors in [6] focused on throughput optimization techniques considering the deployment of small cells in HetNets. The authors in [7] discussed interference management strategies in the context of HetNets. In contrast in [8], the study introduced dynamic interference management techniques for improving the performance of HetNets. Furthermore, the authors in [9] explored strategies to reduce power consumption in heterogeneous networks that have been designed and implemented in a way that is more energy-efficient. In addition, the authors in [10] used dynamic spectrum access techniques, cognitive radio, and interference management strategies, and the overall performance of the network could have been improved. On the other hand, the authors in [11] purpose of this study is to propose novel artificial intelligence (AI)-based algorithms to optimize the performance of ultra-dense HetNets by mitigating interference challenges. In addition, the authors in [12] this work focused on enhancing security protocols and privacy mechanisms in heterogeneous network environments for Internet of Things (IoT) devices. However, all work motioned the trait of the performance of HetNets using many methods; a strategy based just on one criterion or one parameter as management interference, energy-efficient power consumption, or security protocol. Consequently, our work proposed two important parameters: energy efficiency and interference management using AI-based algorithms for optimizing HetNet performance [13].

Recently, many studies have addressed HetNet [14], incorporating two significant key strategies to improve EE and interference mitigation based on the signal-to-interference-to-noise ratio (SINR). In terms of energy efficiency, there are many exciting results in terms of saving battery energy and transmission power in small and macro BSs, as well as significant improvements in network capacity in HetNet in terms of interference reduction. The authors in [15] proposed an algorithm that combined user scheduling and user association for load balancing and EE by maximizing a distributed technique the optimization named the alternating direction method of multipliers. The algorithm proposed in [16] utilizes a utility based on user association, which is primarily focused on subframe interference coordination The authors in [17] proposed many schemes of user association based on weighted average rate using a stochastic geometry approach and on cover probability. The authors in [18] proposed an algorithm of power control with mobility prediction to evaluate radio head switching processes in a heterogeneous cloud radio access network (H-CRAN) to enhance energy efficiency. The authors in [19] presented a joint optimization approach that considers the BS sleep model for ameliorating the energy efficiency of HetNet. The authors in [20] elaborate algorithm for a small-BS on/off switching based on the optimum distance to decrease power consumption in the network. The authors in [21] considered a small-cell sleep methodology based on the traffic load in that BSs are put off for a certain duration when their traffic load reduces under some threshold. The authors in [22] proposed an energy-efficient growth problem based on coverage probability below also a random sleeping strategic sleeping policy and purpose static and dynamic user models in the sleeping policy strategy that only take on considered BS on/off strategies but do not take the user association. The authors in [23] proposed a Markov theory for mobility model based on velocity organization (MM-VC) and studied the characteristics of the phase-type distribution of cell resident time. Also, the sleeping strategy between small cells was useful to decrease the energy consumption of the system. However, it only measured energy consumption and canceled the throughput of the network. Moreover, it did not consider the user association. The authors in [24] proposed strategies that corporate optimization method considering User Association and Small-Cell BS On/Off Strategies for Energy Efficiency. The authors in [25] optimized the Resource Allocation (RA) problem in the IoT based on the heuristic algorithm; this algorithm aims to achieve the goal of optimizing the RA and decreasing the total communication cost between resources and gateways.

## 14.2 PROPOSED WORK

### 14.2.1 System Model

The system model explained in Figure 14.1 contains one Macro BS (Macro Cell) underlaid with Five Small BS (small cell). The number of user mobiles (MUEs) is randomly situated in the coverage area of the Macro Cell, while the MUEs are randomly situated in their associated small cells.

In our model system, each user is identified by several criteria where either one affects the user association. The main objective of this chapter is to improve energy efficiency by reducing and minimizing the wireless interferences and creating an efficient user association considering

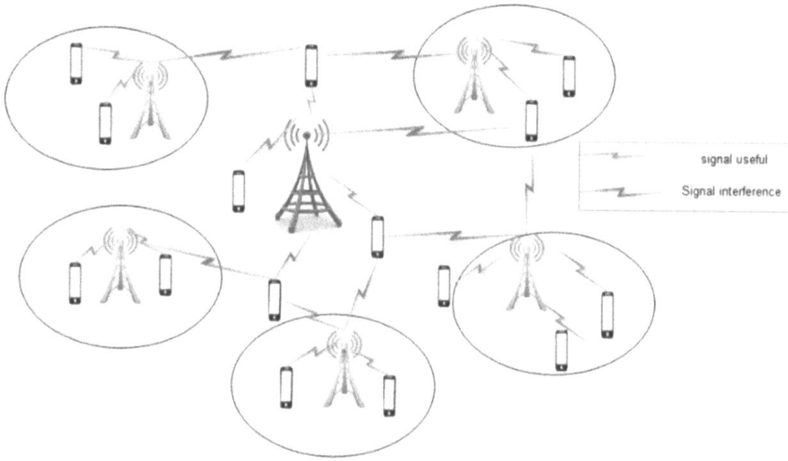


FIGURE 14.1 System model.

various criteria. The SINR of MUE i, associated with its designated MBS k, can be written as [26]:

We have modeled the problem as follows:

$$\boldsymbol{min}\ \boldsymbol{f}(\boldsymbol{u})$$

subject to: $\alpha+\beta+\gamma=1$

$$\alpha,\beta\ and\ \gamma\in[0,1]$$

The objective is to minimize the function f which represents the cost of each user.

We are based on the SINR model of the mobile user MBS at his BS which is expressed mathematically by [27]:

$$\text{SINR}=\frac{P_m^i G_i^{k,m}}{\sum_{j\in I_m} P_j^m G_j^m+\sum_{r\epsilon I_f} P_r^f G_r^f+\sigma^2} \tag{14.1}$$

For the effective energy model is defined as follows [18]:

$$EE_{m,n}=\frac{B\log_2\left(1+Y_{m,n}\right)}{P_n^{sum}} \tag{14.2}$$

$$P_n^{sum}=P_{cu}+\varepsilon P_n^T$$

Où: ε représente le coefficient de puissance qui appartient [0; 1]

Then the method of the resolution proposed by the previous work [28, 29] is as follows, in a HetNet, a user or a set of users moves randomly, which generates a significant consumption also the concern of interference, in order to overcome these problems we have modeled these problems in the form of a cost function of association of mobile user, has this objective we have combined optimal algorithm of users and the method of multicriteria analysis Analytical Hierarchy Process (AHP) in order to associate the user has a BS optimal.

The results show a significant optimization in terms of energy and minimization of interference between wireless devices. Nevertheless, this optimization remains limited, for that we proposed the machine learning method to improve the results obtained from our previous work.

AI technologies provide a range of methods for expanding human intelligence and can generate a wide variety of applications that can be useful to society and companies. There are many examples of the application of AI in our current lives, such as web searches, filtering unwanted messages, image recognition, speech understanding, and self-driving cars.

ICT development can improve the performance and efficiency of the communications infrastructure, systems, and components, create new services, and ensure the best user experience. It also encourages relevant policy contributions and legal, social, and ethical aspects that can help unleash the potential of AI techniques [30] in communication technologies, and promote technical cooperation and digital inclusion.

Communication technologies are expected to become more dynamic and responsive, in order to operate efficiently and at a low cost under difficult conditions. Predictive algorithms, computational analysis, logical thinking, and problem-solving, supported by forward-looking policies and a common set of criteria, can help in this direction.

Our problem situation revolves around the difficulty used in choosing the best and most suitable station with the least power, SINR, and the fastest time. So we are going to use a technology of AI shown in Figure 14.2 to predict the selection of the most suitable station from among the stations near the user, we are going to consider as input to the system both the distance between the user and the station, the SNR that each station produces in relation to each user and the energy needed to create the communication between the user and the station. After we have revised and organized the inputs well, we are getting them into the system for training. The system output will be one of the stations installed in the field to capture users and process their data faster and without any problems.

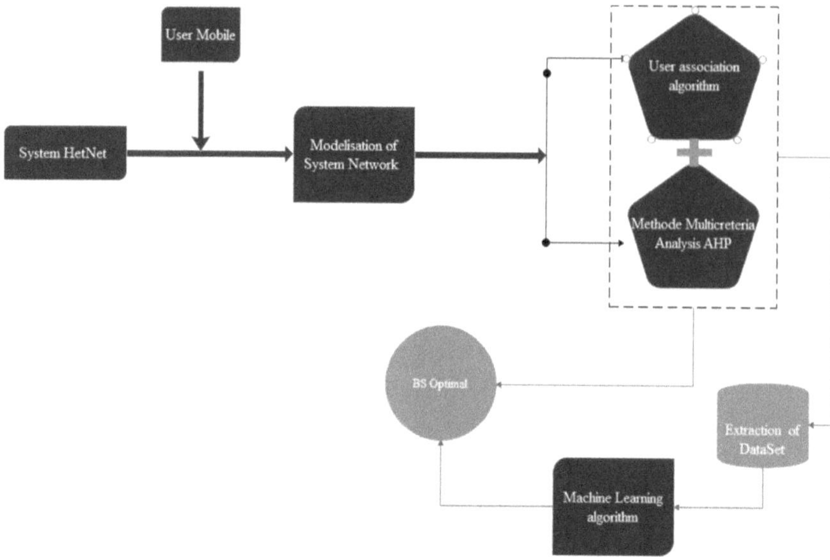


FIGURE 14.2 Methodology of our system model.

***Step 1:***

***For each user j∈{1…, |#ListUsers|}:***

***For each base station k∈{1…, |#ListBS|}:***

***Check distance constraint:***

*If d (j, k) <$R_{BS}$d then:*

***Compute the weights:** α, β, and c using equations (1) and (2).*
***Compute Cost (j, k): Cost (j, k)** ←α·**d** (j, k) + β·$\boldsymbol{E_{tr}}$ (j, k) + c·**SINR** (j, k)*
***End for** (BS loop)*
***End for** (users' loop)*

***Step 2:***

*→Export the data generated by the initial algorithm*

*--->Training a machine learning model to predict cost*
*→Using the model to select the optimal base station for each user.*
*→Implementing an optimization algorithm to improve EE.*
*→Evaluating and refining the model iteratively.*

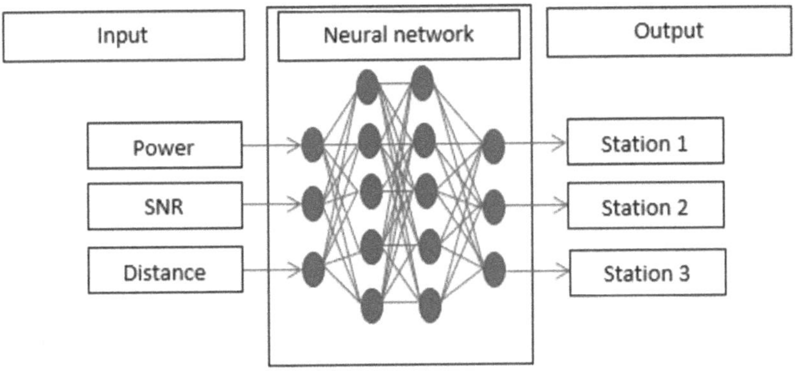


FIGURE 14.3 Description of our method of machine learning.

In this chapter, we present a novel algorithm for user association in HetNet, based on a cost function. We utilized the AHP method to assign significant weights within the cost function, considering three critical factors: SINR, distance, and user power consumption. Driven by the goal of enhancing energy efficiency and optimizing user association, we subsequently exported the data generated by the initial algorithm. This data was then used to select and train a machine learning model as shown in Figure 14.3, which further refined the process by selecting the optimal BS for each user.

## 14.3 RESULTS AND DISCUSSION

It is clear from Figure 14.4 that the top of the error goes down at every training epoch, for train, validation, test, and best. During epoch seven we note that the maximum error value is 2.657E−07, this result suggests that the system has learned well.

The same thing for the gradient shown in Figure 14.5, which is also reduced as increase epochs. In the error histogram shown in Figure 14.4, we also note that the error ratio is between −4.7E−05 and 2.89E−05, we also see that there are very few errors where it is confined to a very small peak which means a system has learned very well.

Figure 14.6 shows all the data that have been trained, where we find that they are on a linear line, as we also note that accuracy is up to 97%, which is a value that is hard to achieve in the implementation. All these results suggest that the system has learned so much and efficiently. And as we know, AI needs very big data to show the results for what they are, but we deliberately take the same data that we worked on in the previous research to see whether or not the AI system is producing good results.

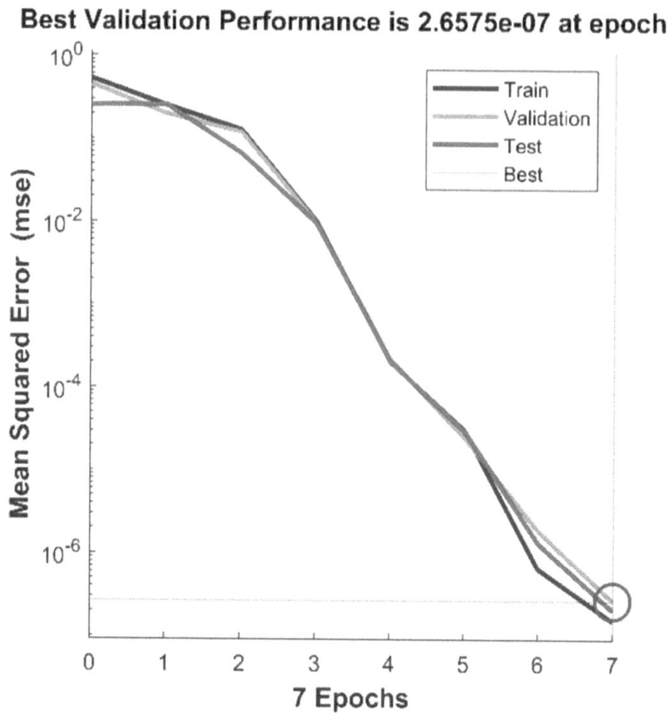


FIGURE 14.4 The performance of the system at epoch.

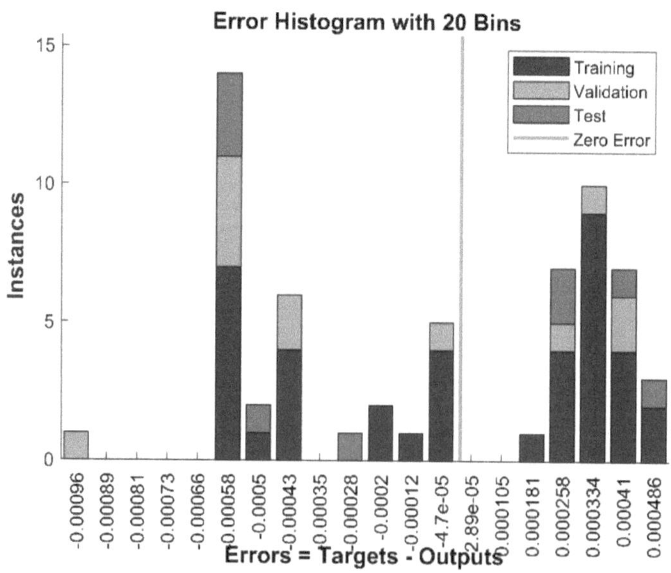


FIGURE 14.5 The performance of the system.

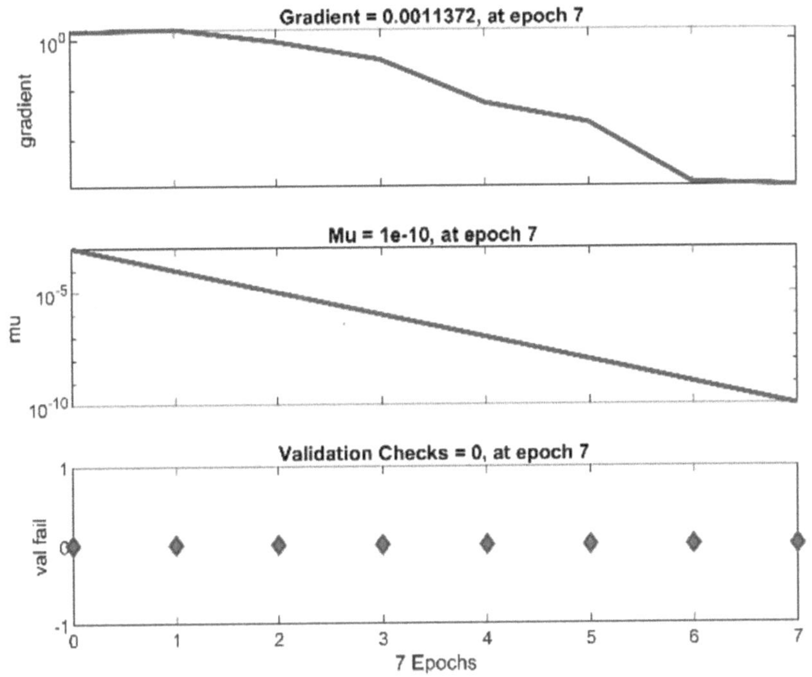


FIGURE 14.6 The gradient system at epoch 7.

After these results, we can conclude that our methodology allows us to improve energy efficiency of BS and mitigate the interference between BS and users, that our system HetNet has learned by way of efficiency to allow each user to select the optimal BS in HetNet. We show in Table 14.1, the comparative of performance results with related work.

TABLE 14.1 The Comparative Study of Result

| Related Work | Problem | Optimization Technique | Main Objective | HetNet | Data Cooperation | Performance EE |
|---|---|---|---|---|---|---|
| [31] | Radio resource and power | Dual decomposition-based method | Minimize the total power consumption | Yes | No | No |
| Our previous work [32] | User association, energy consumption, mitigation interference | AHP theory | Maximize the EE and mitigation of interference | Yes | Yes | Improved the EE with 20% |
| Proposed work | User association, energy consumption, mitigation interference | AHP algorithm and machine learning | Maximize the EE and mitigation of interference | Yes | Yes | Improved the EE with 40% |

## 14.4 CONCLUSIONS

From the results obtained from our previous work, we simultaneously considered the dataset of the SINR, power consumption, and user distance for each user moved randomly in HetNet. This enhanced energy efficiency and mitigation of interference by concurrently preventing recurrent switching between users and SBSs. The simulation results obtained that our proposed approach can enhance network performance in HetNet by 40% compared to 20% of previous work. As a perspective, we propose to combine IoT technologies and machine learning to improve energy efficiency and mitigate interference between users and SBSs.

## REFERENCES

1. Islam, N., Alazab, A., & Agbinya, J. (2019). Energy Efficient and Delay Aware 5G Multi-Tier Network. Remote Sensing. https://doi.org/10.3390/rs11091019
2. Nguyen, D. D., Nguyen, H. X., & White, L. B. (2018). Evaluating Performance of RAT Selection Algorithms for 5G HetNets. IEEE Access. https://doi.org/10.1109/ACCESS.2018.2875469
3. Al Haddad, M., & Bayoumi, M. (2016). Energy-Pivotal Solutions for Green Femtocell Power Control in Hybrid-Dense Deployments. Journal of Green Engineering.
4. Ouamri, M. A., Oteşteanu, M. E., Isar, A., & Azni, M. (2020). Coverage, Handoff and Cost Optimization for 5G Heterogeneous Network. Physical Communication. https://doi.org/10.1016/j.phycom.2020.101037
5. Alhammadi, A., Mardeni, R., Alias, M., Yusoff, Shayea, I., Alriah, S., & Abas, A. (2019). Advanced Handover Self-Optimization Approach for 4G/5G HetNets Using Weighted Fuzzy Logic Control. 1–6. 10.1109/ConTEL.2019.8848507
6. Alwarafy, A., Abdallah, M., Ciftler, B., Al-Fuqaha, A., & Hamdi, M. (2022). The Frontiers of Deep Reinforcement Learning for Resource Management in Future Wireless HetNets: Techniques, Challenges, and Research Directions. IEEE Open Journal of the Communications Society. 3. 1. 10.1109/OJCOMS.2022.3153226.
7. Kose, A., Lee, H., Foh, C., & Shojafar, M. (2024). Multi-Agent Context Learning Strategy for Interference-Aware Beam Allocation in mmWave Vehicular Communications. IEEE Transactions on Intelligent Transportation Systems. 10.1109/TITS.2024.3351488.
8. Iqbal, M., Ansari, E., & Akhtar, S. (2021). Interference Mitigation in HetNets to Improve the QoS Using Q-Learning. IEEE Access. 1. 10.1109/ACCESS.2021.3060480
9. Lavrinenko, V., & Stepanov, M. (2024). Component Analysis for Energy-Efficient Multimedia Networks Utilizing 5G Radio Access Technologies. Technology Audit and Production Reserves. 2. 37–41. 10.15587/2706-5448.2024.301384

10. Nayanam, K., & Sharma, V. (2024). Cognitive Radio Based Enhanced Compressive Spectrum Sensing Technique for 5G Adhoc Networks. International Journal of Engineering Research and. 13. 10.17577/IJERTV13IS020003.
11. Liu, F., & Zhao, S. (2020). Statistical Resource Allocation Based on Cognitive Interference Estimation in Ultra-Dense HetNets. IEEE Access. 1. 10.1109/ACCESS.2020.2987360
12. Hamza, R., & Chikouche, N. (2023). Call for papers: Special Issue on "Privacy-Preserving Cryptography for Emerging Technologies: Applications and Challenges".
13. Airlangga, G., Sihombing, D., Bata, J., & Fitrian Isnawati, A. (2024). Optimizing Femtocell Networks: A Novel Game Theory Based Power Management Model for Enhanced SINR and Energy Efficiency. IEEE Access. 1. 10.1109/ACCESS.2024.3405534
14. Kumar, B. B., Sharma, L., & Wu, S. L. (2019). Online Distributed User Association for Heterogeneous Radio Access Network. Sensors (Switzerland). https://doi.org/10.3390/s19061412
15. Huang, J., Li, J., Chen, Z., & Pan, H. (2018). HICIC: Hybrid Inter-Cell Interference Coordination for Two-Tier Heterogeneous Networks With Non-Uniform Topologies. IEEE Access. https://doi.org/10.1109/ACCESS.2018.2841194
16. Renzo, M., & Guan, P. (2014). Stochastic Geometry Modeling of Coverage and Rate of Cellular Networks Using the Gil-Pelaez Inversion Theorem. Communications Letters, IEEE. 18. 1575–1578. 10.1109/LCOMM.2014.2341251.
17. Park, H., & Lim, Y. Energy-Effective Power Control Algorithm with Mobility Prediction for 5G Heterogeneous Cloud Radio Access Network. Sensors 2018, 18, 2904. https://doi.org/10.3390/s18092904
18. Patil, S., & Bhavikatti, A. (2019). Heterogeneous Network Optimization Using Robust Power-and-Resource Based Algorithm. International Journal of Electrical and Computer Engineering (IJECE). 9. 4226. 10.11591/ijece.v9i5.pp4226-4237
19. Huang, J., Li, J., Chen, Z., & Pan, H. (2018). HICIC: Hybrid inter-cell interference coordination for two Bousia, Alexandra & Antonopoulos, Angelos & Alonso, Luis & Verikoukis, Christos. (2012). "Green" Distance-Aware Base Station Sleeping Algorithm in LTE-Advanced. 10.1109/ICC.2012.6364240
20. Ashraf, I., Boccardi, F., & Ho, L. (2011). SLEEP Mode Techniques for Small Cell Deployments. Communications Magazine, IEEE. 49. 72–79. 10.1109/MCOM.2011.5978418
21. Liu, C., Natarajan, B., & Xia, H. (2015). Small Cell Base Station Sleep Strategies for Energy Efficiency. IEEE Transactions on Vehicular Technology. 65. 1. 10.1109/TVT.2015.2413382
22. Liu, Q., & Li, Y. (2014). Energy conservation schemes based on mobility modeling in small cell networks. 10.1109/URSIGASS.2014.6929302.
23. Gao, J., Ren, Q., Gu, P., & Song, X. (2019). User Association and Small-Cell Base Station On/Off Strategies for Energy Efficiency of Ultradense Networks. Mobile Information Systems. 2019. 1–12. 10.1155/2019/6871378

24. Sangaiah Arun Kumar, Ali Asghar Rahmani Hosseinabadi, Morteza Babazadeh Shareh, Seyed Yaser Bozorgi Rad, Atekeh Zolfagharian, and Naveen Chilamkurti. 2020. "IoT Resource Allocation and Optimization Based on Heuristic Algorithm" Sensors 20, no. 2: 539. https://doi.org/10.3390/s20020539
25. Achki, S., Aziz, L., Gharnati, F., & Ouahman, A. (2020). User Association Strategy for Energy Efficiency and Interference Mitigation of Heterogeneous Networks. Advances in Materials Science and Engineering. 2020. 1–9. 10.1155/2020/7018727
26. Ibrahim, H., Elsawy, H., Nguyen, U., & Alouini, M.-S. (2016). Mobility-Aware Modeling and Analysis of Dense Cellular Networks With C-Plane/U-Plane Split Architecture. IEEE Transactions on Communications. 10.1109/TCOMM.2016.2609905
27. Abdulkafi, A., Tiong, S. K., Chieng, D., Ting, A., Ghaleb, A., & Koh, J. (2013). Modeling of Energy Efficiency in Heterogeneous Network. Research Journal of Applied Sciences, Engineering and Technology. 6. 10.19026/rjaset.6.3623
28. Aziz, L., Achki, S., & El Gourari, A. (2023). Efficient Cluster-Head Selection using BWM-TOPSIS Tool. 1-5. 10.1109/INISTA59065.2023.10310556.
29. Ghazzai, H., Farooq, M. J., Alsharoa, A., Yaacoub, E., Kadri, A., & Alouini, M. S. (2016). Green Networking in Cellular HetNets: A Unified Radio Resource Management Framework With Base Station On/Off Switching. IEEE Transactions on Vehicular Technology. 66. 5879–5893.
30. Sudhakar, K., & Sumaiya, N. (2023). Machine Learning Based Industrial Engineering with 6G Technology. Journal of Pharmaceutical Negative Results. 13. 2022.
31. Aziz, L., El Gourari, A., & Achki, S. (2023). A New Effective Strategy for User Association in Heterogeneous Networks. International Journal of Sensors, Wireless Communications and Control. 13. 10.2174/2210327913666230601153113
32. Aziz, L., & Achki, S. (2021). Optimal User Selection using BWM Tool. 1-4. 10.1145/3454127.3457622

CHAPTER 15

# Legal and Regulatory Challenges in Implementing Blockchain-Enabled Renewable Energy Systems

Karisma Karisma

## 15.1 INTRODUCTION

The depletion of fossil fuel reserves significantly impacts the country's economy. As the cornerstone of global energy supply, it influences energy prices and consequently affects energy availability and affordability. Understandably, scarcity poses a significant threat to energy security by reducing the resilience and flexibility of the energy systems. Moreover, the dependence on fossil fuels is the primary factor driving anthropogenic climate change due to the significant carbon emission levels. Since economic and social stability depends on energy security, the heavy reliance of developed and developing nations on fossil fuels like coal, natural gas, and crude oil leads to widespread disruptions. Therefore, it is crucial to gradually reduce this dependence and shift towards renewable and sustainable energy sources.

There is a need to prevent a path-dependence or lock-in effect from fossil fuel reliance. Decentralised and digitalised energy systems can gradually decrease dependence on fossil fuels, increasing energy security and aligning

DOI: 10.1201/9781003499732-15

with the shift to cleaner energy solutions. Further, the shift away from fossil fuels and renewable energy use represents a promising trend in the energy sector. It empowers individuals, communities, and societies to take charge of energy production, consumption, and trading, promoting sustainability, energy democratisation, and liberalisation in energy systems.

Rising tensions surround the existing energy architecture, which is heavily regulated and monopolised. Conventional centralised and monopolised energy markets, attributable to historical and political factors, are often associated with high bureaucratic hurdles, lack of competition, and inefficiencies. The rigour and rigidity of conventional frameworks pose significant obstacles for energy generation units as large utility companies have (and continue to) dominate energy transmission and distribution. When exploring the concepts of energy democratisation and liberalisation, the author addresses the elements of ownership, sovereignty, and autonomy when allowing more significant participation in Renewable Energy (RE), systems.

### 15.1.1 Ownership

Acquiring ownership of RE infrastructure empowers consumers to become prosumers, generating energy for personal consumption while selling excess energy to other consumers. The plurality, attributes, and material features and traits of Distributed Energy Resources (DERs), make it costly and inefficient to manage in a decentralised manner. In essence, leveraging blockchain to manage the highly heterogeneous nature of DERs through a transparent and tamperproof database can facilitate bidirectional energy trading, enhancing the feasibility and viability of energy systems. Blockchain allows market actors to play distinct roles, conduct various tasks, and create a more heterogeneous atmosphere in futuristic energy systems aligned with decarbonisation and digitalisation goals.

### 15.1.2 Sovereignty

Energy sovereignty allows energy citizens to make informed decisions about their energy mix, use of distributed energy resources, and production, consumption, and trading decisions. This sovereignty facilitates greater integration of renewable energy and decarbonisation while aligning with energy democracy. Further, it gives users choice, control, and flexibility over their energy usage and preferences. The current policy and regulatory landscapes, enacted in siloed ways, lack energy sovereignty and, in many situations, limit its applicability.

### 15.1.3 Autonomy

With autonomy, prosumers and consumers can engage actively and autonomously in energy exchanges unaffected by external intervention from energy companies holding a monopoly and claiming control over the management and operations of energy systems. It is essential to ensure that individuals with the desire or experience in utilising local clean energy and the RE infrastructure partake in peer-to-peer and community-based energy trading models easily, generating environmental and societal benefits and engaging in democratic energy participation.

This chapter delves into the study of the regulatory challenges and jurisdictions in the Global South that confront nations when implementing blockchain-enabled energy trading. We draw upon trust-building and industry-specific challenges to make critical contributions that can gradually decrease dependence on fossil fuels and transition to RE sources. Since global efforts to achieve Sustainable Development Goals align with net-zero commitments, fostering innovation and decentralisation within energy trading systems becomes increasingly critical. In addition, it is essential to revisit and redefine the roles, rights, and obligations conventionally undertaken by incumbent utilities to allow a gradual transition to full-fledged decentralised models. The rigorous legal and regulatory measures, such as significant grid connection fees and unclear policy exemptions, obstruct the prosumer's entry into the national power grid and hinder fair competition in the retail electricity markets. In conventional regimes, network users bear transmission and distribution network charges that do not reflect network costs. In order to enable the smooth integration of blockchain-enabled energy trading systems, national regulatory authorities must redefine tariff methodologies that currently pose significant obstacles to implementation. Encouraging and facilitating the implementation of capacity and dynamic tariffs at the micro and macro levels is essential to motivate proactive customers with diverse energy profiles. The author pursues desktop and doctrinal legal research, acknowledging the dynamic landscape of blockchain-enabled energy trading systems. Further, exploratory research techniques encourage a conceptual and theoretical investigation of the barriers prominent in implementing blockchain-enabled peer-to-peer and community-based models. Given the multidisciplinary nature of the topic, particularly at the intersection of blockchain and law, the analysis incorporates valuable insights from technological advancements to establish a sustainable regulatory framework.

## 15.2 FOMENTING ENERGY SECURITY IN BLOCKCHAIN-ENABLED ENERGY TRADING SYSTEMS

The uninterrupted availability of power supply is crucial in ensuring energy security. Blockchain-enabled energy trading is vital for continuous power supply, reducing the risks of blackouts, downtime, and other disruptive conditions. We argue that blockchain-based energy systems offer optimised, coordinated, and resilient characteristics, facilitating the operation of decentralised energy systems and distributed energy resources. This technology enhances modular energy technology usage and improves production, consumption, and trading. Extending the role of blockchain to energy systems enhances resiliency and robustness, mitigating single points of failure.

Low energy prices ensure that affordability and inclusiveness meet consumers' energy demands. Blockchain-based energy trading systems are promising avenues for facilitating rural electrification initiatives and ensuring isolated communities have affordable access to energy without interruption. Affordable and accessible energy bridges the gap between prominent urban and rural disparities due to smaller generating units. Blockchain eliminates the need for third-party intermediaries, increasing efficiency and reducing costs. The supply of surplus electricity at lower costs with a seamless and secure system proves beneficial to ensuring energy security.

Further, facilitating the use of RE through blockchain-based energy trading systems generates greater public acceptance due to reducing carbon emissions. Fossil fuels are becoming less acceptable due to their numerous negative externalities (Georgarakis et al., 2021). In reflecting public acceptance of RE sources, national and international governance measures, policies, and regulations are means to achieve that end.

It follows that blockchain-enabled Distributed Energy Systems (DESs), will promote the attainment of Sustainable Development Goals and provide promising avenues to achieve energy security, namely the availability, affordability, accessibility, and acceptability. Sustainable Development Goals (SDG), 7 ensures access to "affordable, reliable, sustainable, and modern energy for all." SDG 13 explores the need to "take urgent action to combat climate change and its impacts."

However, implementing technology-centric systems requires institutional frameworks and policies to reduce regulatory challenges.

## 15.3 TOPOLOGIES OF BLOCKCHAIN-ENABLED ENERGY TRADING SYSTEMS

In this section, the author explores the topologies and trajectories of blockchain systems and real-world applications.

This chapter identifies four blockchain types, namely public, private, consortium, and hybrid blockchains in providing a more nuanced understanding of the state-of-the-art. Each has different governance structures, architecture, and degrees of freedom in operations, decision-making, data access, transaction verification, validation, and approval in energy trading systems. We acknowledge that the applicability of a given blockchain architecture is fluid, and different blockchain domains have distinct advantages and disadvantages. Therefore, in line with blockchain's topologies and trajectories, the appropriate blockchain architecture for peer-to-peer and community-based systems should align with distinct business models, use cases, and objectives to offer appreciable contributions to heterogeneous energy trading systems.

*Public blockchain*: A public blockchain is a self-regulating, open, and visible network that is accessible to an unlimited number of participants. It allows for the seamless propagation of blocks and active participation in consensus processes in a decentralised and transparent manner, all without the need for third-party intervention (Ncube et al., 2020; Yang et al., 2020). Public and permissionless platforms will likely be utilised for energy blockchains in the future as a fully distributed and accessible structure implemented on a larger scale due to the features of autonomy and transparency for inter-prosumer trading. One reputable use of the public blockchain network is Bitcoin, a digital currency widely used worldwide. An inherent precept of the public blockchain is that no node exerts significant or disproportionate influence over the other on the network. The public blockchain is also open and accessible to all, including government entities, members and stakeholders of energy communities, corporations and associations, prosumers, distribution system operators, local authorities, small and medium enterprises, prosumers, non-profit organisations, and transactive energy projects. The use of public blockchain depends on specific use cases and business models to minimise risks, costs, and externalities while maximising value and capabilities. This shift is essential for driving the transition from monopolised and centralised energy systems to decentralised ones.

*Private blockchain*: A private blockchain is a permissioned network that imposes restrictions on joining, reading, writing, or auditing data unless authorised. Due to the restricted access, only predetermined nodes can validate and verify transactions, maintaining a level of centralised control over network operations (Kim, 2020). Transactions on private blockchain networks undergo faster processing, thus facilitating transaction efficiency and traceability. As it stands, energy trading platforms, such as Quartierstrom in Switzerland, Power Ledger in Australia, and Brooklyn Microgrid in the United States, adopt a private blockchain to ensure the privacy and security of energy market actors, given that access to the blockchain ecosystem is limited to authorised and identifiable peers (Schneiders & Shipworth, 2018). Unlike fully centralised ledgers, a private blockchain is a partially decentralised network with seamless data transfer features and faster response times.

*Consortium blockchain*: The consortium blockchain shares similar features with a private blockchain in that access is limited. A consortium blockchain is a semi-decentralised network where only pre-selected nodes can participate in the blockchain network. A consortium blockchain is controlled by multiple cooperative organisations working together as joint entities. Participants are required to seek authorisation from the consortia and comply with the blockchain network access rules. A consortium blockchain is particularly advantageous in industries where multiple organisations operate, as it improves transparency and efficiency while reducing transaction verification time. The consortium blockchain can also be used for various energy use cases as an integrative platform to facilitate business processes and transactions, such as localised or microgrid energy trading, due to the high throughput, security, and expeditious propagation of blocks (Bischi et al., 2021).

*Hybrid blockchain*: A hybrid blockchain is a partially decentralised system that combines features from private and public blockchains. The hybrid blockchain represents an ideal solution for energy systems, as it delivers considerable robustness and makes significant contributions to the overall functionality of the energy networks. This blockchain model optimises decentralisation by deciding that certain blockchain transactions are public while securing permission to access other transactions (Jumani et al., 2021).

The utilisation of blockchain is not a mere exaggeration but an invaluable area worthy of research. Through its decentralised ledger, blockchain allows for seamless bidirectional energy trading, enhances the security and efficiency of DERs, supports effective grid management, and enables precise information transfers. The application of blockchain is rapidly proliferating across different business sectors and invoking unprecedented transitions.

## 15.4 CONSENSUS PROTOCOLS AS THE CORE OF BLOCKCHAIN SYSTEMS

As the blockchain ecosystem matures and is grounded in robust and adaptive regimes, we will likely see less resistance to implementing blockchain across industries and geographical landscapes. Broadcasting blockchain transactions to the network enables miners to verify the transactions, thus generating and adding a valid block to the blockchain ledger and eliminating the need for centralised validation (Yadav et al., 2023).

The consensus protocols constitute fundamental blockchain elements as coordination and consistency mechanisms in a distributed setting. Further, the ledger is updated with the help of consensus protocols to maintain a linear, equivocal, and shared ledger of transactions, introducing synchronicity and ensuring the data integrity of the blocks (Sajana et al., 2018). Common consensus protocols include proof-of-work (PoW), proof-of-stake (PoS), and proof-of-authority (PoA).

### 15.4.1 Proof-of-Work (PoW)

PoW involves solving a complex computational puzzle. By solving the puzzle, miners can validate transactions and insert new records onto a blockchain, thus agreeing to a single source of truth. Miners involved in mining compete to find a valid nonce to yield a hash that is less or equal to the target hash (Kirli et al., 2022). They are accorded block rewards by successfully adding new blocks to the chain, resolving the puzzle (Schinckus, 2020).

However, adopting PoW to solve computational puzzles requires significant computational power and energy consumption (Yadav et al., 2023). Contextually, bitcoin mining consumes more energy yearly than Ireland, Portugal, and Denmark combined (Shoker, 2017). It is criticised as a waste of resources as miners compete to mine the block (Chen et al., 2021; Shoker, 2017). The high carbon emission rates from PoW are at odds with the aims of decarbonising energy systems. In integrating and regulating blockchain in various applications and use cases, policymakers must develop a technical understanding of the workability of PoW. While PoW is ideal for establishing consensus, an alternate protocol must be adopted to reduce the environmental impact (Roth et al., 2022).

### 15.4.2 Proof-of-Stake (PoS)

The PoS is an alternative to PoW due to its energy-saving capabilities and reduced processing time (Aggarwal & Kumar, 2021). In generating and validating new blocks, the PoS consensus protocol requires nodes to have a certain number of tokens to be eligible to participate in the network (Baldimtsi et al., 2020; Saleh, 2021). The miners' ability to extend the chain depends on the amount of stake they own and the length of time the tokens have been staked.

Unlike PoW, PoS does not require high computational power to validate transactions and establish trustworthiness. In essence, PoS is an energy-efficient and scalable consensus mechanism that addresses the challenges present under PoW. However, adopting PoS is susceptible to long-range attacks, and nodes can be coerced to participate in malicious transactions (Bouraga, 2021). The blockchain network is prone to adversaries building alternative chains that overtake the main chain (Bouraga, 2021).

The continued adoption of PoW could exhibit a "path dependence" or "lock-in" effect by confining platform development through familiar pathways (Araújo, 2014; Truby et al., 2022). This position suggests why a habitual path of PoW is preferred by developers instead of a more energy-centric PoS mechanism. Platform developers might be reluctant to engage in "path creation," such as adopting PoS as a more suitable alternative (Araújo, 2014). This self-enforcing restriction must be curbed in the bud and warrants regulatory intervention. Through legal and policy tools, policymakers can drive significant transitions towards energy-efficient consensus mechanisms, which is appealing and beneficial to blockchain-enabled energy landscapes. Pragmatically, regulatory regimes must reflect the interest of the public and the environment, which outweighs the commercial interest of businesses, traders, and developers (Truby et al., 2022).

### 15.4.3 Proof-of-Authority (PoA)

The PoA protocol is a variation of the PoS consensus mechanism that blockchain developers can implement. However, unlike the PoW, it is less resource-intensive, making it more energy-efficient and cost-effective (Milunovich, 2022).

The PoA protocol is an energy-efficient consensus mechanism as it does not require the complex computation of puzzles to validate transactions (Khalid et al., 2021). Trusted nodes and authorised accounts can participate in the network and generate new blocks (Rebello et al., 2021). For the PoA consensus protocol, the authorised miner's authority is at stake instead of virtual currencies under the PoS (Reno et al., 2020). The validator can be "voted out" by other approved participants should the validator act maliciously (Rebello et al., 2021). PoA can be applied effectively in permissioned blockchain due to the degree of knowledge and familiarity between the participants on the network (Rebello et al., 2021). The Energy Web Foundation and Grid Singularity leverage blockchain to manage localised energy projects and adopt the PoA consensus mechanisms.

Scholars have transposed other protocols with distinctive features, advantages, and shortcomings as the building blocks of blockchain systems, which include the Practical Byzantine Fault Tolerance, delegated PoS, proof-of-elapsed-time, proof-of-importance, and proof-of-space (Hameed et al., 2022). There is a rising exigency in implementing consensus protocols that are resilient to malicious nodes, energy-efficient, and scalable, considering a steep increase in energy transactions and blocks.

In summary, the author provides a comprehensive overview of blockchain technology in by delineating the blockchain architecture and conceptualising the technical substratum. This section explains the types of blockchains to elucidate the state-of-the-art technology. About the design architecture of blockchain, the author presents the lineage of consensus protocols adopted, each as an improved and comprehensive version of the previous mechanism. Further, it also describes the terminologies commonly applied in the blockchain landscape to provide a better understanding and to stimulate a practical academic discourse. The up-to-date typology of blockchain includes blockchain features that bolster the various phases of blockchain. Further, a structured understanding of the features of blockchain is provided to aid the discussion on blockchain feasibilities and challenges when implemented on a wide scale in the energy sector. This section also allows readers to be completely cognisant of the limitations and barriers that may arise, having devoured the prominent aspects of the blockchain horizon. In furtherance to that, the author addresses the role of blockchain in the energy sector, which is instrumental towards adopting a legal readiness framework that provides a salubrious structure for countries enduring the trough of disillusionment of blockchain technology.

## 15.5 BLOCKCHAIN AT ODDS WITH DATA PROTECTION REGULATIONS

Emerging data protection regulations necessitate a functional understanding of the interactions between blockchain and energy trading systems, so it is crucial to recognise blockchain's potential to ensure decentralisation and autonomy. While the stringency of data protection regulations may pose challenges, the author argues that blockchain can continue to drive technological innovations and enable the operationalisation of energy trading activities with the proper technical and operational measures.

### 15.5.1 Right to Be Forgotten

Under the General Data Protection Regulation (GDPR), data subjects have the right to timely erasure of personal data (European Union, 2016; Holzleitner et al., 2020; Tatar et al., 2020). In private and public blockchain infrastructures, there is an inherent conflict between the immutability feature and the "right to be forgotten" (Godyn et al., 2022; Martins Gonçalves et al., 2023). Transaction data on the ledger cannot be modified, altered, tampered with, or deleted because each block is secured cryptographically using the hash function. The subsequent block includes unique strings of characters of the previous block, forming a continuous link. Modifying or altering one block would necessitate new computational solutions due to the cascading effect on all subsequent blocks, diminishing the blockchain's utility and functionality (Tatar et al., 2020). The immutability feature of blockchain ensures data integrity and trust among blockchain participants once data appends to the ledger. Immutability is a systematic and material feature of blockchain systems in that there are cumulative effects if a malicious actor or an adversary attempts to make changes or assert control. As such, any alterations are readily detectable and evident by other blockchain nodes, making it challenging for malicious actors to compromise the network.

The immutability feature is appealing as it augments integrity but poses issues of fulfilling the data erasure requests, where any infringements to the rights to be forgotten result in fines and penalties. Blockchain developers and platform operators may find themselves at a critical juncture, operating in a place of fear and harbouring considerable uncertainties in developing novel use cases and business models at the expense of technological breakthroughs and advances. Developing a sustained response to these challenges is pertinent to resolve the friction.

### 15.5.2 Right to Rectification

Rectifying data requires the data controller to rectify or erase personal data that is inaccurate, incomplete, or not up-to-date on a decentralised and immutable ledger (European Union, 2016; Li, 2020). In a public blockchain, exercising the data subject's right to rectification entails identifying incorrect or incomplete data and correcting data across numerous nodes on the network (Duarte, 2019). There is a risk of node disconnection, which might complicate the process of creating a fork, making it cumbersome to rectify data effectively (Al-Abdullah et al., 2020; Duarte, 2019; Holzleitner et al., 2020). However, on a private blockchain, coordinating data rectification is more straightforward as it involves a more centralised database than a permissionless blockchain (Duarte, 2019).

The "right to rectification" can be effectively exercised by providing a supplementary statement to resolve the friction between it and blockchain technology (European Union, 2016). Therefore, blockchain participants can add new data blocks to reflect energy participants' latest affairs without deleting inaccurate information. This approach maintains the functionality and sanctity of blockchain while allowing rectification to reflect the status quo. Especially for energy trading platforms, reflecting the accuracy of data is necessary to ensure regulatory compliance and smooth and secure operations of decentralised energy systems.

Since each block is cryptographically linked using hash functions, new computational solutions for all subsequent blocks are necessary for the event of block modifications. Nevertheless, this may result in high computational costs for any ensuing modifications, alterations, and deletions. Instead, file metadata appends without altering the blockchain to ensure transparency and data availability while maintaining the system's integrity.

In rectifying inaccurate data, one could delete the inaccurate data and add information to the chain that provides revised and updated information. However, this solution is only viable if data is stored off-chain in a separate and remote database on centralised servers without jeopardising blockchain networks (Lai & Zhao, 2022; Mekouar et al., 2022).

While the challenges of data protection regulations on blockchain are significant, there are viable technical measures to ensure privacy protection. Off-chain storage, for instance, allows for data rectifications without affecting blockchain operability and functionality. It is essential to recognise that blockchain can coexist with data protection regulations, and by implementing various technical and operational measures, we can ensure blockchain technology's continued success and security.

## 15.6 BLOCKCHAIN SECURITY THREATS AND VULNERABILITIES

This section is purposefully descriptive, elucidating security vulnerabilities and threats surrounding blockchain systems and highlighting the qualitative and quantitative impact on critical infrastructures due to the lack of oversight and accountability. With the widespread security challenges, the prospects of blockchain in the energy sector is nebulous and obscure. Energy security relates to and links decentralised power systems and effective energy policies.

A resilient energy system, supported by diverse market actors and infrastructures, demands institutional solidity and governance frameworks to mitigate common threats. As economies grow, energy supply shortages and resource overuse become pressing issues. Decentralised, digitalised energy systems are vital in leveraging renewable energy, reducing fossil fuel dependency, and boosting energy efficiency and security. Blockchain technology plays a significant role here, providing high reliability and fault tolerance, maintaining data integrity, and eliminating single points of failure. However, while blockchain enables greater energy trading participation, it also presents numerous cybersecurity challenges that could disrupt connectivity and automation.

Nations highly value energy security. Blockchain technology promotes energy availability, affordability, and accessibility, but security concerns can undermine these benefits. Cyberattacks on blockchain-enabled energy systems can disrupt production and supply, creating economic challenges. Reliance on blockchain for prosumer-centric activities increases the risk of sophisticated cyberattacks, potentially leading to local energy market disruptions. Cyberattacks threaten the development of equitable energy services, posing challenges to universal electrification and sustainable supply. Distributed denial of service (DDoS) and 51% attacks are significant threats that can interrupt energy supply and hinder economic growth.

DDoS attacks overwhelm networks with disruptive traffic, causing service disruptions (Jahnavi, 2021; Wani et al., 2021).

Although decentralised blockchain platforms are generally resistant, they are not immune. For example, in 2021, Polygon and Solana networks suffered DDoS attacks, causing congestion and delays (Hulubas, 2022; Thurman, 2022). Binance Exchange faced a similar attack in 2020, resulting in performance disruptions (Trajcevski, 2020). Ethereum network and BitMEXalso experienced DDoS attacks, leading to stability issues and revenue losses. A 51% attack occurs when attackers gain majority control of mining power, allowing them to rewrite transaction histories and block new transactions, which leads to double spending and revenue losses. Ethereum Classic and Monacoin faced such attacks in 2018 and 2019, resulting in significant losses (Fadilpašić, 2020). While no significant cybersecurity attacks have yet occurred in blockchain-enabled energy trading, the potential for such incidents necessitates robust security measures to protect trading platforms and ensure smooth operations. Increased blockchain trading volumes demand enhanced cybersecurity.

Cyber risks in blockchain systems encompass vulnerabilities, threats, and impacts. The International Organization for Standardization and the U.S. National Institute of Standards and Technology define "risk" as the probability of a security threat exploiting a vulnerability (ISO/IEC 27005:2011, 2011). The author's trajectory is that the adverse impact of such events can be taken as a starting point to aid the reader's understanding.

### 15.6.1 Vulnerabilities

Technical, organisational, and systemic vulnerabilities compromise blockchain's inherent cyberattack resistance (Lesniewska et al., 2019). A brief explanation of these vulnerabilities are as follows:

i. Technical Vulnerabilities: Misconfigurations, system design flaws, security weaknesses in code, and inadequate cryptographic implementations can lead to attacks (Lesniewska et al., 2019).

ii. Organisational Vulnerabilities: Weak security policies, lack of skilled blockchain developers, and insufficient training contribute to vulnerabilities.

iii. Systemic Vulnerabilities: Interconnected systems can spread the impact of an attack, destabilising multiple systems (Limongi & Galderisi, 2021).

Blockchain architecture includes application, data, and network layers, each susceptible to specific vulnerabilities. For example, the 2016 decentralised autonomous organisation (DAO) attack impacted the Ethereum platform where malicious actors aligned their interests to the detriment of their victims (DuPont, 2017; Samreen & Alalfi, 2020). This section assesses the vulnerabilities associated with the blockchain architecture layers towards energy policy goals, which could exacerbate trust, compromising democratic participation in the energy market. A brief explanation of these layers are as follows:

i. Application Layer: This layer comprises decentralised applications (dApps), smart contracts, and user interfaces. It is vulnerable to DDoS attacks, user wallet breaches, and network disruptions, which can impact local energy transactions (Hameed et al., 2022).

ii. Data Layer: This layer is the blockchain's core data structure, maintaining transactional data. Privacy leakages and private key thefts are significant risks here (Deepa et al., 2022).

iii. Network Layer: This layer facilitates communication among nodes but is prone to attacks like DDoS, routing, and Sybil attacks, affecting network performance and data integrity (Ul Hassan et al., 2020).

### 15.6.2 Threats

Malicious actors can exploit blockchain software to launch new attacks or augment existing ones. Internal and external attacks categorise these threats, summarised as follows:

i. Internal Attacks: Malicious nodes within the system can launch DDoS attacks or gain consensus power for 51% of attacks, compromising system integrity (Marijan & Lal, 2022; Sai et al., 2021).

ii. External Attacks: These include denial of service, double-spending attacks, and user-wallet breaches, resulting in significant disruptions (Harika et al., 2019; Le et al., 2019).

Proper governance and coordinated policies are crucial to mitigate these threats. Developers must implement validating mechanisms, while platform operators should conduct red-teaming exercises to assess and improve system resilience.

### 15.6.3 Impact

Cyberattacks on blockchain systems have significant societal and technological impacts, potentially undermining the viability and effectiveness of blockchain in the energy sector. Qualitative impacts include denial of network access, system inefficiencies, and reputational damage. Quantitative impacts involve revenue losses from network downtime, malicious transaction alterations, and recovery costs (Guru et al., 2021; Jahnavi, 2021; Kausar et al., 2021; Raza et al., 2021; Wani et al., 2021; Xu et al., 2021b).

The complexity and novelty of blockchain technology raise cybersecurity concerns, threatening energy security and the continued operation of energy infrastructures. Effective legal, institutional, and regulatory frameworks are essential to address these challenges and ensure a resilient energy system.

While blockchain technology offers enhanced security capabilities, it is not entirely immune to cyberattacks. The design and operation of blockchain networks must prioritise reliability to prevent disruptions in energy supply. Critical factors include the number of nodes, authorisation requirements, consensus mechanisms, and encryption strength.

## 15.7 INDUSTRY-SPECIFIC CHALLENGES OF BLOCKCHAIN-ENABLED ENERGY TRADING SYSTEMS

### 15.7.1 Monopolisation of Public Distribution Networks and Lack of Bidirectional Access to the Grid

The monopolisation of public distribution networks poses significant obstacles for new entrants with generation units, hindering market liberalisation. The development of conventional frameworks is centred around large, monopolised power stations, preventing jurisdictions in both the Global North and South from fully embracing and integrating prosumer-centric initiatives. There is a pressing need to reduce reliance on heavily regulated and monopolised markets and pave the way for more agile, responsive, and resilient technology-driven systems, recognising new market actors as pivotal to facilitate blockchain-enabled energy systems (Amenta et al., 2021; Bukar et al., 2023). Blockchain-enabled energy systems represent a bottom-up approach that contrasts the dominant top-down model. Unlike monopolised systems where energy is generated and transmitted through high-voltage transmission lines to end-users, technology-driven systems utilise bidirectional energy flows and distributed energy resources, thereby transforming the use of network grids.

In monopolised systems, a lock-in effect occurs due to the sustained dependence on an incumbent provider, making switching difficult. The lock-in stifles competition and disincentives to switch to more innovative-centric energy production, consumption, and transaction (Stein, 2016). Further, existing capital infrastructural investments in fossil fuels can impede the adoption of renewable energy sources. Reliance on fossil fuels and the resulting entanglement restrict the funding for the transition to decentralised energy trading. This situation arises because overcoming the dominance of vertically integrated monopolies is a complex task that requires implementing robust regulatory and policy frameworks to break free from the current system. However, these frameworks still favour established and incumbent players, hindering new entrants' crucial need for non-discriminatory and equitable network access (Kallies, 2016).

These limiting factors do not make the transition fairer, inclusive, just, and transparent.

We delve into the lack of non-discriminatory prosumer access to the power grid, which generates a bottleneck, limiting the exchange and surplus of energy via public distribution networks. In microgrid landscapes, peers can easily trade energy without needing third parties to manage transactions (Zhou & Wu, 2021). Such initiatives reduce energy poverty through decentralised systems, particularly in rural areas. As acknowledged above, access to secure, affordable, and reliable energy sources causes social and economic ramifications in developed and developing nations.

As marginalised communities are less likely to invest in individualised distributed energy resources, shared infrastructure such as sensors, smart meters, appliances, and IoT-enabled devices ensures access to reliable energy generation on microgrid systems. Blockchain utilisation allows for complete decentralisation when deployed on microgrid infrastructures. Brooklyn microgrid is a monumental real-world microgrid-based peer-to-peer energy trading system. Developing a blockchain-based platform in Brooklyn (New York) comprises multiple framework components, including microgrid setup, grid connection, information systems, market mechanism, pricing mechanism, energy management trading system, and regulation (Mengelkamp et al., 2018).

However, microgrids have higher investment costs than traditional grid infrastructures and may raise scalability issues (Soshinskaya et al., 2014). Since it is not practical cost-wise to construct an entirely new distribution network, any decision to implement a microgrid to enable a blockchain-based peer-to-peer energy trading system should be on a case-by-case basis after a cost-based analysis. Adopting existing network infrastructures is necessary for blockchain-enabled energy trading systems, as constructing new distribution grids or microgrids solely for peer-to-peer energy trading is not a practical and feasible approach. Given that the utilisation of microgrids may not be entirely practical, open network access for the transportation of electricity is pertinent.

### 15.7.2 Lack of Cost-Reflective Tariffs and Dynamic Tariffs

Many jurisdictions in the Global North and South do not advance efficient and cost-reflective tariff designs considering the augmentation of distributed energy resources and the impact on energy networks. An updated network tariff methodology should incentivise active consumers and ensure the fair allocation of costs that do not unfairly burden passive consumers (Ashour et al., 2021). The current grid tariff methodology does not alleviate grid congestion or promote the dynamically efficient conduct of prosumers engaging in blockchain-enabled peer-to-peer energy trading systems (Felice et al., 2022). Regulators must carefully consider the technical variables when overseeing energy markets in blockchain environments, especially regarding electricity tariffs. There is a compelling need for updated tariffs that facilitate bidirectional energy trading systems and guarantee grid flexibility and reliability. In a residential setting, blockchain-enabled energy trading is highly appealing due to the autonomy and control it provides prosumers when buying and selling energy with nearby peers. Blockchain is utilised in grid management to monitor and facilitate optimal operations and performance, promoting effective and reliable interactions between network participants and stakeholders (Hua et al., 2022). These systems alleviate grid congestion, crucial for transactive energy systems, and reduce tariffs, ultimately increasing profit maximisation. Moreover, the tariff data stored on the blockchain platform enables enhanced visibility and accessibility for all participants in the energy trading sector, painting a promising picture of the future of energy markets (Aman et al., 2024; Dinesha & Balachandra, 2022).

Conventional tariffs present hindrances and disincentivise prosumers from participating in individual and collective prosumerism while implementing cost-reflective tariff designs that consider price and economic signals (Felice et al., 2022). Therefore, ensuring the successful integration of renewable energy assets and distributed energy resources is crucial in addressing network congestion issues caused by network augmentation and peak demand.

Dynamic tariffs effectively convey price and economic signals by considering geographical and temporal granularities (Abdelmotteleb et al., 2022). Such tariffs provide incentives to market participants when engaging in transactive energy systems. Scholars recognise the potential of dynamic tariffs in blockchain-enabled energy systems as they facilitate active trading while considering network load profiles for effective grid utilisation (Brown et al., 2020; De Clercq et al., 2020). Dynamic tariffs allow prosumers to effectively respond to energy market conditions in light of increased utilisation of DERs, community-owned assets, and metering infrastructure to support peer-to-peer models. A proper governance structure can ensure the development of dynamic tariffs to optimise such resources and infrastructure.

Cost-reflective and dynamic tariffs, including time-of-use (ToU), critical peak pricing, and real-time pricing, can progressively enhance engagement in revenue maximisation. ToU tariffs, for instance, are not just economically effective network tariffs, but they also allow peers to formulate trading strategies by accounting for peak and non-peak hours, providing a clear path towards a more efficient and cost-effective energy market (Torriti & Yunusov, 2020). Critical peak pricing, on the other hand, is a tariff structure that charges higher rates during periods of high demand, encouraging consumers to reduce their usage at these times (Gyamfi et al., 2013). Real-time pricing, as the name suggests, is a tariff structure that varies the price of electricity based on the current market price, allowing consumers to adjust their usage to take advantage of lower prices (Xu et al., 2021a).

Regions can gradually implement ToU tariffs before transitioning to more specialised and complex dynamic regimes to prevent governance gridlocks and achieve effective results. This approach will help to avoid unexpected and abrupt tariff changes, ensuring a smoother transition for all stakeholders involved. However, regulators must consider the gestation period, which refers to the time it takes to introduce cost-reflective tariffs and for the widespread implementation of DESs to occur. Many countries that cater to flexibility and technology-driven energy markets have begun implementing tariff methodologies that consider distributed generation and the diversity of network users with varying load profiles. In its working paper, the U.K. Office of Gas and Electricity Markets (Ofgem) identifies five designs for distribution charges, including agreed capacity, actual capacity, volumetric ToU, dynamic charging, and critical peak rebates to provide accurate price signals to consumers and are cost-reflective (Ofgem, 2022). However, Ofgem is still assessing the feasibility of such tariff designs by weighing up the costs and benefits. Ofgem recently published the Decision and Direction on the Access and Forward-Looking Charges Significant Code Review to ensure energy networks' efficiency, flexibility, and cost-reflectivity. However, Ofgem acknowledges the need to enact legislation to implement the charging decisions and directions, expecting reforms to be in place by 2023 (Ofgem, 2022). Rule 6.21 of the National Energy Rules in Australia allows distribution network service providers to adopt cost-reflective tariffs. According to the Australian Energy Regulator (AER), this includes shifting from a "single rate usage tariff" to a rate that varies based on peak and off-peak times. Further, the tariff reform will allow for the widespread deployment of distributed energy resources while encouraging more efficient use of energy networks with lower costs to consumers (Australian Energy Regulator, n.d.).

Similarly, in France, according to L341-2 Energy Code "Code de l'énergie", tariffs for the transmission and distribution networks are transparent and non-discriminate to reflect and recover the costs incurred by the network operator ("French Energy Code," 2011). Further, under L341-3 Energy Code "Code de l'énergie", the Energy Regulation Commission (CRE) sets the tariff methodology for public transmission and distribution networks. In doing so, the CRE takes into consideration the energy policy guidelines ("French Energy Code," 2011). According to the Deliberation of the French Energy Regulatory Commission, the CRE develops tariffs by adhering to guiding principles: "Pricing must reflect the costs generated by each user category independently of their final use of the electricity" (Commission De Regulation De L'Energie, 2021). Developing tariff signals that accurately reflect grid costs is pertinent to achieving long-term optimisation of grid investments as they encourage network users to align their energy behaviour efficiently to meet grid requirements. The CRE considers that grid costs are generally subject to variations in time, localisation, and congestion. However, a perfectly cost-reflective tariff would be overly complex, and the costs associated with implementing it would outweigh the benefits. Accordingly, the CRE states that the tariff structure must achieve the "right balance between the reflection of investment and operating decisions by all participants (producers, customers and storage) on grid costs and the readability of tariffs" (Commission De Regulation De L'Energie, 2021). Governance regimes must strategically develop tariff designs that account for the complexities of transitioning to renewable energy and accurately reflect the cost impact on network users. This includes implementing capacity-based and dynamic tariffs to address these challenges.

### 15.7.3 Absence of Mandates Regarding the Installation of Smart Meters

Smart meters comprising metering and communication devices are crucial in digitalising electricity systems (European Data Protection Supervisor, 2022). They are recognised as enablers in blockchain-enabled energy trading systems and offer various benefits to utilities, network operators, government and regulatory bodies, service providers, and energy consumers. Smart meters ensure accurate measurement and exchange of production and consumption data between stakeholders to ensure smoother peer-to-peer interactions, effective integration of DERs on the grid, and efficiency in load management (European Commission, 2019; European Data Protection Supervisor, 2022; Guimarães, 2017).

Energy readings gathered by smart meters in short intervals encourage energy savings and remove financial burdens. Smart meters allow the imposition of real-time pricing plans and dynamic tariffs, constituting a progressive structure in light of transactive energy systems. At present, there is an aggregated representation and breakdown of energy data based on the consumption of individual appliances (Hennig et al., 2022; Vijayapriya et al., 2010). Smart meters can classify and determine individual appliance usage with greater precision and accuracy, as each appliance has unique usage patterns. Smart meters can measure, record, and transmit smart meter data for energy planning and managing a specific household, particularly when engaging in distributed energy trading (International Renewable Energy Agency, 2019, 2020).

By transmitting instantaneous energy production and consumption values through smart meter installations, energy consumers and interested third parties can adjust their energy behaviour patterns while improving energy efficiency. They also actively integrate various energy appliances to enable individual load management, billing, forecasting, and interoperability. Smart meters have tremendous potential to facilitate continuous communication processes to establish secure and reliable communication and transmission channels. In what follows, such equipment empowers prosumers and consumers to make cost-effective consumption decisions and enhance energy security while meeting environmental targets.

Japan is leading the adoption of smart meters in the Asia-Pacific region. The Ministry of Economy, Trade and Industry's mandate of energy sector liberalisation and the fourth Strategic Energy Plan required all households and businesses to install smart meters (Agency for Natural Resources and Energy, 2014). Switzerland provides an enabling framework in that Article 17a, paragraph 1 of Electricity Supply Act "Stromversorgungsgesetz" (StromVG) defines smart meters as intelligent metering systems, referring to it as "a measuring device for recording electrical energy that supports bidirectional data transmission and records the actual energy flow and its progression over time ("Electricity Supply Act "Stromversorgungsgesetz"," 2007). Article 17a, paragraph 2 allows the Federal Council to set guidelines for adopting intelligent metering systems. The Federal Council shall consider globally recognised standards and recommendations from reputable professional organisations when developing the standards. It can oblige network operators to facilitate the installation of intelligent metering systems within a specific time frame. Further, the Federal Council can ascertain the technical requirements smart meters have to adhere to and the properties, equipment, and functionalities they must possess concerning the "transmission of measurement data," "support of tariff systems," and "support of other services and applications" ("Electricity Supply Act "Stromversorgungsgesetz"," 2007). The definition of technical requirements of intelligent metering systems under Article 8a Electricity Supply Ordinance "Stromversorgungsverordnung" (StromVV) require system interfaces to support bidirectional communication, amongst others. Under Article 31e, paragraph 1, Switzerland has set an objective to replace 80% of current metering systems with intelligent metering devices by 2027 ("Electricity Supply Ordinance "Stromversorgungsverordnung"," 2008). Further, based on Article 31e, paragraph 2, within the period specified in paragraph 1, grid operators are given the autonomy to determine which end consumers and generators they want to equip with intelligent metering systems and when. However, only end consumers must be equipped by the grid operator when they exercise their right to network access and generators when they connect a new generation facility to the grid ("Electricity Supply Ordinance "Stromversorgungsverordnung"," 2008). State and territorial regulations govern Smart meter deployment in Australia. While smart meters are mandatory in Victoria, New South Wales, the Australian Capital Territory, South Australia, and Queensland have a lower smart meter rollout rate (Jones, 2021). Australian Energy Market Commission acknowledges that the existing arrangements do not support the full rollout of smart meters, and changes to regulatory frameworks are necessary to increase smart meter deployment (Australian Energy Market Commission, 2021).

Smart meter rollouts enable more accurate and informative billing, thus promoting simplicity in bidirectional energy trading systems. Many countries have existing silos in their energy regulations, which include a notable absence of mandates regarding the installation of smart meters. The lack of smart meters generates negative externalities and causes legal gaps in the strategic implementation of dynamic pricing (Office of Gas and Electricity Markets, 2019a, 2019b). The full deployment of smart meters is a proactive catalyst for tariff simplicity, dependent on regulatory agility. Such deployment begs whether legislative frameworks mandate the rollout of smart meters, considering the improvement of prosumers' engagement and energy literacy through real-time records.

## 15.8 CONCLUSION

When navigating the regulatory landscape of energy technology, it is essential to recognise the intricate dynamics inherent in transitioning from traditional energy systems to those that bolster energy democratisation and security while actively engaging citizens in the energy supply chain. The diminishing availability of fossil fuels, traditionally employed for electricity generation, poses a substantial threat to energy security and has far-reaching implications for developed and developing economies. Blockchains' technical features and trajectories precipitate and radically transform energy systems. Through its decentralised ledger, blockchain allows for seamless bidirectional energy trading, enhances the security and efficiency of DERs, supports effective grid management, and enables real-time information transfers. Drawing on the topologies of individual and collective prosumerism, legal and regulatory frameworks in many nations remain fragmented, impeding the adoption of blockchain energy applications.

This chapter makes the following findings. Firstly, due to blockchain immutability, deleting transaction data impairs the utility of blockchain systems. As such, the rights to be forgotten and rectification is at odds with blockchain-enabled energy trading systems since energy data is replicated and distributed across all full nodes perpetually. Several approaches are employable to minimise friction, including leveraging off-chain storage and pruning transactions on blockchain ledgers.[1] Secondly, using blockchain technology in energy operations has led to a surge in cyberattacks, causing disruptions to fully decentralised and community-based energy models. The rise of blockchain attacks, ranging from transaction malleability to routing, time jacking, and Sybil attacks, has instigated widespread apprehension regarding revenue losses and potential implications on data integrity. Thirdly, achieving complete decentralisation requires suitable legal and regulatory instruments, which should commence with precise delineation of roles, rights, and responsibilities of market actors and removing barriers to entry and allowing structural transitions from monopolised to liberalised structures. Fourthly, the rigorous legal and regulatory measures, such as significant grid connection fees and unclear policy exemptions obstruct the prosumer's entry into the national power grid and hinder fair competition in the retail electricity markets. Finally, appropriate network tariff structures play a critical role in enabling the successful implementation of peer-to-peer energy trading systems. The current conventional tariff designs fail to communicate accurate price and economic signals to accommodate the dynamic nature of the energy sector. It is crucial to prioritise prosumer-centric activities, clearly define stakeholder roles and responsibilities, including prosumers and energy communities, and incorporate supportive measures to ensure the sustainability of renewable energy across diverse geographical and temporal contexts.

## NOTE

1. Also available are applying encryption, utilising chameleon hash, relying on consensus-based voting, adopting a block matrix, and safeguarding private keys.

## REFERENCES


Abdelmotteleb, I., Fumagalli, E., & Gibescu, M. (2022). Assessing Customer Engagement in Electricity Distribution-Level Flexibility Product Provision: The Norwegian Case. *Sustainable Energy, Grids and Networks*, *29*, 100564. https://doi.org/10.1016/j.segan.2021.100564

Agency for Natural Resources and Energy. (2014, 1 April 2014). *Japan 4th Strategic Energy Plan.* https://www.enecho.meti.go.jp/en/category/others/basic_plan/pdf/4th_strategic_energy_plan.pdf

Aggarwal, S., & Kumar, N. (2021). Cryptographic Consensus Mechanisms. In S. Aggarwal, N. Kumar, & P. Raj (Eds.), *Advances in Computers* (Vol. 121, pp. 211–226). Elsevier.

Al-Abdullah, M., Alsmadi, I., AlAbdullah, R., & Farkas, B. (2020). Designing Privacy-Friendly Data Repositories: A Framework for a Blockchain that follows the GDPR. *Digital Policy, Regulation and Governance*, *22*(5–6), 389–411. https://doi.org/10.1108/DPRG-04-2020-0050

Aman, A. H. M., Shaari, N., Bashi, Z. S. A., Iftikhar, S., Bawazeer, S., Osman, S. H., & Hasan, N. S. (2024). A Review of Residential Blockchain Internet of Things Energy Systems: Resources, Storage and Challenges. *Energy Reports*, *11*, 1225–1241.

Amenta, C., Sanseverino, E. R., & Stagnaro, C. (2021). Regulating Blockchain for Sustainability? The Critical Relationship between Digital Innovation, Regulation, and Electricity Governance. *Energy Research & Social Science*, *76*, 102060, Article 102060. https://doi.org/10.1016/j.erss.2021.102060

Araújo, K. (2014). The Emerging Field of Energy Transitions: Progress, Challenges, and Opportunities. *Energy Research & Social Science*, *1*, 112–121. https://doi.org/10.1016/j.erss.2014.03.002

Ashour, N. A., Bhuiyan, R., Bichler, M., Buhl, H. U., Fridgen, G., Fugger, C., Gretschko, V., Hanny, L., Knörr, J., Neuhoff, K., Neumann, C., Ott, M., Richstein, J. C., Rinck, M., Röhrich, F., Schöpf, M., Sitzmann, A., Wagner, J., & Weibelzahl, M. (2021). *Electricity Market Design 2030-2050: Moving Towards Implementation.* https://synergie-projekt.de/

Australian Energy Market Commission. (2021, 16 September 2021). *Review of the Regulatory Framework for Metering Services.* https://www.aemc.gov.au/sites/default/files/2021-09/EMO0040%20Metering%20Review%20Directions%20paper%20FINAL.pdf

Australian Energy Regulator. (n.d.). *Network Tariff Reform*. https://www.aer.gov.au/networks-pipelines/network-tariff-reform

Baldimtsi, F., Madathil, V., Scafuro, A., & Zhou, L. (2020, 22 June 2020). *Anonymous Lottery in the Proof-of-Stake Setting*. 2020 IEEE 33rd Computer Security Foundations Symposium (CSF), Boston, MA, USA.

Bischi, A., Basile, M., Poli, D., Vallati, C., Miliani, F., Caposciutti, G., Marracci, M., Dini, G., & Desideri, U. (2021). Enabling Low-Voltage, Peer-to-Peer, Quasi-Real-Time Electricity Markets through Consortium Blockchains. *Applied Energy*, *288*, 116365. https://doi.org/10.1016/j.apenergy.2020.116365

Bouraga, S. (2021). A Taxonomy of Blockchain Consensus Protocols: A Survey and Classification Framework. *Expert Systems with Applications*, *168*, 114384. https://doi.org/10.1016/j.eswa.2020.114384

Brown, D., Ehrtmann, M., Holstenkamp, L., Hall, S., & Davis, M. (2020). *Policies for Prosumer Business Models in the EU* (Prosumers for the Energy Union: Mainstreaming Active Participation of Citizens in the Energy Transition, Issue). https://proseu.eu/resource/policies-prosumer-business-models-eu

Bukar, A. L., Hamza, M. F., Ayub, S., Abobaker, A. K., Modu, B., Mohseni, S., Brent, A. C., Ogbonnaya, C., Mustapha, K., & Idakwo, H. O. (2023). Peer-to-Peer Electricity Trading: A Systematic Review on Current Developments and Perspectives. *Renewable Energy Focus*, *44*, 317–333. https://doi.org/10.1016/j.ref.2023.01.008

Chen, S., Shen, Z., Zhang, L., Yan, Z., Li, C., Zhang, N., & Wu, J. (2021). A Trusted Energy Trading Framework by Marrying Blockchain and Optimization. *Advances in Applied Energy*, *2*, 100029. https://doi.org/10.1016/j.adapen.2021.100029

Commission De Regulation De L'Energie. (2021). *Deliberation of the French Energy Regulatory Commission of 21 January 2021 Deciding on the Tariffs for the Use of Public Transmission Electricity Grids (TURPE 6 HTB)* (Deliberation N°2021-12 Issue). https://assets.rte-france.com/prod/public/2021-07/TURPE6_tariff_decision.pdf

De Clercq, S., Schwabeneder, D., Corinaldesi, C., & Fleischhacker, A. (2020). Emerging Aggregator Business Models in European Electricity Markets. In F. Sioshansi (Ed.), *Behind and Beyond the Meter: Digitalization, Aggregation, Optimization, Monetization* (pp. 285–303). Academic Press.

Deepa, N., Pham, Q.-V., Nguyen, D. C., Bhattacharya, S., Prabadevi, B., Gadekallu, T. R., Maddikunta, P. K. R., Fang, F., & Pathirana, P. N. (2022). A Survey on Blockchain for Big Data: Approaches, Opportunities, and Future Directions. *Future Generation Computer Systems*, *131*, 209–226. https://doi.org/10.1016/j.future.2022.01.017

Dinesha, D. L., & Balachandra, P. (2022). Conceptualization of Blockchain Enabled Interconnected Smart Microgrids. *Renewable and Sustainable Energy Reviews*, *168*, 112848.

Duarte, D. (2019). An Introduction to Blockchain Technology from a Legal Perspective and Its Tensions with the GDPR. *Cyberlaw Journal of the Cyberlaw Research Centre of the University of Lisbon School of Law*, 1–58. https://papers.ssrn.com/sol3/papers.cfm?abstract_id=3545331

DuPont, Q. (2017). Experiments in Algorithmic Governance: A History and Ethnography of "The DAO," a Failed Decentralized Autonomous Organization. In M. Campbell-Verduyn (Ed.), *Bitcoin and Beyond* (1st ed., pp. 157–177). Routledge. https://doi.org/10.4324/9781315211909

Electricity Supply Act "Stromversorgungsgesetz", (2007). https://www.fedlex.admin.ch/eli/cc/2007/418/de

Electricity Supply Ordinance "Stromversorgungsverordnung", (2008).

European Commission. (2019, 19 October 2019). *Smart Grids and Meters*. https://energy.ec.europa.eu/topics/markets-and-consumers/smart-grids-and-meters_en

European Data Protection Supervisor. (2022, 24 August 2022). *EDPS Formal Comments on the Draft Commission Implementing Regulation on Interoperability Requirements and Non-Discriminatory and Transparent Procedures for Access to Metering and Consumption Data*. https://edps.europa.eu/system/files/2022-09/22-08-24_access-metering-and-consumption-data_en.pdf

European Union. (2016). Regulation (EU) 2016/679 of the European Parliament and of the Council of 27 April 2016 on the Protection of Natural Persons with regard to the Processing of Personal Data and on the Free Movement of such Data, and Repealing Directive 95/46/EC (General Data Protection Regulation). *Official Journal of the European Union Law, 119*(1), 59.

Fadilpašić, S. (2020, 17 March 2020). *BitMEX Explains the Attack to Doubting Customers; Refunds BTC 40*. Retrieved February 10, 2023 from https://cryptonews.com/news/bitmex-explains-the-attack-to-doubting-customers-refunds-btc-6048.htm

Felice, A., Rakocevic, L., Peeters, L., Messagie, M., Coosemans, T., & Ramirez Camargo, L. (2022). Renewable Energy Communities: Do They Have a Business Case in Flanders? *Applied Energy, 322*, 119419. https://doi.org/10.1016/j.apenergy.2022.119419

French Energy Code, (2011).

Georgarakis, E., Bauwens, T., Pronk, A.-M., & AlSkaif, T. (2021). Keep It Green, Simple and Socially fair: A Choice Experiment on Prosumers' Preferences for Peer-to-Peer Electricity Trading in the Netherlands. *Energy Policy, 159*, 112615. https://doi.org/10.1016/j.enpol.2021.112615

Godyn, M., Kedziora, M., Ren, Y., Liu, Y., & Song, H. H. (2022). Analysis of Solutions for a Blockchain Compliance with GDPR. *Scientific Reports, 12*(1), 15021.

Guimarães, L. N. (2017). The Dichotomy between Smart Metering and the Protection of Consumer's Personal Data in Brazilian Law. *Brazilian Journal of Public Policy, 7*(3), 275. https://doi.org/10.5102/rbpp.v7i3.4898

Guru, D., Perumal, S., & Varadarajan, V. (2021). Approaches towards Blockchain Innovation: A Survey and Future Directions. *Electronics, 10*(10), 1219. https://doi.org/10.3390/electronics10101219

Gyamfi, S., Krumdieck, S., & Urmee, T. (2013). Residential Peak Electricity Demand Response—Highlights of Some Behavioural Issues. *Renewable and Sustainable Energy Reviews, 25*, 71–77.

Hameed, K., Barika, M., Garg, S., Amin, M. B., & Kang, B. (2022). A Taxonomy Study on Securing Blockchain-Based Industrial Applications: An Overview, Application Perspectives, Requirements, Attacks, Countermeasures, and Open Issues. *Journal of Industrial Information Integration, 26*, 100312. https://doi.org/10.1016/j.jii.2021.100312

Harika, M., Rahmadika, S., & Ramdania, D. R. (2019). Blockchain Technology for Managing an Architectural Model of Decentralized Medical Record. *Journal of Physics: Conference Series, 1402*(7), 077027.

Hennig, R. J., Ribó-Pérez, D., de Vries, L. J., & Tindemans, S. H. (2022). What Is a Good Distribution Network Tariff?—Developing Indicators for Performance Assessment. *Applied Energy, 318*, 119186. https://doi.org/10.1016/j.apenergy.2022.119186

Holzleitner, M.-T., Burgstaller, K., Cejka, S., & Veseli, A. (2020). Electricity Trading via Blockchain in an Energy Community from a Data Protection Point of View. *European Energy & Climate Journal, 9*(2), 33–43. https://ezproxy.um.edu.my/login?url=https://search.ebscohost.com/login.aspx?direct=true&db=edb&AN=145712706&site=eds-live

Hua, W., Chen, Y., Qadrdan, M., Jiang, J., Sun, H., & Wu, J. (2022). Applications of Blockchain and Artificial Intelligence Technologies for Enabling Prosumers in Smart Grids: A Review. *Renewable and Sustainable Energy Reviews, 161*, 112308. https://doi.org/10.1016/j.rser.2022.112308

Hulubas, A. (2022, 5 January 2022). *Solana Network Goes Through Another DDoS Attack*. Retrieved February 10, 2023 from https://cryptorobin.com/solana-network-goes-through-another-ddos-attack/

International Renewable Energy Agency. (2019). *Net Billing Schemes* (Innovation Landscape Brief, Issue). https://www.irena.org/-/media/Files/IRENA/Agency/Publication/2019/Feb/IRENA_Net_billing_2019.pdf?la=en&hash=DD239111CB0649A9A9018BAE77B9AC06B9EA0D25

International Renewable Energy Agency. (2020). *Peer-to-Peer Electricity Trading* (Innovation Landscape Brief, Issue). https://www.irena.org/-/media/Files/IRENA/Agency/Publication/2020/Jul/IRENA_Peer-to-peer_electricity_trading_2020.pdf?la=en&hash=AB7E0CFED5A51CFD75EB2CC1FB635B64329BB439

ISO/IEC 27005:2011. (2011, June 2011). *Information Technology—Security Techniques—Information Security Risk Management*. https://www.iso.org/obp/ui/#iso:std:iso-iec:27005:ed-2:v1:en

Jahnavi, K. (2021). The Blockchain Technology and Attacks on It. *Turkish Journal of Computer and Mathematics Education (TURCOMAT), 12*(13), 571–581.

Jones, J. S. (2021). *Australia: New Directions Proposed to Optimise Smart Meter Rollout*. Retrieved February 8, 2022 from https://www.smart-energy.com/industry-sectors/smart-meters/australia-new-directions-proposed-to-optimise-smart-meter-rollout/

Jumani, A. K., Laghari, A. A., & Khan, A. A. (2021). Blockchain and Big Data: Supportive Aid for Daily Life. In S. David, R. S. Anand, V. Jeyakrishnan, & M. Niranjanamurthy (Eds.), *Security Issues and Privacy Concerns in Industry 4.0 Applications* (pp. 141–178). Wiley.

Kallies, A. (2016). A Barrier for Australia's Climate Commitments?: Law, the Electricity Market and Transitioning the Stationary Electricity Sector. *University of New South Wales Law Journal, 39*(4), 1547–1582.

Kausar, F., Senan, F. M., Asif, H. M., & Raahemifar, K. (2021). 6G Technology and Taxonomy of Attacks on Blockchain Technology. *Alexandria Engineering Journal, 61*(6), 4295–4306. https://doi.org/10.1016/j.aej.2021.09.051

Khalid, R., Malik, M. W., Alghamdi, T. A., & Javaid, N. (2021). A Consortium Blockchain Based Energy Trading Scheme for Electric Vehicles in Smart Cities. *Journal of Information Security and Applications, 63*, 102998. https://doi.org/10.1016/j.jisa.2021.102998

Kim, J.-w. (2020). Blockchain Technology and Its Applications: Case Studies. *Journal of System and Management Sciences, 10*(1), 83–93.

Kirli, D., Couraud, B., Robu, V., Salgado-Bravo, M., Norbu, S., Andoni, M., Antonopoulos, I., Negrete-Pincetic, M., Flynn, D., & Kiprakis, A. (2022). Smart Contracts in Energy Systems: A Systematic Review of Fundamental Approaches and Implementations. *Renewable and Sustainable Energy Reviews, 158*, 112013. https://doi.org/10.1016/j.rser.2021.112013

Lai, R., & Zhao, G. (2022). Blockchain for Achieving Accountable Outsourcing Computations in Edge Computing. *Computer Communications, 200*, 17–29. https://doi.org/10.1016/j.comcom.2022.12.024

Le, D.-N., Kumar, R., Mishra, B. K., Khari, M., & Chatterjee, J. M. (2019). *Cyber Security in Parallel and Distributed Computing: Concepts, Techniques, Applications and Case Studies*. John Wiley & Sons.

Lesniewska, F., Ani, U., Carr, M., & Watson, J. (2019). In the Eye of a Storm: Governance of Emerging Technologies in UK Ports Post Brexit. *Living in the Internet of Things*, 1–9. https://doi.org/10.1049/cp.2019.0165

Li, Z. (2020, 10 June 2020). *Standardisation of Blockchain and Distributed Ledger Technologies—A Legal Voice from the Data Protection Law Perspective.* EURAS Annual Standardisation Conference–Standards for Digital Transformation: Blockchain and Innovation. https://papers.ssrn.com/sol3/papers.cfm?abstract_id=3892298

Limongi, G., & Galderisi, A. (2021). Twenty Years of European and International Research on Vulnerability: A Multi-Faceted Concept for Better Dealing with Evolving Risk Landscapes. *International Journal of Disaster Risk Reduction, 63*, 102451. https://doi.org/10.1016/j.ijdrr.2021.102451

Marijan, D., & Lal, C. (2022). Blockchain Verification and Validation: Techniques, Challenges, and Research Directions. *Computer Science Review, 45*, 100492. https://doi.org/10.1016/j.cosrev.2022.100492

Martins Gonçalves, R., Mira da Silva, M., & Rupino da Cunha, P. (2023). Implementing GDPR-Compliant Surveys Using Blockchain. *Future Internet, 15*(4), 143.

Mekouar, L., Iraqi, Y., Damaj, I., & Naous, T. (2022). A Survey on Blockchain-Based Recommender Systems: Integration Architecture and Taxonomy. *Computer Communications, 187*, 1–19. https://doi.org/10.1016/j.comcom.2022.01.020

Mengelkamp, E., Gärttner, J., Rock, K., Kessler, S., Orsini, L., & Weinhardt, C. (2018). Designing Microgrid Energy Markets: A Case Study: The Brooklyn Microgrid. *Applied Energy, 210*, 870–880. https://doi.org/10.1016/j.apenergy.2017.06.054

Milunovich, G. (2022). Assessing the Connectedness between Proof of Work and Proof of Stake/Other Digital Coins. *Economics Letters, 211*, 110243.

Ncube, T., Dlodlo, N., & Terzoli, A. (2020, 25 November 2020). *Private Blockchain Networks: A Solution for Data Privacy.* International Multidisciplinary Information Technology and Engineering Conference (IMITEC), Sol Plaatje University in Kimberley, South Africa. https://ieeexplore.ieee.org/document/9334132

Office of Gas and Electricity Markets. (2019a, 2019). *Smart Meter Transition and the Data Communications Company (DCC).* Retrieved February 7, 2023 from https://www.ofgem.gov.uk/energy-policy-and-regulation/policy-and-regulatory-programmes/smart-meter-transition-and-data-communications-company-dcc

Office of Gas and Electricity Markets. (2019b, 21 February 2019). *Supply Licence Guide: Smart Metering.* Retrieved February, 7 2023 from https://www.ofgem.gov.uk/sites/default/files/docs/2019/02/licence_guide_smart_metering_0.pdf

Ofgem. (2022, 3 May 2022). *Access and Forward-Looking Charges Significant Code Review: Final Decision.* https://www.ofgem.gov.uk/sites/default/files/2022-05/Access%20SCR%20-%20Final%20Decision.pdf

Raza, Z., Haq, I. u., Muneeb, M., & Shafiq, O. (2021). Energy Efficient Multiprocessing Solo Mining Algorithms for Public Blockchain Systems. *Scientific Programming, 2021*, 1–13.

Rebello, G. A. F., Camilo, G. F., Guimarães, L. C. B., de Souza, L. A. C., Thomaz, G. A., & Duarte, O. C. M. B. (2021). A Security and Performance Analysis of Proof-Based Consensus Protocols. *Annals of Telecommunications, 77*, 517–537. https://doi.org/10.1007/s12243-021-00896-2

Reno, S., Bhuiyan, M. R. I., Ahmed, M., & Monyeem, M. (2020). Implementation of Private Blockchain in Smart Card Management System. *Baiust Academic Journal, 1*(1), 12–20. https://www.researchgate.net/publication/352900860_Implementation_of_Private_Blockchain_in_Smart_Card_Management_System/citations

Roth, T., Utz, M., Baumgarte, F., Rieger, A., Sedlmeir, J., & Strüker, J. (2022). Electricity Powered by Blockchain: A Review with a European Perspective. *Applied Energy, 325*, 119799. https://doi.org/10.1016/j.apenergy.2022.119799

Sai, A. R., Buckley, J., Fitzgerald, B., & Gear, A. L. (2021). Taxonomy of Centralization in Public Blockchain Systems: A Systematic Literature Review. *Information Processing & Management, 58*(4), 102584. https://doi.org/10.1016/j.ipm.2021.102584

Sajana, P., Sindhu, M., & Sethumadhavan, M. (2018). On Blockchain Applications: Hyperledger Fabric and Ethereum. *International Journal of Pure and Applied Mathematics, 118*(18), 2965–2970.

Saleh, F. (2021). Blockchain without Waste: Proof-of-Stake. *The Review of Financial Studies, 34*(3), 1156–1190.

Samreen, N. F., & Alalfi, M. H. (2020, 18 February 2020). *Reentrancy Vulnerability Identification in Ethereum Smart Contracts.* International Workshop on Blockchain Oriented Software Engineering (IWBOSE).

Schinckus, C. (2020). The Good, the Bad and the Ugly: An Overview of the Sustainability of Blockchain Technology. *Energy Research & Social Science, 69*, 101614. https://doi.org/10.1016/j.erss.2020.101614

Schneiders, A., & Shipworth, D. (2018). *Energy Cooperatives: A Missing Piece of the Peer-to-Peer Energy Regulation Puzzle?* [Preprint]. University College London Energy Institute.

Shoker, A. (2017, 30 October 2017). *Sustainable Blockchains through Proof of Exercise.* 2017 IEEE 16th International Symposium on Network Computing and Applications (NCA), Cambridge, MA, USA.

Soshinskaya, M., Crijns-Graus, W. H., Guerrero, J. M., & Vasquez, J. C. (2014). Microgrids: Experiences, Barriers and Success Factors. *Renewable and Sustainable Energy Reviews, 40*, 659–672.

Stein, A. L. (2016). Breaking Energy path Dependencies. *Brooklyn Law Review, 82*, 559.

Tatar, U., Gokce, Y., & Nussbaum, B. (2020). Law Versus Technology: Blockchain, GDPR, and Tough Tradeoffs. *Computer Law and Security Review, 38*, Article 105454. https://doi.org/10.1016/j.clsr.2020.105454

Thurman, A. (2022, 12 May 2023). *Polygon under Accidental Attack from Swarm of Sunflower Farmers.* Retrieved February 10, 2023 from https://www.coindesk.com/tech/2022/01/06/polygon-under-accidental-attack-from-swarm-of-sunflower-farmers/

Torriti, J., & Yunusov, T. (2020). It's Only a Matter of Time: Flexibility, Activities and Time of Use Tariffs in the United Kingdom. *Energy Research & Social Science, 69*, 101697.

Trajcevski, M. (2020, 30 April 2020). *Binance Suffered Series of DDoS Attacks.* https://dailycoin.com/binance-suffered-series-of-ddos-attacks/

Truby, J., Brown, R. D., Dahdal, A., & Ibrahim, I. (2022). Blockchain, Climate Damage, and Death: Policy Interventions to Reduce the Carbon Emissions, Mortality, and Net-Zero Implications of Non-Fungible Tokens and Bitcoin. *Energy Research & Social Science, 88*, 102499. https://doi.org/10.1016/j.erss.2022.102499

Ul Hassan, M., Rehmani, M. H., & Chen, J. (2020). Differential Privacy in Blockchain Technology: A Futuristic Approach. *Journal of Parallel and Distributed Computing, 145*, 50–74. https://doi.org/10.1016/j.jpdc.2020.06.003

Vijayapriya, P., Bapna, G., & Kothari, D. P. (2010). Smart Tariff for Smart Meters in Smart Grid. *International Journal of Engineering and Technology, 2*(5), 310–315.

Wani, S., Imthiyas, M., Almohamedh, H., M Alhamed, K., Almotairi, S., & Gulzar, Y. (2021). Distributed Denial of Service (DDoS) Mitigation Using Blockchain—A Comprehensive Insight. *Symmetry, 13*(2), 227. https://doi.org/10.3390/sym13020227

Xu, B., Wang, J., Guo, M., Lu, J., Li, G., & Han, L. (2021a). A Hybrid Demand Response Mechanism Based on Real-Time Incentive and Real-Time Pricing. *Energy, 231*, 120940. https://doi.org/10.1016/j.energy.2021.120940

Xu, C., Qu, Y., Luan, T. H., Eklund, P. W., Xiang, Y., & Gao, L. (2021b). A Light-Weight and Attack-Proof Bidirectional Blockchain Paradigm for Internet of Things. *IEEE Internet of Things Journal, 9*(6), 4371–4384.

Yadav, A. K., Singh, K., Amin, A. H., Almutairi, L., Alsenani, T. R., & Ahmadian, A. (2023). A Comparative Study on Consensus Mechanism with Security Threats and Future Scopes: Blockchain. *Computer Communications, 201*, 102–115. https://doi.org/10.1016/j.comcom.2023.01.018

Yang, R., Wakefield, R., Lyu, S., Jayasuriya, S., Han, F., Yi, X., Yang, X., Amarasinghe, G., & Chen, S. (2020). Public and Private Blockchain in Construction Business Process and Information Integration. *Automation in Construction, 118*, 103276. https://doi.org/10.1016/j.autcon.2020.103276

Zhou, Y., & Wu, J. (2021). Peer-to-peer Energy Trading in Microgrids and Local Energy Systems. In N. Jenkins (Ed.), *Microgrids and Local Energy Systems*. IntechOpen. https://doi.org/10.5772/intechopen.99437

# Index

Note: **Bold** and *italic* page numbers refer to **tables** and *figures*.

## Q



## R



## S

FSC
www.fsc.org
MIX
Papier | Fördert gute Waldnutzung
FSC® C083411

Kolibri 360 GmbH
Ferdinand-Jühlke-Straße 7
99095 Erfurt, Deutschland
produktsicherheit@kolibri360.de